AF693987

ELK FARMING HANDBOOK

WRITTEN, COMPILED AND EDITED BY

IAN THORLEIFSON
TONY PEARSE AND
BRUCE FRIEDEL

WITH MAJOR ASSISTANCE FROM

MARLENE ABBOTT
ABBOTT RICHARDS GRAPHIC DESIGN INC.
SAINT JOHN, NEW BRUNSWICK

Copyright © 2000 by the authors
All Rights Reserved
No Reproduction Without Written Permission
Printed in Canada
ISBN 0-9687367-3-4

Two of the authors, Ian Thorleifson and Bruce Friedel, with some good friends.

Table of Contents

TABLE OF CONTENTS (cont'd)

Chapter 1 **Background**

INTRODUCTION TO ELK FARMING

Elk farming is based on a remarkable partnership with nature. Wild elk were historically distributed throughout the areas of North America where agriculture is now successful. It is only **natural** that they would continue to thrive in those areas. Nature has provided us with animals exceptionally well suited to our farms, and in return, we, the farmers provide the resources to enhance their natural patterns of growth and reproduction, and guide natural selection towards our production goals.

This discussion will introduce you to the philosophy, practice and business of elk farming. To help you remember the key concepts I have highlighted the **key words** in each subject area.

In addition to their historical and environmental suitability, elk have several other natural advantages. They are by nature **easily domesticated** and **adaptable** to a great variety of farmlands and management strategies. People are amazed at how easily wild elk may be herded and captured, and once captured, how quickly they adapt to their new environment. With patience and persistence, those wild elk will eat from your hand in a few months. Farmed elk breed very successfully, averaging over 90% calving rates, and their calves grow very quickly - almost half their mother's weight and ready to thrive on their own in three to four months. They may be farmed in bush pastures or on highly productive intensive grazing operations, on huge wide-open ranches or virtual feedlots, and from the subarctic to the Deep South.

Another tremendous advantage elk farming has over both long- established farming operations and most of the "new and non- traditional" agventures is that it is **profitable**, with **well established markets.** Elk and other deer have been farmed in Asia for over 2000 years. The demand for antler products for pharmaceutical use in Asia is even older. Venison is a traditional and highly desired food item in European markets, and elk (wapiti) meat is a scarce and superior type of venison. Elk trophies are one of the most sought after goals for hunters, with demand far exceeding the supply.

With these advantages, it won't be long before you are progressing to the next stage of your inquiry, which should be **farm selection**. The key word in finding the right farm is **balance**. Elk farming is profitable, and will return handsomely to investments you put in. But should you buy the most productive (and expensive) cropland and put high fences around it? You'll be in the business of marketing livestock, and a high traffic location would be valuable (and more expensive). But the traffic may also bring with it unwanted visitors. How do you choose? The answer is to balance all of these desires and concerns. First, decide who you are. Are you an outgoing, gregarious type of person who communicates well with people, especially strangers? Or are you more quiet, preferring your own company? Think about how you will fit into the group that you are planning to sell to, and adjust your target market according to who you work with the best. Maybe you aren't a showperson, but you can still sell high quality stock. A very valuable saying is **"Where you buy is where you sell"** and the more you think about it, the more it rings true. If you pay the "big bucks" at Select Sales, the odds are that the people who frequent those sales will pay the same prices to buy from you, especially if they become your friends and business associates. If you buy a few head on farm each year, those people will likely be back at your place next year, buying a breeding bull.

Once you've decided where your personality, skills, pocketbook and background will best fit, then start building a farm that also fits. Be honest with yourself. The decisions you make at this stage will influence how successful you are, because what you are doing is creating an **identity**. Once that identity has been created, you have to stick with it, or make changes very slowly. People have to remember you and what you stand for and what you have to offer. That identity is your Market Profile.

Now back to your farm. You've already considered **production potential versus cost,** and **access versus aesthetics**. You also need to consider more basic considerations. A heavy clay soil may have great production potential, but it is difficult to work with when you get that long, wet spring. **Drainage** must be good for elk to thrive. This drainage may come with medium textured soils, and will also be improved with rolling (not steep) terrain. **Water** - good clean drinking water, is also essential for any livestock operation, although elk need less per head than cattle. These are the basic considerations for selecting your farm. The next step is **farm layout**.

The key concept to bear in mind in laying out an elk farm is **flow**. Think of a pool of water. Now try moving that pool around your farm. This is the way an elk herd works. An elk herd - because you will almost always be moving a herd. Elk are strongly herd oriented, and they become very uncomfortable if you try to move them by themselves, especially in a pasture situation. Water never moves well through a square corner. It hits that corner and piles up and splashes. Elk always move best through and around **curves and corners**. A square corner is alright in the far corners of the farm, but it does not work in high traffic areas.

You need to particularly consider two things in each pasture, **water and shelter**. Water may be in dugouts, but a clean water fountain of some type is preferable. Dugouts can present dangers in terms of water quality, drowning and falling through thin ice. Your water needs will be most pressing in the summertime, so aboveground waterlines are adequate with proper maintenance. Adult elk, especially bred cows, do fine with just clean snow in the wintertime, especially if you are feeding high moisture haylage or silage, but young growing stock do much better with clean water all year around.

When building gates and alleyways, always consider adequate **access**. If you decide to break up and replant a pasture, can you get in there with the tillage and seeding equipment you will be using? How wide a haybine will you be using to harvest surplus forage in July? Do you need to get through that gate with a 60-foot highway tractor and trailer?

Handling elk is in many ways more challenging than handling other types of livestock. They are relatively large, excitable and very athletic. But handling elk does not need to be difficult, if you do adequate **planning**. Always remember that elk have a **herd mentality**. This has advantages and disadvantages - if one goes, the rest will usually follow, so if you convince one to go where you want, the job is mostly done. But if one goes where you do not want them, then the whole herd might be right behind. Avoid trying to move just one elk, and always try to keep the herd calm. The key words for moving elk are **posture, pressure and patience**. Your posture tells elk a lot about who you are, your confidence level and your intent. You must appear confident whenever you are among elk, or they may decide to challenge you. If you want elk to move, stand up tall and approach with your eyes focussed right on them. If you want to walk by and not disturb them, lower your shoulders and head and avert your gaze. Elk have a **flight zone**, an area around them that varies in size. If you step into that flight zone, they respond. Elk that are calm and comfortable with your presence may have a very small flight zone - maybe a few meters around them. Wild or frightened elk may have a flight zone two kilometers across! You will use that flight zone by applying pressure to the edges. Elk will move away from the point where you apply pressure. If you press on the right edge, they will move left, and so on. The more you apply, the more they move. If you apply too much and too quickly, they may respond by fleeing very quickly. This is where patience becomes so important. Lack of patience leads to wrecks, with injured animals, damaged facilities and nothing accomplished. If the elk are too worked up, leave them alone for a few hours or days, then try again. Windy days are particularly difficult. The wind interferes with the ability of the elk to smell and hear, and they become agitated and much more likely to react more quickly. Be patient, wait another day.

The last very important word in handling is **seasonality**. Elk are adapted to life in temperate climates, so they live by a tight seasonal schedule. They calve in the spring, breed in the fall. Consequently, their moods and reactions vary with the seasons. A bull that was your best friend in the spring could willingly crush your ribs in autumn. Mid- winter is always the best time for handling elk.

An essential part of efficiently handling elk is properly designed and built **handling facilities**. The key word is **smooth**. Smooth entry, smooth corners, smooth handling. After twenty years of experience, the industry has also included three more essential "S" words, **sweeps, slides and squeeze**. Sweeps and slides refers to the style of construction that has been proven to be the best by far. The elk flow through the rounded curves and around the corners described above, and into a roofed area made up of a series of steel and wooden cubes, or modules. Each side of these modules is a gate. The gate between modules is a slider gate, the side of each is a sweep gate. To move elk from one to another, you slide open the divider gate, then sweep the elk into the next module. At no time is the human in the pen with the elk, and the elk can see all around, making them much more calm and easy to handle. The elk move through one module after another until they reach the squeeze. A squeeze may be a "cow box" with two sweep gates forming a rectangular box about the length and width of an adult cow's body, with sliding access doors on the sides and small swinging doors on both ends.

Or, it may be operated with small winches. The best so far are the very carefully designed hydraulic squeezes. Once a farm has more than two or three bulls, a hydraulic squeeze is almost mandatory.

One more "S" word can be added here - **safe** loading & transport. Well- designed loadouts and trailers are essential if injuries and stress are to be minimized.

Mention of **stress** leads us to the next important topic - animal health. Stress, and the reduction of stress, is the single most important consideration in maintaining the health and productivity of your elk. Elk are exceptionally healthy animals with very tough immune systems and remarkable abilities to recover from injury. Stress reduces the effectiveness of their immune systems, and increases the possibilities of mistakes and consequent injuries. The most successful strategy for avoiding stress is **good management** in every area of your elk farming operation. It would not be appropriate for me to talk about all that is required for health management on your farm. A well-established Veterinary Client Relationship is essential in creating a management plan that will ensure optimum health and production. Your veterinarian will help outline the concerns that need addressing and the strategies and timing that will work best on your elk operation.

One last pair of key words is **identification and inventory.** Without an efficient and current ID and inventory system, you cannot deliver proper health management, and you cannot assure your customers of the safety and quality of the products you are providing.

When we take elk away from their wild environment and put them behind high fences, we accept a significant responsibility. We have restricted their ability to range in search of the right combinations of plants and other nutrient sources to deliver the nutrition they require. On the other hand, a well designed nutrition program can not only compensate the elk for their limited range, but can actually provide a superior diet along with protection from exposure to the elements and predators that goes along with trying to survive and prosper in the wild. To create and deliver this well designed program, we need to regularly **test and balance.** The compositions of your soil, water, forage, browse and supplementary feed all need to be considered and compared to the needs of your elk at different times of the year. Your local feed company nutritionist can be an excellent source of help. Keep in mind the kind of animal an elk is. Elk are "concentrate feeders", skimming the best from a huge variety of plants. They both **graze** grasses, legumes and other herbaceous plants, and **browse** the twigs, leaves and bark of woody shrubs and trees. The key words to remember in providing graze and browse for elk are **"YTRG" - young, tender, rapidly growing.** Pasture "up to their bellies" is of little value for elk. They will eat a tremendous variety of plants, but only when those plants are "YTRG". This is especially important during spring and summer, when cows are lactating, calves are growing, and bulls are growing their antlers. This brings back the key word I've used before: **"seasonality"**. Nutrient requirements vary considerably with age, gender and season of the year. Bulls virtually give up eating during the peak of the rut, the breeding season, and cows need twice as much during lactation than during the winter.

Winter can be a very lean time for wild elk. On farms, we make it much easier for them by providing **supplements**. These may be simply stored forages like hay, haylage or silage, but may also include minerals or vitamins that are shown by analysis to be lacking in the regular feed, or energy and protein supplements when pasture starts to deteriorate in late summer. Supplementary feed will "flush" cows before the rut, and will work as treats to keep your elk friendly and trusting. A bit of oats every day reduces their flight zone, and makes them much easier to handle and to show to customers.

Speaking of the rut and breeding, the key word once again is **seasonality**. Elk are keyed to the seasons. They breed when the pressures of birth and lactation subside, and calve just before the spring flush of nutrition becomes available, perfectly timed. Your choice of bulls should not be haphazard. Rather, you should develop a breeding plan with stated **goals**, and stick with it. You will improve your herd much more quickly, and your customers will learn what your program offers, and come back for more. Don't skimp on your breeding bull. The **bull = 50%** of all your calves, so do your best to buy not only profile but also proven performance.

Performance refers not just to breeding performance, which is essential, but also to the products other than calves that elk provide for us. The most unique product is Elk Velvet Antler, or EVA. No other source exists for EVA other than our elk, and research and testimonial evidence continues to build that supports the claims for its effectiveness as a nutraceutical and dietary supplement. The key word for farmers is **optimum** production. Elk can be encouraged to produce more EVA by excessive feeding programs, but this practice can shorten their productive lifespans. Is it really worth it? Single trait selection in breeding programs, in this case for EVA production alone, is also a dangerous road to travel. The other key words here are **food quality.** We have to remember that we are not only producing a food product, we are producing a health food product! That means no drug residues, minimal stress, and handling practices that are consistent with the health foods market.

The other health food that we produce is elk (wapiti) meat. This is not just "venison" - that's what you get from regular hunting, from all other deer, from New Zealand. This is a **distinct** and more desirable product. To deliver the exceptional quality consistent with that marketing approach, we go back to the key words I used in describing pasture - **"YTRG"** - elk need to be young, tender and rapidly growing to yield superior quality. Elk are exceptional as compared to other meat producers, in that females can yield tender and acceptable quality meat at almost any stage of their lives if they have just completed a rapid growth (weight gain) phase.

Elk bulls that have passed the age for premium meat production don't go down in value - they may actually increase substantially if their antlers are right. Trophy elk bulls are one of the most sought after species in the world. Most jurisdictions with huntable wild elk populations sell ten times more hunting licenses than are actually filled. Elk hunting ranches can provide a **quality experience,** a recreational and educational opportunity that is pure **magic** for the hunter and his companions.

The magic of elk extends beyond their production capabilities. These majestic creatures have thrilled and intrigued hunters and other lovers of nature who have encountered them in their wild habitats, and the qualities that excite these passions continue to be expressed in their fenced environments. Just seeing and working with them provides great pleasure and satisfaction.

To summarize, elk farming is a **partnership with nature**. We need to keep it that way. Never be totally satisfied with your herd or your operation - always continue to **study and improve**. Keep a few copies of the **Elk Farming Handbook** around to remind yourself and to provide to your new customers.

Enjoy Elk Farming!

Historical - Global Perspective

The past 20 years have seen a dramatic pace of development in the elk farming industry in North America. Before the 1970's, there were very few commercial elk farms in operation. "Game farms" raised and sold live animals, but these operations were more like zoos rather than commercial farms. The impetus to begin commercial development came from contact with Asian buyers of velvet and other products, and from the pioneering efforts of farmers from around the world, but particularly New Zealand farmers. North American farmers realized that a large unsatisfied market existed for elk and the products derived from those animals. The challenge was to establish management systems and political environments that would allow efficient and profitable production.

Management systems for farmed elk in North America were not difficult to develop. Elk were native to much of the southern half of the continent and proved to be highly domesticable and easily adapted to the farm environment. The New Zealanders readily shared husbandry skills and knowledge, with further information provided by Europeans and South Africans. Most of these first elk farmers were already accomplished stockmen who then adapted the techniques learned in other livestock industries to this new strategy.

The first commercial farmers in the 1980's obtained livestock from the wild when permits could be issued by the wildlife agencies, or purchased from zoos and game farms. Markets were readily found for velvet antler, primarily in the Asian community, and for aged bulls on the established hunting farms in the United States. The industry strengthened, and the market for breeding stock boomed. This good fortune, unfortunately, precipitated the most difficult times the industry has faced. Science had not kept pace with the development of the industry. Large numbers of live animals were moving all across the country with minimal control or testing. Techniques used for detection of various diseases and parasites in other species were directly transferred to cervids without assessment of their efficacy or accuracy. As a result, Tuberculosis was diagnosed in farmed elk in the late 1980's. The discovery of these health problems was sufficient bad news to the industry - but the reaction of opposition groups was much worse.

The development of commercial elk farms in North America had been strongly criticized by some wildlife, hunting and environmental advocacy groups. Canada shares with the United States a history of exploitation of native wildlife that endangered the survival of many of those species. By the early 1900's, uncontrolled commercial hunting had reduced the wild bison and elk to remnant populations in Western North America. Conservation groups, led by Theodore Roosevelt among others, demanded systems of wildlife conservation be put in place. These protection systems, built on a philosophical base of public ownership of wildlife, are a source of great pride to many North Americans. To their credit, these systems have served wildlife well. Populations of most native North American species are healthy, and provide viewing and hunting opportunities to rival any other part of our world. Many passionate supporters of these systems saw any commercialization of wildlife as a threat to the survival of the protection systems and to the wildlife species themselves. Although many proponents of conservation could accept commercial outfitting for trophy hunters, they opposed any form of wildlife ranching. They predicted the introduction of dangerous diseases and parasites, genetic pollution of existing stocks with imported and escaped farm animals, increases in illegal harvest as legal pathways for production and sale were developed, and decreased profile and support for wild populations as farmed populations increased. The appearance of Tuberculosis confirmed, in their minds, all of their predictions. This perception led to a period of active lobbying in some jurisdictions for the complete eradication of all farmed cervids and an end to the elk farming industry.

In the face of this opposition, prices and demand for breeding stock dropped substantially in the late 1980's and early 1990's. The industry was hurt, but pioneering elk farmers were not at all prepared to quit. They quickly realized that national organizations were needed to focus their own lobbying, research and industry development activities. In 1990, Rush Johnson, an elk farmer from Missouri, organized a meeting in Denver of 17 keen supporters of elk farming. That core group then recruited several more supporters, and each of them contributed $2,000.00 of their personal funds to establish the North American Elk Breeders Association. The first convention of NAEBA was held in Kansas City in the spring of 1991, with about 450 people in attendance - less than one year after establishment! NAEBA is now a powerful and well-respected promotion and lobby group, with over 1,400 members in 1997.

In the spring of 1992, representatives from all across Canada met in Saskatchewan to form the Canadian Venison Council. This organization became the vehicle to carry the messages of the Canadian deer and elk farmers to the federal government. CVC representatives met with the Canadian groups who were in opposition to the industry at the time to discuss their concerns and to develop possible resolutions. They met with animal health professionals and supported research efforts aimed at resolving the current difficulties. They also accepted, along with all other Canadian farmers, the forced destruction of thousands of their livestock as part of the measures taken to rid the Canadian herd of Tuberculosis.

Although this was a difficult time for our industry, the net result was a strengthening of every aspect of management, and a hardening of the resolve of North American Elk farmers to see the industry succeed. Industry-supported research continued, as it does today, towards improvement of disease and parasite detection and management techniques. With these ongoing programs, and the development and broad acceptance of the Code of Practice for the Care and Handling of Farmed Cervids has come acceptance and support of the elk farming industry throughout most of North America.

Elk Farming in North America is strongly based on the market for velvet antler as a food supplement and nutriceutical. Production of unprocessed antler is estimated to have surpassed the 150,000 pound mark in 1996. Good markets also exist for meat and hunting. These strong markets, plus the limited availability of breeding stock, has pushed the prices of livestock to unprecedented levels. These levels will probably stabilize, then the averages will slowly decline as availability of animals increases. With increasing sophistication in knowledge of productive capacities and efficiencies, the spread of prices paid will likely increase, with more valuable genetics increasing in price.

Current estimates place the number of farmed elk in Canada at about 50,000, and in the United States at about 80,000. These numbers are increasing steadily each year.

Elk and deer farming industries have been established in many other countries. Several species are raised, varying with the climate and target markets. Population numbers are shown in the following table.

Approximate Populations of Farmed Elk and Deer 1999

Country	Number	Breed and Use	1998 Velvet Antler Production (tonnes - green)
New Zealand	1,400,000	Mainly red deer for venison and antlers	450
China	1,000,000	Red, elk, sika for antlers	200
Russia	400,000	Elk, red, sika for venison and antlers	180
United States	250,000	Elk, red, fallow & others for venison and antlers	50
Australia	180,000	Elk, red, rusa, fallow for venison and antlers	10
Germany	150,000	Red and fallow for venison	
Korea	112,000	Elk, red, sika for antlers	40
Canada	115,000	Elk, red, fallow, whitetails for antlers and venison	52
Mauritius	60,000	Rusa for venison	45
England	50,000	Red and fallow for venison	
Eire (Ireland)	38,000	Red for venison	
Taiwan	36,000	Sika, sambar, red deer for velvet	
Sweden	35,000	Red and fallow for venison	15
Denmark	30,000	Red and fallow for venison	
France	30,000	Red and fallow for venison	
New Caledonia	20,000	Rusa for venison and velvet antler	
Vietnam	15,000	Sika for antlers	1
Malaysia	15,000	Red, fallow, rusa for venison and antlers	5
Thailand	5,000	Sambar, red for venison and antler	5
Norway	1,000	Red for venison	2

Table 1. Official Estimates - does not include about two million extensively herded reindeer in Canada, Alaska, Siberia, and Scandinavia (reindeer velvet production 5 - 40 tonnes/year)

The Products of Farmed Elk

As in all livestock based farming systems the major goal, at some point in the industry establishment, is to convert on-farm management of the animals into a product based return. A huge variety of elk products are recognized. These come from the heritage of man's interaction with elk through hunting for necessity or sport, and from the traditional Asian use of the animal as a medicinal resource. Modern elk farming adds to that interaction by producing with consistency and the highest quality standards, a safe, nutritionally desirable range of marketable food, fashion and pharmacological resources, with excellent returns.

VELVET ANTLER

Elk are most known, admired and distinguished for their annual growth and display of antler. Their form and role in competition and breeding has a fascination in wildlife appreciation and in hunting lore. Antlers are recognized as major features of individual superiority. In elk farming terms, antler is also seen as an animal resource of great medicinal and pharmacological significance. Antler is a valuable revenue source that commands attention to management and welfare for quality production of a unique edible and medicinal product.

Ian and a buddy

The term velvet antler refers to the whole antler humanely harvested from the males at a particular stage in growth. It refers to the blood filled cartilaginous matrix, skin and the richly vascular growing points prior to the onset of antler hardening and calcification. In this form the velvet antler is processed through traditional cooking and air drying techniques learned from masters in Korea, China and Russia. In the Orient, processed antler fulfills a spectrum of health and physical needs. Whole dried, sliced, powdered, or as an extracted form, it is used as an ingredient in preventative tonics, curatives and valuable dietary supplements. Velvet is often used with other traditionally recognized herbs and marine life and animal derivatives. Because of a largely western misunderstanding of traditional oriental medicine, the philosophy of prevention and disease cure and the balance of the yin and yang life forces, it is occasionally dismissed as a potent panacea or ridiculed in terms of its powers and function.

As the westernized farmed elk industry expands, there is growing use of velvet antler capsules and extracts as health and lifestyle supplements. More people are convinced of its significant effects in promoting well-being and relief for conditions affecting aging, circulation and some pain symptoms particularly in arthritis and in muscle and skeletal damage or disease.

Antler is a renewable product, grown annually. Its characteristics vary with elk bloodlines and with age, nutritional status and genetic background. Successful farming for antler production requires a full understanding of growth patterns of the individual and of the species. Farmers must make a commitment to animal handling, veterinary attention and welfare principles, as well as hygiene and care for quality production. Returns related to product weight, style and quality can be very high on a per animal basis, and superior antler growth is highly heritable. North American elk are regarded as superior producers of top quality elk velvet antler, or EVA.

Using the tools of selection and breeding, elk farmers have the option of antler harvest or may choose promotion, through the aesthetic appeal of bulls in full antler display. Such superior herd sires may also contribute to elk farming diversification as trophy animals. For farm management, however, the balance between animal well-being, risk, quality meat production, and safety, requires that hard antler be removed. It is in elk farming's favour that removal of the velvet antler allows an additional revenue stream and the quality production of a significant medicinal product.

FARM RAISED ELK MEAT - "WAPITI"

Farmed elk produce a healthy, tender and tasty meat. Its leanness, low calorie value, and high protein and mineral values create a nutritional profile that fits with modern culinary trends and consumer demand.

The product is extremely versatile with an end use in traditional cuisine, or modern cooking with minimal preparation and cooking time. It is increasingly being used in many traditional North American preparations - (sausage, added value specialty meals, jerky and smoked meat), as well as being highly acceptable in ground meat outlets and in many ethnic preparations. Limited production tends to focus elk at the white table cloth hospitality trade although increasingly, venison is in demand at retail outlets. There is considerable research devoted to microwave ready added value preparation and appropriate upmarket retail presentation and promotion.

The unique nature of wapiti meat and the special relationship that elk farmers quickly develop with this emerging industry leave room for individual marketing through farm gate sales and a personal involvement with promotion. Elk recipes and presentation combine well with other unique food products that characterize a region or culture.

Post and antemortem meat inspection can ensure that wapiti retains its farmed origins of healthy, quality production.

The eating qualities of wapiti meat are largely unaffected by the great variety of diets and foods that farmed elk are offered. Most importantly, from an economic viewpoint, farmed elk retain an efficiency of production that naturally converts plant protein to red meat protein with qualities far in excess of traditional livestock breeds that have been selected and managed for meat production. Carcass yields and meat : fat : bone ratios are substantially higher than other livestock species, yet elk are just as easily handled for slaughter and processing.

There is a major need for consumer education in the differences between the quality and presentation of farm raised wapiti meat, to that of wild shot and hunting season harvest. International trends in successful marketing and in good returns to the producer are rapidly creating a unique situation where demand outstrips supply in contrast to many other farmed products. The industry challenge is to provide year round availability and consistency of quality to maintain premium returns above the commodity price levels of traditional meats.

Elk farming produces meat that is a healthy and tasty alternative food item, rather than a variation of common meat and protein sources and subject to commodity trading.

CO-PRODUCTS

SKINS

The suppleness and strength of elk leather is rapidly being recognized by fashion designers for both men and women's clothing and is regarded amongst the finest of the world's natural skin products. Pelts from farmed elk offer a product of uniform size and thickness and grain character so important to fashion and design. Elk leather grain is distinctly different from sheep and cattle and produces the finest nappa leather, and also can be split and shaved to create the buckskin suede of North American heritage. While farmed elk can produce a higher and more uniform type of leather, it does require strict adherence to quality management to avoid damage that may result from the intensity of farming and lack of consideration to the value of this important product.

MEDICINAL CO-PRODUCTS

The Chinese, Koreans and other Asian groups recognize deer as a symbol of good luck, health and longevity and treat parts of the animal as unique and valuable for their curative and preventative properties in many medicinal preparations.

The lower leg sinews, the tails of red deer and elk, and the pizzle (penis, testes and haitch bone) are all recovered at slaughter and further processed using traditional techniques that are as complex and secretive as the many functions and forms in which they are used. Most co-products used are soaked in wine or strong alcohol or consumed in soups and eaten. The common benefit is to promote energy flow and rejuvenative powers. In time, aided by growing Western acceptance they will provide additional revenue streams contributing to the extended utilization and profitability of these remarkable animals.

OTHER PRODUCTS

Small and very specialized markets can also be found for the edible by-products (tongue, heart, liver and kidneys), and also hard antler and cast antler stubs for bone carving, knife handles, belt buckles, etc., and for the recovered eye teeth particularly of older females. Uses of dried blood, fetus and placental material and urine are also well known but relatively specialized. They are not yet significant features of elk farming enterprise except in the traditional oriental countries.

As the theory and practice of elk farming has developed, several foundation principles have become obvious. This handbook deals with most of these principles in detail, but it is important to review the generic features of the elk we successfully farm. Elk are sophisticated ruminants, very efficient converters of forage and browse. They are also strongly social herding animals, traits which make them well suited to a farming situation. If well-fed and comfortable, they are amazingly free of disease and stress. A key feature of this unique deer species is the close correlation between life cycles and seasonality.

SEASONALITY

In temperate environments, elk have a strong seasonal pattern of growth. Maximum gain of body tissue occurs in spring, summer and fall. Weight may be lost, particularly by mature bulls during autumn, (the rut or breeding season) and little or no weight gain takes place over winter. This is classically illustrated in Fig 1.

Extensive experimental work now shows that deer growth cycles can be defined as patterns of weight change which occur under proper nutrition and can be influenced by changes in daylight length.

The increasing daylight length of spring and summer stimulates appetite and growth, and decreasing days of autumn and winter reduce appetite and growth.

The cycle of active reproduction is the opposite, or reciprocal of the growth cycle.

Further work has since shown that physiological systems the deer might employ during the seasonal extremes, (i.e. a raised metabolic rate during summer compared with winter) are a direct consequence of the food intake cycle.

For on-farm management of growth and production we need to be aware that elk have:

- a yearly cycle of growth potential best expressed in the young animal,
- a seasonal cycle of food intake - increased in summer, but restricted in winter,
- an all year round requirement for quality feed,
- an ability to digest feed better in summer rather than winter - so high quality, highly digestible feeds with adequate fibre content are critical in winter.

The annual growth curves show there is a strong seasonal change in liveweight. This is shown in adult bulls as the ability to lay down stores of body fat and mobilize it during the autumn breeding season. This is particularly true in males greater than 2 years of age. Breeding bulls can lose up to 30% body weight during this time of rut. Clearly this has a major implication for lean venison

Modern North American Elk Farm Scene

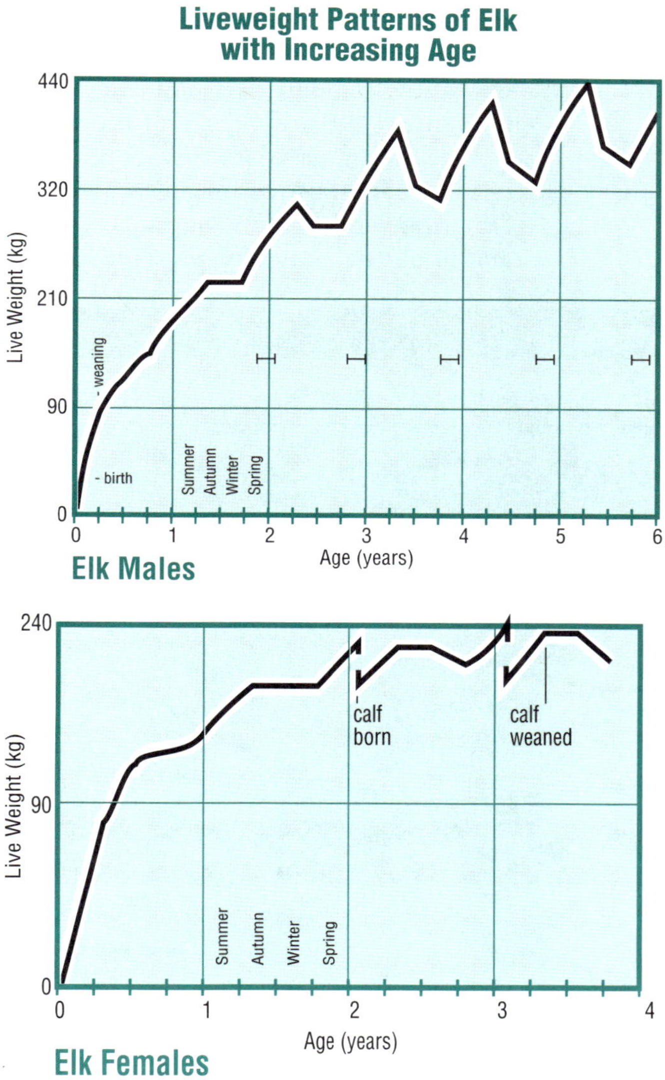

Fig. 1

production. A slaughter target prior to winter at 1 or 2 years of age is ideal for young males. Slaughter for older bulls is appropriate only during post breeding season, once condition and fitness have recovered if over fatness is to be avoided. Surprisingly, elk females yield excellent quality meat at any time in their life. Farm management production targets on which to base feeding strategies and management then become:

June - September (100 days)
- summer growth and lactation
- best time for growth and slaughter

September - December (100 days)
- weaning, breeding
- potential for growth
- slaughter options

December - March (100 days)
- winter - supplemental feeding
- quality feed with good digestibility and adequate fibre.
- limited growth

March - May (65 days)
- spring
- rapid growth of young stock and antlers

It is no accident that these seasonal features of elk biology have a parallel fit with the seasonal task of feeding elk to seasonal requirements that are predictable and repeatable.

Figs. 2, 3, 4 (Adapted from Haigh & Hudson 1993) summarize the seasonal influences of elk physiology on growth rates and reproductive status for females, young males, and adult bulls.

The hormone melatonin is the primary hormone produced in response to changes in daylight length that leads to the increase or decrease in the production of all of the hormones involved in reproduction, sperm production in males and seasonal oestrus in females. Changes of secondary sex characteristics, (coat colour, neck and mane development, roaring behaviour, etc.) are included.

SIGNIFICANT FEATURES FOR MANAGEMENT

1) Calving season and the onset of breeding are very predictably timed to allow a spring calf the opportunity to grow quickly to a target weaning weight by winter. Good lactation can be enhanced by good pasture feeding in this time.

2) The young stock can be targeted to maximize growth during autumn and winter by maximum feeding in smaller pens, which also allows familiarization and quieting of the calves.

3) Spring growth potential must be maximized by responding with quality feeding at the appropriate time even if spring conditions and pasture growth are affected by bad weather and poor growing conditions.

4) There is a direct effect on quality production, animal well being and health status by under feeding during any of the critical times of the year. Successful elk farming means matching feed supply to animal requirements.

On the whole, the seasonal patterns of agriculture in North America with high quality surplus feed grown in spring and summer and readily stored for autumn and winter supplementation, provide a sound basis for successful elk farming.

ELK AS FARMED ANIMALS

Deer species that are successfully farmed share a number of common features that contribute to that success. Normally, deer that adapt well to traditional or modern farming systems show the following features:

- They are largely a herd animal, social in nature and behaviour and are gregarious. An elk herd is controlled by experienced dominant females who alert the herd and lead in the education and experience of calves and herd movement and acceptance of routines.
- Deer are multi-plant species browsers and grazers. The most commonly farmed deer are "intermediate" feeders, opportunistic, adaptable and in part selective over a wide range of plants

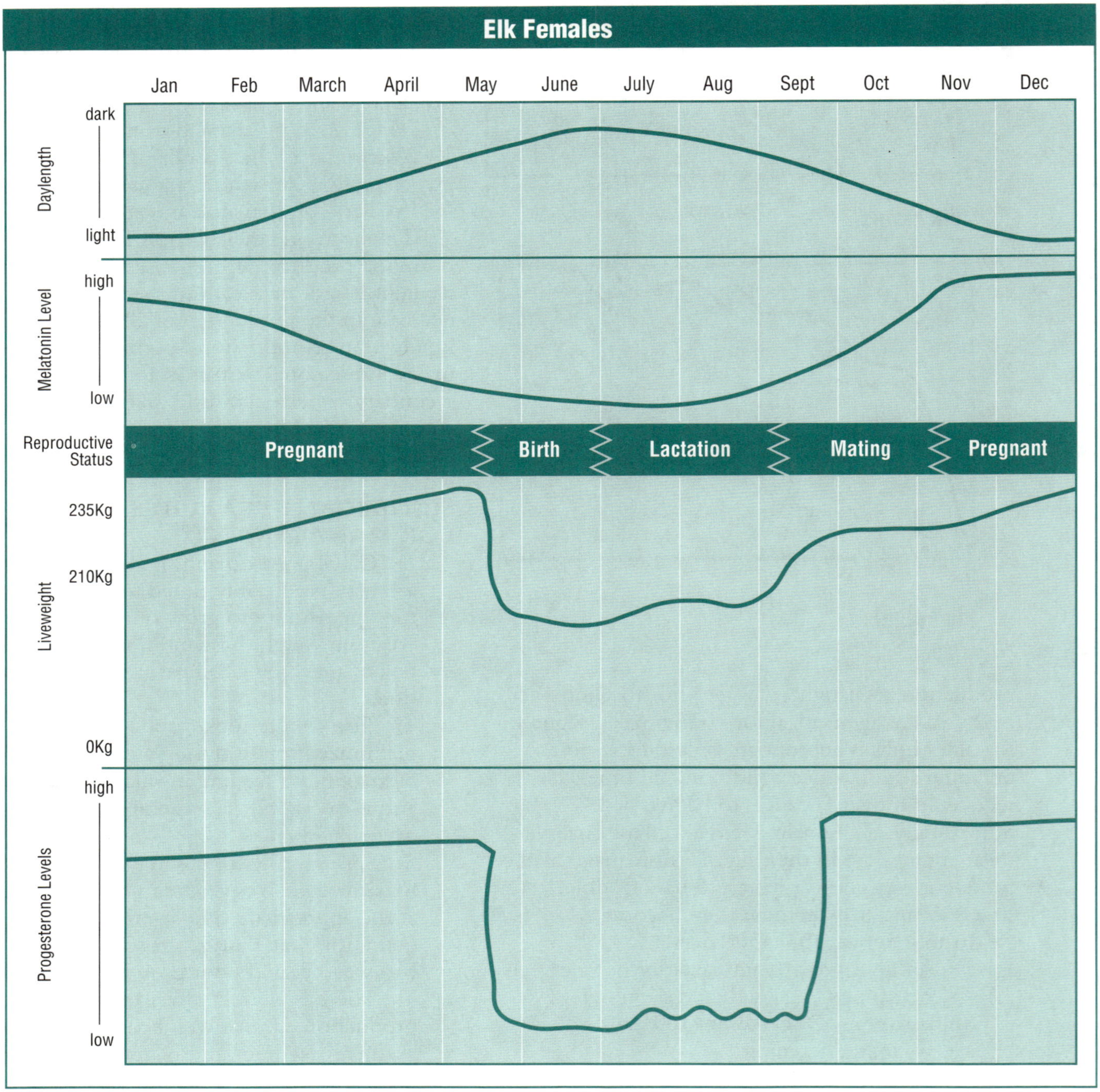

Fig. 2

and pastures. Elk have proven to be highly adaptable to a variety of farm environments.

• The behaviour of elk, its polygamous nature and productivity mean there are powerful tools available for selection and improvement. A dominant or superior male can mate many females in the season allowing a high female : male ratio to be established in a farmed herd. In turn large numbers of surplus males are available for production end points, (velvet, meat, or hunting) allowing truly superior animals to be retained for breeding.

• Farmed elk have great longevity compared to traditional livestock with a 12 - 15 year breeding life a realistic target. Herds can quickly expand in numbers or there is the potential to supply surplus breeding females to a growing industry. With a low replacement rate required, selection for superiority - mothering ability, growth rate, antler genetics

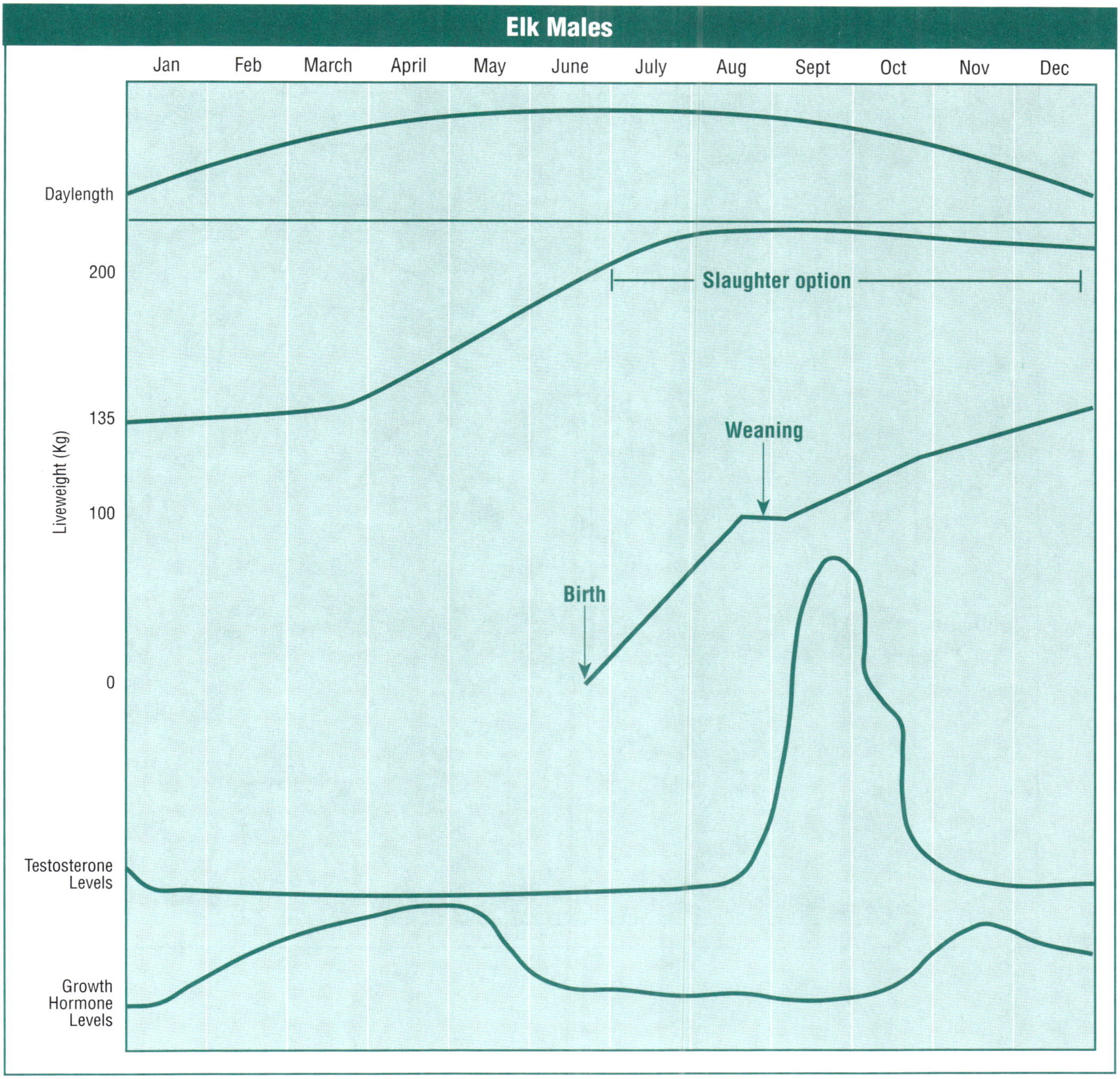

Fig. 3

can have rapid returns in herd improvement and in return on investment.

• Elk under good nutrition thrive, are healthy and carry natural resistance to many diseases. As they have evolved as wildlife their relationship with man on farms shows little dependence on intervention, provided the basics of exercise, food, water and shelter are present.

• Elk have proven to be highly adaptable, responsive to good routine and management and can be handled easily and systematically with appropriate facilities and equipment. Most elk farmers share a very close relationship and interaction with their animals, without compromising the spirit and natural behaviour of the animals.

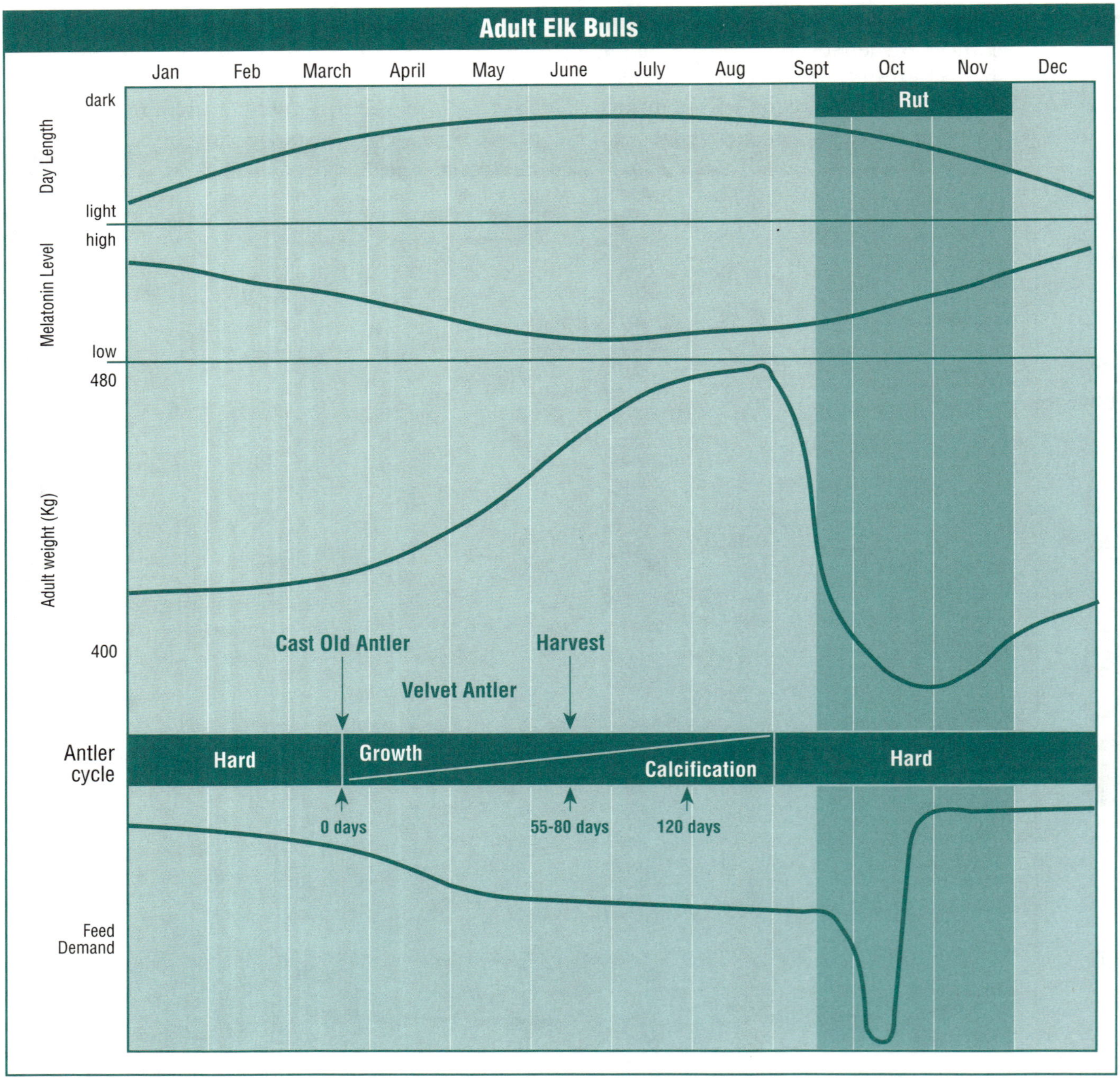

Fig. 4

CHARACTERISTICS OF NORTH AMERICAN ELK

5 yr. old Elk Bull

Males - Bulls	420 - 600 kg
Females - Cows	220 - 320 kg
Young: Calves	
Birthweight: Males	18 - 28 kg
Females	16 - 26 kg
Sexual maturity (highly dependent on nutritional status)	15 - 27 months
Gestation length	247 ± 10 days
Oestrus cycle	21 days
Breeding season	Sept - Nov
Calving - single, (twins, rare)	May - July

Young Elk Cows

Typical Elk Antler Formation

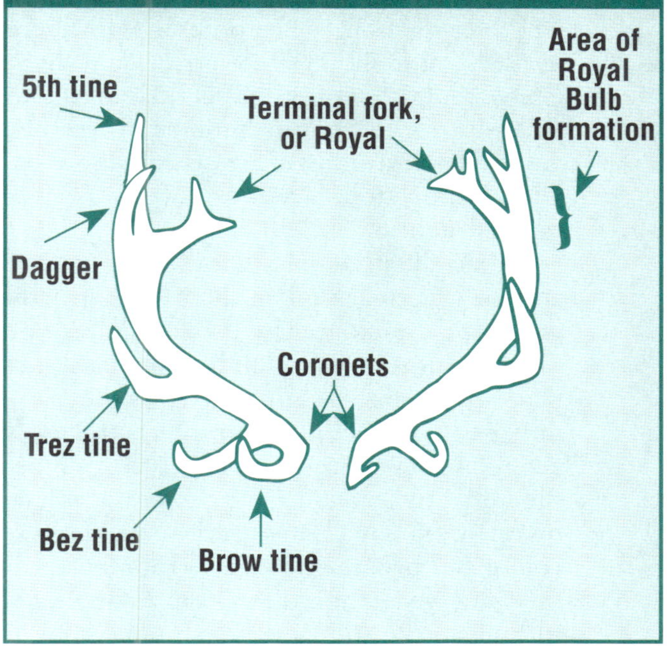

A Wildlife Perspective

The North American elk, or wapiti, is the largest form of the species *Cervus elaphus*. "Elk" is the name by which most North Americans know this majestic deer. "Wapiti", meaning "white rump", is the Shawnee Indian name and the common name preferred by scientists, because the animal known as an "elk" (alg) in Europe is not an elk at all but a close relative of the North American moose. Other breeds of the species, all smaller and known as "red deer", "maral" or other local names, are found throughout the northern hemisphere: in Scotland and continental Europe, North Africa, and Asia.

Appearance

In general appearance elk are obviously kin to the well known white-tailed deer. However, elk are much larger. Among North American deer, they are second in size only to the moose.

The colour of the elk's coat ranges from reddish brown in summer to dark brown in winter. Although it looks white from a distance, on closer inspection the rump colour is ivory to orange. In contrast to the rump, the head and neck are dark. Elk have long, blackish hair on the neck that is referred to as a mane. Their hooves are rounded and their tracks may be confused with those of yearling cattle in range country. Their droppings (scats), like those of other deer, are in the form of pellets in winter, but in summer, when the animals are on new green forage, resemble those of cattle. Closer inspection, however, reveals traces of a pellet structure.

Male elk are notable for their impressively large antlers. It is amazing that these large structures are grown new each year by the animals in a period of a few months in spring and summer. Antlers look particularly large in summer when they are encased in velvet - a covering that protects the growing antlers. In later summer, the velvet is rubbed from the fully grown antlers and the boney structure is revealed. Newly cleaned antlers are light grey in colour but become stained by rubbing and thrashing vegetation during the excitement of the rutting season. (Rutting is an annual state of sexual excitement in the male deer.)

The elk is highly vocal for an ungulate. A person close to a group of elk can hear frequent grunts and squeals as they keep in touch with each other. When alarmed the cows give sharp barks to warn the rest of the group. The whistling roar of rutting bulls is a spine-tingling sound on a frosty autumn morning.

Diet

Elk are plant eaters. There are few plants that occur on their range that they do not eat in certain areas under certain conditions. In winter they eat grasses when they can obtain them. However, when the snow gets deep, they readily eat twigs of woody species, even conifers like Douglas fir. In spring, grasses and sedges are favourite foods. As broad-leaved herbaceous plants spring up in early summer, elk include a high proportion of them in their diet. They also consume shrub and tree twigs and leaves. Early summer is the time of year when a wide variety of nutritious food is available for elk; it is also the time when elk cows are providing milk for their newborn calves.

As summer passes, the herbaceous plants dry out and elk turn again to dry grasses and browse. When the frosty nights of autumn arrive, leaves begin to fall in trembling aspen forests on the western ranges of the elk. Elk include dry leaves in their diet and continue to eat them until they are buried by snow. When winter comes, elk diets are controlled largely by snow. Elk dig craters in loose snow to expose dry grass and leaves, but when the snow gets too deep or too hard they must shift their feeding largely to woody twigs. In mountainous terrain elk must leave areas of deep snow cover and seek locations such as valley bottoms where snow cover is shallow or absent. In areas where deep snow seldom occurs, they may frequent high or low elevation ranges at any time of the year.

Life History

Elk are sociable animals. They are seldom found without other elk nearby. The herd lifestyle is characteristic of animals that live in open country. However, elk populations today occupy forest or parkland regions, where small groups averaging six or seven animals are common.

Elk are long-lived animals: males survive to an average of 14 years, whereas females live and breed as long as 24 years. Although they may travel widely, each elk is strongly attached to certain localities within its home range.

The annual cycle of the elk begins in spring with release from the snows and food shortages of winter. This is the time of calving and increasing of the herds. Calving usually occurs in areas with which the cow is very familiar. Some cows may seek the same area to calve in year after year. Others give birth to their calves in whatever part of their home range they happen to be in when the time comes. Whenever calving occurs, the cows split off from other elk and seek seclusion and cover a few days before giving birth. Elk hide their calves for 10 days or more after they are born. The calves are genetically programmed to remain quiet and concealed as a defence against predators. Later mother and offspring join others in cow/calf bands on the summer range. Beginning in August, the quiet summer life of the elk comes to an end with the start of the rut. The bulls, which have passed a lazy summer in small groups while their antlers grew large and heavy, now move into the cow/calf group and establish harems of cows. In the process there is considerable fighting among the bulls. Large bulls eventually get control of as many as 20 or 30 cows and drive other males to the fringes of the herds. This does not mean, however, that the young males are totally left out of the breeding. While the large harem masters are running off intruders or rounding up straying females on one side of their group, a young bull may sneak in and mate with a female on the other side.

Following the turmoil of the rut, the bull elk leave the females and move to good foraging areas to recoup their losses in weight and condition before winter. Some go back up the mountains to spend a few more weeks on the nutritious pastures of the alpine zone before snow forces them down. Elk usually, but not always, wait for the coming of snow to move down to the valleys. There is considerable overlap between the winter ranges of bulls and cows. As bulls are larger and more powerful they can travel and dig through deep snow more readily than the cows, and by doing so they are able to have foraging areas to themselves.

Distribution

The present population of wild elk in North America is a reflection of the changes made by settlement and large scale crop agriculture. Elk once ranged from New York State and southern Ontario to Mexico, and all through the Prairies, Parklands and Mountains. Now they are associated with the wildlands and wildlife reserves. It is very difficult to estimate how low elk populations reached in the early 1900's, nor how high they may have been prior to that date, but there are now an estimated 600,000 wild elk in all of North America. The largest herds are found in Colorado, Wyoming, Montana, Washington, Alberta and British Columbia - the States and Provinces with the greatest areas of suitable habitat.

Distribution of Elk in the Wild

Limits of Population

The principal limiting factor on the number of elk in North America has been loss of habitat to agriculture. Fortunately, extensive areas do remain to the elk. Hunting serves to keep elk numbers within the carrying capacity of the ranges. In parks elk numbers are sometimes reduced by capture and transplant of surplus animals.

Aside from humankind the most important predator of the elk is the wolf. In spite of their size and power, elk are readily killed by wolves. The distribution of elk in Canada overlaps with wolf distribution, so most elk herds are culled to some extent by wolves. Black bears also kill considerable numbers of elk. Recent studies have demonstrated that in some areas black bears may kill as many as 50% of the elk calves. This predation occurs during the first two or three weeks of the calf's life. Once calves become strong enough to keep up with their mothers, and mother and calf rejoin the rest of the herd, most predation ceases. However, grizzly bears may kill an occasional adult elk. Coyotes take some calves, and cougars, which share the elk's range from the Rocky Mountains west, take elk of all ages.

Where predation and hunting do not keep them low, elk numbers usually increase until they are limited by lack of food. At high population levels, elk can have a significant impact on their range and on their food plants by grazing, browsing, and trampling of vegetation. During severe winters or droughts, significant numbers of elk may starve or become predisposed to disease. The managers of many wild elk populations that are not in parks aim to keep elk numbers well below the maximum dictated by food resources so that elk will be less likely to experience die-offs.

Relationship to People

Elk are highly esteemed by hunters and lovers of wildness. They are one of North America's major big game species, with tens of thousands harvested by hunters each year. In wild areas where they are subjected to hunting pressure, they become very wary and difficult to capture, but in parks where they are not hunted by humans, they quickly habituate to the presence of people. Banff and Jasper National Parks are well known for their very tame elk, so tame as to pose danger to unwary or foolish people. The National Elk Refuge near Jackson Hole, Wyoming, provides safe haven and supplemental feed for as many as 10,000 elk each winter. These elk are a great attraction for tourists, drawing thousands of elk lovers there each winter.

In mountain areas during winter, elk share valley bottoms with major transportation corridors. This leads to many elk-vehicle collisions, with disastrous results to the elk and to humans and their automobiles. This costly hazard has been controlled in many areas, by the construction of a system of fences, cattleguard gates, and underpasses along highways.

The readiness with which elk can be habituated to people and the value of products derived from them have recently aroused considerable interest in domestication and ranching of the animals. One of the most valuable elk products is their antlers. Since ancient times, Oriental people have believed that medicinal preparations from elk antlers that have been removed while still in velvet are useful for diverse therapeutic purposes. Thus Oriental medicine consumes large quantities of elk antler at a high price. Antlers are surgically removed when they have reached maximum size but before they harden; then they are dried, sorted by grade, and shipped to Asian markets.

In many areas elk and domestic cattle share the same ranges. Because both eat the same foods and the presence of cattle brings human activity, there is some conflict between the two species. In mountain areas where elk concentrate in valleys that are also important winter range for cattle there is competition for scarce forage and disturbance of elk at a time when they are under stress due to severe weather. Such situations call for close cooperation between ranchers and wildlife managers to keep problems under control.

The future welfare of elk in general depends on cooperation between wildlife authorities and all land managers, including forest industries, oil and mining companies, park managers, and Indian bands, as well as ranchers.

Farming Suitability

Elk are well adapted to the North American climate and integrate readily into farming management systems. Dietary needs are easily satisfied with a requirement for fibre in their diet via browse or quality hay supplementation. Current farming emphasis is on the supply of breeding stock and velvet antler production. Selection for genetic improvement in body size, growth rate and antler production has been rapid with truly remarkable antlers being produced by the leading breeders.

Elk have proven extremely responsive to on-farm situations as long as due regard is given to their size and speed that the animals can move under intensive handling conditions. Elk tend to be the most pasture-quiet of all farmed deer species and rarely track fence lines, jump at fences or create pasture damage through wallowing such as red deer are more prone to do.

Elk genes are being used successfully to advance the productivity and efficiency of farmed red deer, and also in upgrade programmes of red/elk crossbred females.

Traditionally the North American species is recognized as four subspecies or breeds, relating to geographical origin, body size and antler shape. Terms include Manitoban, Rocky Mountain, Roosevelt and Tule elk. In wildlife relocations there has been much transplanting of subspecies and interbreeding. The North American Elk Breeders Association (NAEBA) is committed to blood typing and DNA analysis of farmed elk to identify parentage and pedigree and to define productive traits and genetics for an improved breeding registry.

ELK ON A BEEF FARM

Many new elk farmers come to our industry from a background in the beef business. Many more beef farmers are watching our industry grow, and considering the possibility of adding an elk component to their beef farm. For those with an understanding of cattle production, the following discussion will allow you to think about elk farming as a comparison.

Elk and high wire are changing our agricultural landscape. In just over twenty years, the pioneers and forward thinkers have built a thriving industry.

As of 1999, in the United States and Canada, there are over 2,000 elk farms stocked with over 130,000 head of elk.

So, why raise elk? Is there a place for them on your beef operation? To make that diversification move a success, here are some things you should know. Both beef cattle and elk work well on farms for some of the same reasons:

- They are ruminants, able to eat vegetation that people cannot live on, and efficiently turn it into good nutritious food;
- They are social animals with strong herding instincts, traits that make them easy to manage in a farm situation;
- They are polygamous, meaning that we need only one excellent bull to breed many cows;
- They fit well in our climate.

The most important difference between cattle and elk is summed up in the word "seasonality". Elk are still much closer to nature. They have strong seasonal cycles in breeding activity, feed requirements and general management requirements. Elk bulls are only capable of breeding from about August to January, and the cows cycle only during that same period. In order to grow antlers, elk bulls must have very low levels of testosterone in their systems. In January, their testicles begin to shrink, and their entire reproductive systems slow right down. During velvet antler season, the bulls are kittens to handle. During the rut, the breeding season, they can be monsters. This temperament change is entirely related to hormone levels. Cattle, on the other hand, breed all year around, and their temperament is little affected by the seasons.

Feed requirements for cattle vary more with temperature and lactation status than with the actual seasons. Elk, conversely, show strong seasonal swings in feed requirements. Bulls pack on the weight in summer, then show little interest in feed and lose up to 40% of their bodyweight during the rut. They begin eating seriously again in November, then slow down again in late winter. As spring greenup occurs the bulls go back to putting on weight. Elk cows maintain more consistent weights than elk bulls, staying much the same over winter as the weight of bodyfat accumulated in the fall is replaced by the weight of the growing calf. Cows then lose pounds to the heavy demands of producing rich, concentrated milk all summer and recover those lost pounds on the hard, rich feed of autumn. Elk calves grow very quickly.

An elk cow averaging 550 pounds liveweight will raise a 210 pound calf in 100 days. A beef cow averaging 1,100 pounds will wean a 550 pound calf in 200 days. Cattle and elk eat essentially the same feeds. An elk will browse on leaves and twigs more than most cattle, and cattle will use more coarse feed, especially coarse hay or straw. This reflects the need elk have for higher mineral levels, especially copper, which are found in higher concentrations in browse. Cattle can utilize coarser feed largely because of the relative size of a cow's mouth, teeth and rumen. Basically, good alfalfa/ grass pastures are the best choice for both elk and cattle in most of Canada and the northern U.S. Elk have virtually no problems with bloat, due to the fact that their rumen releases gases and foam more often than in cattle, thereby avoiding the buildups that cause bloat. However, elk are more susceptible to grain overload than cattle. Grain overload causes rumen acidosis in elk, which can be fatal or may damage the rumen walls and cause tendonitis and lameness.

Rotational grazing systems work well for both elk and cattle, using either permanent fences or temporary electric subdivisions. In winter, both elk and cattle are fed similar feeds. Elk cows are supplemented with grain and pellets just before and during the breeding season (late August to early November) then fed hay to maintain their weight through to calving in May and June. Stockpiling pasture or swath grazing can work for both, effectively extending the grazing season.

Breeding and calving have many similarities in cattle and elk. Elk cows are exposed to a bull in ratios of 5:1 for a yearling up to 50:1 for an older, experienced bull. Elk breed for the first time as yearlings, at about 16 months of age, and these yearlings show conception rates of about 85% with good management. Conception rates for adult cows should be slightly higher, at about 95%. Semen collection and artificial insemination is becoming commonplace for elk, although embryo collection and transfer is much less successful than for cattle. A good AI program, utilizing CIDR's (controlled internal drug release devices which are inserted vaginally to program the cycle) and cervical AI should yield at least 60% conception. A well designed breeding program for both cattle and elk will have similar goals, such as structural soundness, manageable temperament, and efficient production, with the addition of a significantly different product of elk - their antlers.

Elk calve only in the springtime, just as the food supply booms, and nurse their calves when that feed is least expensive. Cattle were probably that way long ago, and our selective breeding has changed them to be able to calve at any time of the year. This is a big advantage for some reasons, but you have to wonder when you're carrying those calves in out of a blizzard. Calving difficulties are very rare in well managed elk herds. Selection for larger and larger bulls, and luxury feeding programs can lead to difficulties. An assisted birth of an elk calf generally results in a bottle fed calf, as elk cows are usually very stressed by the procedure.

Herd health management for cattle and elk is somewhat similar with some key differences. Elk are susceptible to most diseases that trouble cattle, but less so. As mentioned before, elk are much closer to nature than cattle. In nature, individuals or bloodlines that tend to succumb to disease are quickly wiped out. Even if the disease does not kill directly, it weakens the animal to the point that predators or exposure does the job. Thus the weak or susceptible genetic makeup is removed from the population. With domesticated livestock, we develop vaccines and antibiotics to fight disease and keep our animals healthy and breeding. This is effective in the short term, but it perpetuates the susceptibility to that disease. Eventually, we might breed that same susceptibility into farmed elk. For now, elk are less prone to most diseases, especially the respiratory complex. Most elk farmers vaccinate for Chlostridial disease. The need for other vaccines, such as for scours, rabies, leptospirosis and others, depends on the farm location and history. Parasite control is a necessity for elk as it is with cattle. The intensity of control required for both depends on stocking density and herd history. Winter ticks are a concern with elk, but are rarely seen on cattle. Other parasite concerns and control strategies are very similar.

Good management is the key component of a herd health program for both cattle and elk. Livestock, which are well fed, have access to comfortable resting and shelter areas, and live without stress will rarely become ill.

The most significant difference in herd health

management for cattle and elk is the requirement for whole herd tuberculosis testing of elk. At present this must be done every three years. This is required because so few elk are slaughtered. A large portion of the population of cattle are slaughtered on a regular and continuous basis, allowing inspection of the carcasses for abnormalities or disease, particularly any signs of TB. Most elk are too valuable alive to make slaughter for meat a viable option, making regular testing a necessity. The test site for TB is slightly different - beside the tail for cattle - on a shaved patch on the neck for elk. The farmed elk and cattle herds in all of Canada and most of the United States are currently considered to be free of Tuberculosis, Brucellosis and all other dangerous communicable diseases.

Handling facilities and fencing requirements for elk are more robust and substantial than what is required for cattle on most farms. Elk are comparatively more athletic and excitable. The well constructed steel facilities of a modern auction mart are not a lot different from those required for elk, but most small cattle operations can manage safely with facilities which cost one third to one half of what facilities for a small elk operation would require. Perimeter fences for an elk farm must be eight foot tall page wire, with handling facility walls at ten feet tall. Handling elk and cattle employs very similar techniques, as long as you consider the elk to be the same as the wildest, longest-legged exotic cattle beast you can remember from the 1980's, except a lot smarter! They both handle as herd animals, following a leader, and prefer some company at all times. While cattle work well in facilities with solid walls, elk prefer to be able to see all around them. To accommodate this, elk facilities are built with horizontal plank walls and spaces between the planks. Good stockmanship is the key. Curves and corners in the facility, strategic placement of gates and walls, and the right amount of pressure, posture and patience is what is required. Stressed cattle and stressed elk both need time to settle before handling. The preferred method is to avoid stressing them.

Two things set elk apart from cattle - suitability and products. God made elk for our climate and environment. They are not exotic, not imported from anywhere. When it's thirty below, with a stiff breeze blowing, elk are just as likely to be laying in the snow on an exposed hillside as to be seeking shelter. Their annual cycle is totally in tune with the changing seasons, and one of the most noticeable aspects of that annual cycle is the growth and shedding of the antlers on the bulls. Antlers are the most unique trait of all members of the deer family. As the most rapidly growing living tissue (other than mushrooms), antlers have an exceptional chemical makeup. Practitioners of traditional Asian medicine, who use different portions of the antler as food supplements, recognized this long ago. Users of these medicines feel that antler helps them with energy, growth and blood circulation, and alleviates symptoms of arthritis and pre-menstrual syndrome. A huge and growing market exists for these products. Such a market makes velvet antler production one of the most profitable agricultural enterprises today. Hard antlers are also a very valuable product, especially for hunters who are thrilled with the opportunity to hunt and harvest a spectacular bull.

Are elk the right diversification choice for your beef farm? Potential profits are presently very attractive. Farmers in other provinces have already shown us the path. Capital requirements are high, and management skills learned from cattle farming will need to be fine-tuned. Before you make the decision, be sure you read the best sources of information, and visit a few well established farms.

CROSSBREEDING
THE MOST PRODUCTIVE OPTION FOR RED DEER FARMERS

An Elk is a big Red Deer. A Red Deer is a smaller Elk. Acceptance of these statements opens up a whole new world of possibilities for red deer and elk farmers. Crossbreeding is a standard practice in most sectors of agriculture. The advantages of crossbreeding are summed up in the term heterosis, or "hybrid vigour". Think of beef, pork or corn production. All of the mainstream commercial producers depend on hybrid strains. This concept is now being accepted by venison and velvet producers as well. Fallow deer farmers realized several years ago that, if they were to be more efficient producers of venison, then Mesopotamian, or Persian fallow genetics crossed with the standard Europeans was a better way to go. Red Deer offer even more options. Instead of just the two or three different strains (European, Mesopotamian, improved European such as Danish) of Fallow Deer, Red Deer producers have close to thirty different breeds of genetic stock to choose from. And, there is tremendous diversity between these breeds - all the way from Scottish Red females, at 180 pounds liveweight, to North American Elk bulls at 1,200 pounds. This realization of the potential benefits of heterosis "arrived" in New Zealand several years ago. Now it is sweeping, like a wave, over Eastern North America. Although good statistics are not available, it is possible that 40% of the Red Deer females bred in 1996 in North America were exposed to Elk bulls.

Why would a Red Deer farmer crossbreed his females? Is this a mule or beefalo type of thing? Experience would suggest not. Elk/Red crossbreeding is more like the Hereford/Simmental cross. Enhanced growth rates (venison), better velvet antler production, with excellent reproductive characteristics and temperament, all add up to a very desirable alternative. In fact, for velvet and venison production, it is the best alternative from an efficiency point of view. Take a look at beef production. The standard for efficiency has emerged in the medium sized British-breed dam crossed with the fast growing Continental sire. The same principles apply in Red Deer crossbreeding. For maximum performance, we draw on not just any Red females, nor just any Elk males. We choose sound, healthy productive Reds in the 225 lb. + weight range as adults, and Elk/Wapiti males with proven ability to aggressively breed Red females, and the ability to produce exceptional gains in their offspring. Liveweight gain is the primary goal in the first two years of the crossbred deer's life. After that, velvet antler production is the target. Again, crossbreds have shown impressive abilities in this area. Most data, to date, originate from New Zealand, as shown in the following tables:

300+ lb. F1 Crossbred Yearling

First, consider liveweight and venison production:

	Breed Type				
	Red	Super Red	Elk/Red	Elk	Fallow
Liveweight (av.) Age 12-15 mos.	230 lbs	266 lbs	320 lbs	450 lbs	105 lbs
Carcass weight	131 lbs	153 lbs	186 lbs	252 lbs	62 lbs
Dressing %	57%	57.5%	58%	56%	59%
Range %	52-59%	53-59%	56-59%	51-58%	56-60%
Liveweight 24-27 mos.	310 lbs	365 lbs	420 lbs	660 lbs	140 lbs
Carcass weight	180 lbs	210 lbs	243 lbs	375 lbs	83 lbs

Table 2 Dressing %: Ratio of liveweight to hot carcass weight. To estimate optimal slaughter weights, it is necessary to know accurate liveweight to carcass weight ratios as represented in this Table.

In this table, Elk/Red refers to an F1 - the first cross of a pure Elk and pure Red. A Super Red is an F2 - the second cross, usually a pure Red female with an F1 male, or an F1 female bred to a pure Red male for her first calf, (25% Elk, 75% Red). Note that for all of the first three categories, Red, Super Red and Elk/Red Crossbred, the dam in each case is usually pure Red.

Carcass weight and dressing percentage are the first key data when assessing value for venison production. However, we are all aware that the choice cuts sell much more easily and for much higher prices than the rest of the carcass. Table 3 shows comparative yield statistics.

Although not shown in this table, experience has shown that dressing percentage and comparative yield advantage increases for 26 month crossbreds, with no loss in meat quality.

Stag Carcass Composition - Averages			
	Red Deer		Elk / Red Crossbreed
	15 month	26 month	11 month
Liveweight (lbs)	229	286	255
Hot carcass weight (lbs)	128	163	151
Cold carcass weight (lbs)	124	159	147
Dressing %	56	57	59
Typical Carcass Yields	**lbs (%CW)**	**lbs (% CW)**	**lbs (%CW)**
Saddle	18.4 (14.5)	24 (15)	26 (18)
Hind legs	47.7 (38.5)	61.8 (39)	58.8 (40
Shoulder	25.3 (21)	30.1 (19)	28.4 (20)
Neck	18.7 (15)	25.5 (16)	20.5 (14)
Ribs	12.5 (10)	17.6 (11)	12.1 (8)
G.R. (tissue depth 11/12th ribs) -fatness indicator	6 mm	10 mm	4.7 mm

Table 3 Stag Carcass Composition

The next consideration in assessing value is the actual composition of those choice cuts, as shown in Table 4.

Lean, Fat & Bone Content of Primal Cuts (Tissue as % Primal Cuts)						
Primal Cuts	**Lean**		**Fat**		**Bone**	
	Red	Elk/Red	Red	Elk/Red	Red	Elk/Red
Saddle	69	74	7.3	3.9	24	22
Shoulder	74	76	6.2	4.6	16	16
Hind Leg	77	79	5.2	3.5	21	21
Neck	67	70	6.8	5.7	26	24
Ribs	68	75	12.3	5.5	20	19
Total Yield	**72.7**	**76**	**7.0**	**4.7**	**20**	**19**

Table 4 After Drew 1989

These data show a consistent and considerable advantage of crossbreds for venison production. The crossbreds show higher yields for loin and hindquarters. These choice cuts from the crossbreds also show a more desirable composition of lean meat vs. fat and bone compared to red stags at over twice the age. The red stags' tendency to lay down fat in the rib section indicates the potential for over-fatness in even relatively young stags. Elk and crossbreds are always leaner at the same age, or weight for age.

Production Trends in Farmed Deer Antler Growth				
	Breed Type			
	Red	F2-25% Elk	F1-50% Elk/Red	Elk
Velvet antler production (lbs, adjusted to 65 days, average)				
3 years old	4.2	5.5	6.8	12.1
6 years old	6.4	8.4	12.1	16.5
Casting Dates (expect variation +/- 3-9 days)				
3 years old	11 April	10 April	8 April	5 April
6 years old	28 March	20 March	16 March	14 March
Days until harvest (individual variation +/- 5-10 days)				
3 year olds	56	59	64	69
6 year olds	69	69	75	75
Liveweight (lbs, average)				
3 year olds	330	451	561	704
6 year olds	429	594	726	968
Relative antler yield (Red Deer at 1.00, unadjusted)				
3 year olds	1.00	1.3	1.6	2.9
6 year olds	1.00	1.3	1.9	2.6
Relative antler yield (Red Deer at 1.00, adjusted*)				
3 year olds	1.00	1.02	1.08	1.63
6 year olds	1.00	1.01	1.28	1.52

* Antler yield adjusted for relative liveweight and metabolic efficiency. Research has shown that, in cold climates, an increase in LW of 100% results in an increased food intake of only 68%, a fact which allows us to adjust for both LW and efficiency.

Table 5 Antler Growth

Now, let's take a look at velvet antler production - Table 5 clearly shows the increase in velvet antler production as we crossbreed from "pure" Red to "pure" Elk. The point must be made, however, that liveweight increases with the increase in antler production. This, of course, increases feed requirements and, ultimately, increases the cost of producing that antler. This is unquestionably true, but two factors work to make Elk antler a more profitable product:

Metabolic efficiency

Extensive research work has shown that, in cold climates, a larger body size is more efficient in terms of energy use. In fact, the estimate for Red Deer and Elk is that an increase in liveweight of 100% results in an increased food requirement of only 68%! This allows us to adjust Relative Antler Yield for both liveweight and metabolic efficiency. As shown in Table 5, Crossbreds and Elk maintain a substantial advantage over Reds.

Market value of antler
Buyers have consistently paid more for larger Elk antlers than for smaller Red antlers. Future market prospects, with steady competition from New Zealand Reds and possibly from Canadian, Alaskan, Siberian and Scandinavian Reindeer, will maintain the relative scarcity and desirability of larger antlers with exceptional quality.

These two factors give another strong boost to the idea of crossbreeding Elk over Reds.

F1 Crossbred Females

MANAGEMENT CONSIDERATIONS

Nothing is ever really free. So far, we've established a clear advantage in venison and velvet production to Crossbreds. Why shouldn't all Red females be exposed to Elk bulls?

The first answer is obvious. Crossbreeding requires a steady supply of replacement Red females, preferably top-quality females. Considering the life expectancy of a Red female (15 - 20 years), perhaps the practice of breeding all first-calvers to the best Red stags will supply enough replacements.

The next, and more complex answer is the management skills required to produce crossbreds. Red stags are very aggressive breeders, even as yearlings. With this attitude, and no natural constraints to breeding, calving rates of 95% and weaning rates over 90% are easily achievable with good husbandry of a Red over Red herd. On the other hand, elk bulls are less aggressive breeders. A yearling, or even a two year old, may be intimidated by a nearby adult bull or stag, to the point where he will lose interest in breeding. In addition, there is a substantial size difference. With Red Deer females, a shorter-legged Elk bull will generally be more successful.

This discussion of body size leads into the question of sire selection. Some Elk bulls just simply don't like Red Deer females. Although I've stated that Reds and Elk are the same creatures, the animals themselves might take exception to that. Mix a herd of Red, Crossbred and Elk females, and you'll find that they segregate into "type groups".

2 yr F1 beside a 4 yr Red

Whether the signals are visual, smell or behaviour based, they can definitely tell the difference. Elk bulls need to "learn" to breed Reds. Some catch on quickly, and do as good a job as any Red. Some do a lack-lustre type of job, and some turn up their noses and walk away, forever! The result may be a much lower conception rate. Many crossbreeders follow up the Elk bull with a clean-up Red stag, possibly resulting in an extended calving season, and lower weaning weights.

Dystocia, or calving difficulty, is also more common with crossbreeding. Generally, the problems result from too big a calf in too small a birth canal. These problems can be virtually eliminated by selecting for bulls that produce long, slim calves, and managing nutrition and condition of the Red females. The most successful feeding strategy is to wean early, "flush" the Reds with excellent nutrition before and during the breeding season, then let the bred females "coast" during the winter on a medium quality forage ration with appropriate mineral and vitamin supplementation. After calving, these mamas need more special attention. Raising a crossbred calf that will weigh 50 - 70% of her liveweight by 100 days requires a furious lactation pace. Good quality milk is expensive to produce, so these females need the very best pasture available, plus energy, mineral and vitamin supplementation. Their Crossbred calves also need access to this top-quality diet, as soon as they'll eat it and as much as they want, to reduce pressure on their mothers, and to allow them to grow to their full, vigorous, genetic potential.

Is crossbreeding for you? It's a management-intensive operation that can yield remarkable returns. There are many different options to consider in profitably marketing your crossbred calves. Elk venison and velvet antler are only two. Two more very exciting options involve upgrading females to pure elk, and use of crossbred females as recipients for pure elk embryos. There's no more productive option for a Red Deer farmer!

3 yr Crossbred Male

Chapter 1 ***Building an Elk Farm***

Farm Selection

Farmed elk are truly so adaptable that the first question to ask in the process of selecting an elk farm is "where would I like to live?"

Very few areas are totally unsuitable for elk farming, but many areas would be more economically feasible. In particular, the preferred option would be a location in an established agricultural area with access to productive soils, supplies and services.

Other desirable characteristics include selecting

- coarser soil types - tending to silt/sand textures, rather than clay. This allows better drainage and workability;
- soil pH 6 - 7, with a balance of soil nutrients. This can be built up but natural soil conditions such as these are ideal.
- gently sloping or rolling land also improves drainage, and gives a variety of options for location of buildings and facilities. Variety also gives the elk a choice of sites for feeding, resting or other activities. A sheltered south-facing slope provides an ideal wintering location.
- a predictable supply of good drinking water may be sourced from wells, streams or ponds.
- areas of trees provide shade, shelter and browse. These are particularly desirable during calving season, in summer heat, and in winter.
- existing buildings that may be converted to serve as winter shelter and handling facilities, or feed and equipment storage areas.
- quieter locations, away from school yards, industrial activities or other sources of abrupt noise and movement.

The last point to remember is "Start Small - Think Big." Plan for expansion in your farming operation, both in pasture area and in handling area.

Farm Layout

In planning the layout of an elk farm a useful analogy is to think of an elk herd as flowing water. Elk move as a unit flowing across pastures, through gateways, down slopes, around corners. They bottle up in square or dead-end corners and under such pressure can be forced through holes or weaknesses in or under the fence or handling areas.

To properly plan a layout, start with a map, and hike or ride through the farm, with "flow" in mind. Plan with the idea that your farm will be a "one-person" operation. The two keys to farm layout are raceways and handling facilities. Raceways join all the paddocks together, allowing elk to move easily from any one paddock to another, or into the handling yards. Handling yards should always be as central as possible. Raceway widths may range from 5 to 10 meters (15 to 36 feet), with larger widths accommodating larger herds of elk and larger farm equipment. A one person operation would favour narrower raceways. Pasture size should vary, with smaller areas near the handling yards and take account of local geography (trees, watercourses, access and security).

Overall, for efficient pasture and stock management, the more pasture subdivision the better. A typical small farm base layout is shown in Fig. 1.

However, where boundary fences must be built to very high standards and expense, it is often wise to consider fencing a larger area than the stock numbers may initially need and develop this area over time with less expensive yet effective subdivision fencing. These considerations can often be recognized in advance by taking time in planning layout using farm maps, aerial photographs and taking time for a farm walk-about in the planning stages.

The handling yards should be on well drained, gently sloping ground. You need access to the yards with a truck and trailer for loading and transport. From the yards, the central raceway extends out to the pastures. You may use one or several raceways, depending on your land. Often, during stock movement, or a variety of on-farm activities, the raceways themselves can be used for sorting or holding areas. A specially designed "keyhole" can be effective in providing rest areas for stock, the ability to sort or draft groups, or simply as vehicle turning areas or for feeding stations. Pastures feed into the raceways at keyholes (Fig. 2). Keyholes are essential for one-person operations.

Layout of pastures must consider

- availability of drinking water. Each pasture must have a water supply. Springs or ponds work well, except the elk tend to muddy them with wallowing and playing. This activity produces environmental and public relations problems in some circumstances. A pressurized system of pipes and water bowls is the most appropriate for spring and summer use but must be drained in October for overwintering.
- access to shade, shelter and escape from insects during the various seasons of the year. Trees, sheds and open windswept hilltops provide this protection.
- access for vehicles and machinery. Widths of gates and raceways should be a minimum 12' up to 16' to allow most equipment easy passage. Ground levelling and gravel cover may also be necessary for easy access and maintenance.
- prevailing winds, for both shelter and snow accumulation concerns.

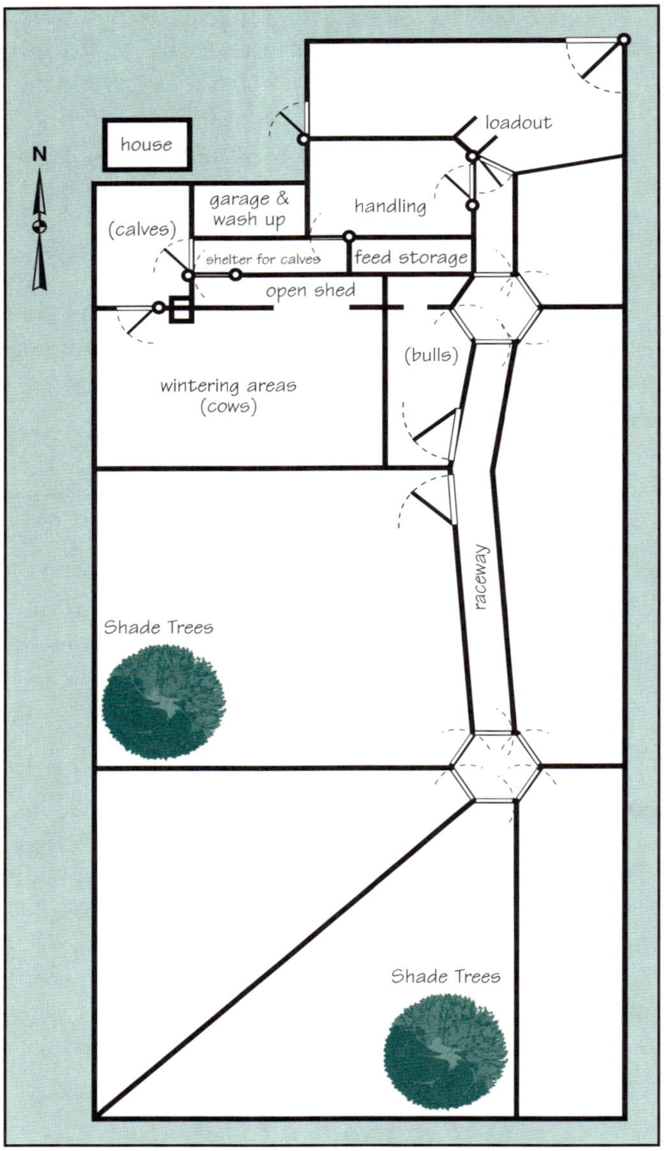

Fig. 1 Sample Farm Layout For a Small Breeding Unit

Overwintering areas should include shelter in the form of open sheds, roofed shelters or good tree cover to keep your elk as dry as possible. Generally, cold and snow don't bother elk, but wet conditions, particularly with strong wind and freezing temperatures, are hard on them and access to shelter is critical.

Hard ground and good drainage are important to allow good access for feeding equipment. Feed storage facilities should be located as near as possible to the overwintering areas. For a one-person operation, the ability to drop feed through or over a fence saves a lot of scrambling at the gate!

As mentioned in the farm selection section, keep further expansion in mind. "Start Small, Think Big".

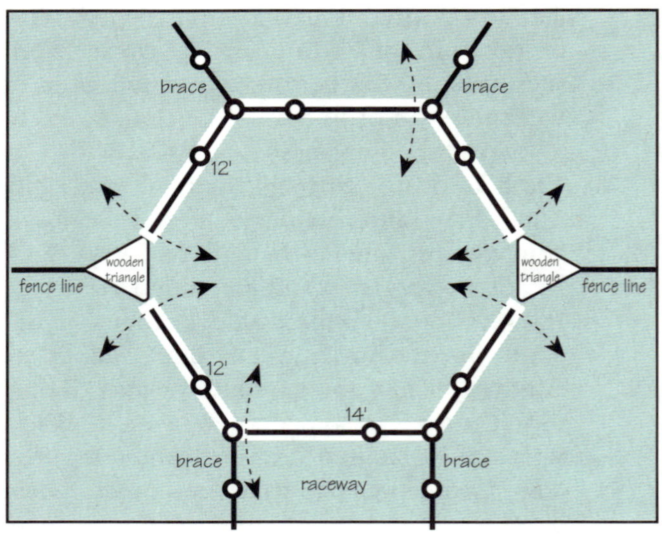

Fig. 2 Standard 6-Gate Keyhole

FENCING

INTRODUCTION

The investment in elk fence in relation to overall establishment of a farm is relatively high (15 - 25% of budget) and critical for stock security and farm management. The boundary fence exists as much to contain farm stock and prevent escape, as to keep predators, native deer species and the inquisitive out. In most provinces there are established regulations for post spacing, fence height, construction and materials and maintenance requirements. The standard for elk is 8' tall (20 - 96 - 6) pagewire.

Posts are recommended at 6 - 8 m spacings (20') but this distance may vary. Even on straight, level ground, spacing should not be more than 9 m (30').

Minimum standards have been determined in consultation with wire manufacturers, elk farmers and relevant environmental and wildlife agencies for the benefit of all. The detailed dimensions are as much for sound fence construction as for satisfying official constraints and there is no advantage in trying to short cut for cost sake. It has been proven often that trying to upgrade existing fence, or failing to plan prior to fencing is a costly and short term exercise.

The critical foundation of any elk farm operation is the fence and it pays to consult with existing farmers, specialist fencing contractors and the companies that supply product for assistance with farm layout, choice of material and construction techniques. Some specific regulations exist - contact local agriculture staff for direction.

MATERIALS

Modern elk farm fencing has been revolutionized by the development of the hi-tensile woven wire netting fence. The strength and integrity of the netting is maintained by the high tensile nature of the wire giving high breaking strain loading and a construction that allows contraction and expansion under the extremes of climate without reducing wire properties or fence life.

Netting is given function and strength with the development of a special locking tie that joins horizontal wire and vertical stays. This gives rigidity to the construction at a single point, but allows the fence elasticity to repel animal pressure, and the whole netting area acts as a unit. Livestock can bounce off it, but generally cannot distort netting spacing, not crawl through between wires, or in a well constructed and properly maintained fence, crawl under it.

The dimensions of wire spacings have been carefully planned and tested so that they are closer at the base and mid fence where the main pressures are applied.

This construction repels many predators and will withstand shock from stock pressure either in groups in laneways or single animals charging the fence for any reason. The spacings with a 150 mm vertical stay will stop calves from slipping through fence to find outside cover or shelter, and the product is ideal for game farm fencing in every situation possible. Legislators in some US States have adopted special concession in their typical rules for such fence. These recognize the superior properties of the high tensile netting which allows cost savings in posts used, through increased allowable distance between them, and gauge and fastening of other wires. These recommendations have all been tested over some years and farmers are advised that additional shortcutting or trying to save on construction and costs is a false economy in the long term.

The final feature is that the netting is manufactured to high specifications of production including heavy zinc galvanizing to ensure long life and durability. Hi-tensile netting fence has a record of remaining rust and corrosion free for many years even in quite hostile coastal areas.

Sound maintenance-free fencing can be relatively easily constructed with attention to a few basic principles, remembering that the integrity of any fence is only as great as the weakest point.

A shortcut in cost savings or erection technique can be enormously costly through accident or escape of animals, or a simple deterioration in fence quality. Conditions in much of North America are extreme with heat, humidity, snow drift, ice storms and frost heave, and conditions of spring thaw all creating their own pressure on post rigidity, alignment and wire elasticity and strength.

Today's materials are well proven in all these situations but the following are key elements in design and function.

FEATURES

The structures that make corners, ends of fence lines, gateways, or angles in fence lines are the most critical aspects of fence construction. These constructions carry all the weight and tension in a fence. Because of the height of netting and the combination of strain in high tensile horizontal line wire forces, the twisting or torque of wire tied around the anchor post, and the pressure of change in direction in fence lines, extreme forces can be applied.

Anchor posts

or strainers and their associated brace must be built from material that is strong enough to avoid breakage from wire pull and have strength for the counteracting force of the brace. They must be deep in the ground - 40% of the length to provide a solid anchor. For a 2.4 m fence (8') a 3.3 (14') 200 mm diameter (8" minimum) wooden post is needed, or a large diameter (150 - 200 mm) structural tubing steel pipe. For a 2 m fence, anchor posts of 3.8 m (12') length are sufficient but similar minimum diameters are needed.

Braces

The choice of brace, and brace construction variations are illustrated in figures prepared by Bill Daugherty, with Langley Wire Industries and Ray Nixdorf, of Saskatchewan Agriculture. (pages 41 - 43).

The brace assembly is an integral part of the anchor post construction. Size and strength must be in proportion to the size of anchor post or strainer selected. The positioning of the brace on the strainer is critical to allow forces and tensions to be distributed uniformly.

The recommendation is to use a horizontal brace assembly, or a diagonal brace that acts against a well placed stay block. The stay block or the secondary strainer post is the most critical part in a brace assembly of either construction. Pages 39 to 43 of this handbook illustrate brace construction techniques.

A horizontal brace can be positioned at the top of the post and tensioned diagonally by a wire twitch to the base of the anchor post, but for neatness and ease of construction it is often set down about 20% of the fence height.

The brace should be at least twice the length of the fence height, remembering the more line wires in the netting, the greater the tension on the whole assembly. If the brace is too low on an anchor post it can exert a strong leverage effect particularly on corners or acute angles and in time lever the anchor post upwards. Braces must also neatly fit mortise joints prepared for them in wooden posts to counteract the twisting effect of tensioned wires. Many contractors will also prepare a footing for the anchor posts to lock the base post in the ground.

Note that braces must be placed exactly in line with the direction of line wires or netting to avoid any rotation or twisting. Even a slight angle will weaken the whole assembly.

SITE SELECTION, CONSTRUCTION KEYPOINTS

Site and contour Anchor posts, braces and line posts should stand vertically to the base ground contour. Strainer shape should also be straight. If wooden posts have a slight curve, they should be vertical above ground and any curve in the post positioned towards the stay to avoid twisting when wire tension is applied - i.e. construction in contour.

The fence must be tight and follow the ground surface. This requires anchor posts to be virtually immovable to hold the netting or line wires at the correct tension. Even a slight twist or upwards movement in a strainer will cause a major reaction in fence tension and therefore function. If a strainer assembly moves, then attempting to retighten wires or netting will only increase the problem at the strainer.

Posts must be placed at exact changes in contour to ensure the wire stays parallel to the ground. In very uneven terrain it is often worthwhile to level the fence site in advance to remove minor humps and hollows. Care must be taken, however, not to create a water run off that could create potential erosion or soft areas filling hollows that in time can settle leaving gaps under fences.

Fence lines should be as long and straight as possible to reduce costs of additional angle braces and allow more effective tensioning of netting and wire. Netting rolls are joined or spliced at a convenient central point in a fence line to allow an even pull from both end points and consistent tension along the whole fence. Netting is supplied in 100 m rolls (330') or 200 m (660'), so fence layout can be mapped accordingly.

Netting attachment

Once the fence is correctly tensioned, the netting is attached to posts and pulled down into hollows or lifted over high ground always parallel to the contour. With netting it is not necessary to staple every horizontal wire except for top and bottom. In stapling, galvanized angle cut barbed staples should be used with staples placed on an angle to reduce the staple width to just above the gauge of the wire.

Notes

- Staples should never be driven hard into posts and onto the wire as this can damage the galvanizing and create weakness in time. Neither should they be left too far from the post as this weakens the lateral strength of the netting.
- Staples should be placed on the side away from the horizontal pull when the tightlock and vertical stay wires of netting meet. This allows the expansion and contraction of line wires and the constant tension of fence throughout its length. This is particularly important when first tensioning the wire. It is not useful to try maintain tension by stapling at the tightlock junction as inevitable wear and damage results.
- Staples in the end anchor post should never be used to hold the wire in position by hammering them into the post. Any hammer bruise on wire or impact of staple on wire at the post will weaken it in time through damage of galvanized zinc coating.
- Where line wire slides around angle braces a staple should be placed flat against the post to allow the wire to slide over it without biting into the wood. This staple can be fixed in position with another staple leaving sufficient room for movement during tensioning.
- In sharp dips or rises, wire should be double stapled as far apart as practical to lessen the angle and position tension wear on the wire.
- In thinner line posts, staples should be staggered in position down the post to avoid the chance of the post splitting down the line of staples.
- Hi-tensile wire is not difficult to work with but does require some experience in efficient knot tying and handling, particularly around strainer posts. There are a number of tools (pliers, twitches) available to help handling. It is better to strip off the tightlock fence knot at the tying end and have slightly more length to work with, rather than tying off wires around the knot and avoid uneven tensions and pressures in the netting, or on the strainer post.
- Fence ends can be tied with standard knots, or by an inexpensive wire crimp. Knot techniques should employ some variation of a locking knot and the ends cut close to the line wire and not left projecting into deer pasture. Poorly trimmed fence ends can cause severe eye, pelt or carcass damage to deer that often run up and down fence lines if agitated. Good neat construction is essential in laneways where the animals often group and put most pressure on netting areas.

All cut ends, loose wire, cast off netting ends or old fence must be removed as the last job in fence construction. There are too many reports of livestock suffering mechanical damage to feet and limbs from material left lying around.

Loops of high tensile wire have caused death by strangulation or entanglement and small wire pieces have been ingested causing fatal injury. Farmed elk are intensively inquisitive and if there is a potential trouble point they will invariably find it.

GATES AND GATE POSTS

Gates are an important investment and given the amount of use and time involved in opening and closing they should be properly hinged with a secure locking chain and hook and be an appropriate width and size for the task. Gates tied up with wire or string and not hinged are frustrating, timewasters and can be a disaster in the making if they are needed in a hurry in laneways or yard situations. As in anchor post construction, the gate and its post must stand in line with contour, or at 90% to the ground surface. Hinges should be as far apart as possible and positioned so that the gate can open back hard against a fence line.

In the snow and ice conditions of North America it may be more prudent to raise the gate above the 100 mm from the ground usually recommended, to 150 - 200 mm. During spring, summer and fall an attached heavy gauge steel mesh from the base of the gate to ground level should be fitted. This can be lifted and tied to the gate as conditions of snow dictate in winter. Elk will always investigate gate openings and calves can crawl under 200 - 250 mm gaps.

The opening end should also be supported when closed by a wedge or block at the base of the locking post to support the weight and prevent hinge damage and sagging.

Any chain or locking device should pass around the post and gate and be stapled in position for easy use as the gate is closed and be as tight as practical function allows. There are a number of spring loaded locking clamps available and these should pull the gate against the locking posts with some tension, but not at the expense of being difficult to use.

It is recommended that the services of an experienced fencing contractor, familiar with the high tensile product be considered. There is a wealth of knowledge available to be learned from these specialists, especially considering the importance of the perimeter fence and the value of the stock investment and costs in fencing. Invariably short cuts in material selection or construction techniques create long term loss and significantly affect good farm management.

FENCE SITING AND LAYOUT

To preserve the integrity of perimeter fence two major considerations apply - i.e. risk of fence washout and water damage, and risk of fence collapse from falling trees, or more rarely collision from vehicles leaving roadways etc.

Elk, flowing water, or standing swampy conditions and elk fence are bad combinations.

In swampy conditions, or through soft ground depressions and conditions, deep trenching along the fenceline can rapidly form. Conditions are very much worse during winter feeding, thaw conditions, and during mating season.

Males in particular can also create wallows in such areas all adding to the risk of undermining the fence and eroding solid soil around posts. The conditions can get so bad that parts of the fence may become buried and eventually the bottom wires and stay knots can corrode and break, adding further to loss of fence tension.

Ideally swampy conditions should be avoided altogether, or if a boundary fence must pass across them, isolate the area with less expensive internal fence, plant suitable trees or leave it unused.

At worst, heavy tracking areas in soft ground can be back filled with rock making it uncomfortable for walking, but this is a drastic cure for a problem best identified at the time of construction. See "frost heave" next page.

Crossing permanent flowing water, or areas prone to flash flooding is risky from a security viewpoint. Even though netting fence is extremely strong it will capture any debris carried by water and create tremendous strain on anchor posts. If there is no choice, there are two common approaches.

Floodgate control

For small creeks a floodgate that can swing in the direction of flow as debris and pressure build up is workable. Gates can be built of 15 mm pipe laced with 150 mm square steel reinforcing mesh cut to shape and suspended by a steel cable running at the base of the fence.

The fence itself should be supported by a typical inline strainer assembly or brace at each side of the bank. The fence can end at each brace and the creek crossing netted independently or carry over the span at the normal height.

If the span is large (greater than 12 - 15 mm), it too can be supported from a cable running from the base of the anchor brace and across the top of the two anchor posts. The fence is then clipped to both steel cables and the flood gate swung underneath.

In more serious situations a small steel bridge or large diameter pipe can be built to support a large flood gate, or even allow vehicle access.

Obviously the flood gate is a point of weakness and should be kept free of debris.

Ideally a small concrete weir or support of the sides will prevent erosion during flood turbulence and is the first place to check after bad weather.

Sacrifice fence

In difficult areas a sacrifice fence, plain wire ideally, or low netting plus a top up plain wire can extend away from the perimeter fence, crossing the water at a low flat point and then angling back to the perimeter. The perimeter itself can be made secure with netting and tie downs. In times of flood the sacrifice fence will collect debris up to a point and any chance of flood compromising the main fence is reduced.

The wire and netting should be stapled with smaller non-barbed staples onto posts below the direction of flow so that in extreme cases the fence will pull out of the posts with added weight and can be easily cleared and resurrected afterwards.

If netting in the perimeter fence across the water course is attached lightly with low tensile lacing wire it too can let go under pressure and avoid extensive damage.

However, in either case avoidance of problems with water is a more reliable option.

Some success with sacrifice fence has been had by making them all electric and using avoidance of the area to reinforce security, but these can be high maintenance and should only be considered if other electric fence uses are considered.

SPECIFIC FENCE PROBLEMS

Frost heave

Can be minimized by avoiding wet areas where water stands in spring and autumn. The most successful technique for fencing through such areas is to construct a berm of soil and run fencing on top of that. The berm will drain and remain drier than surrounding wet areas. Deep driving posts may also limit heaving, but many areas are wet due to the shallow depth of soil above bedrock, which becomes the limiting factor.

Snow accumulation on fences must be considered during construction. Deep drifts will both damage the fence by weighing it down, and will allow stock to escape over top of it. In some winters, regular snowblowing is the only solution.

Tree and mechanical damage

Hi-tensile netting properly supported is enormously strong but not totally invincible, but even large trees falling on the fence during storms or by accident will not destroy the fence. Because the vertical stays are of milder steel than the horizontal wires, the pressure of a tree falling tends to tear out staples and buckle the netting to ground level under the weight. Once the tree is removed the netting can be stood up, vertical stays straightened and staples or any damaged posts replaced. There are few reports of elk escaping in such situations, provided fence inspection and maintenance after storm is followed.

Elk perceive this as a major disturbance in their own security and will avoid such areas, at least for awhile. However, if fencing is through woods it is prudent to clear a pathway in advance (8 - 10 m wide) and remove any dead or suspect trees in advance.

Such clearance can allow a 3 - 4 m corridor around the perimeter for quick and easy inspection of the perimeter. In fact, many layouts in areas where land area is not limited develop a drive around perimeter road that aids in stock inspection without disturbance, and helps in security checks. These perimeter access ways are useful for public farm tours, or for inspecting velvetting or calving stock without disturbance or the need to continually open and close gates.

In wooded areas these can be successfully seeded with shade tolerant grasses and clover to provide a small but effective barrier between farmed elk, predators or native deer, as well as making fence construction and maintenance so much easier.

Rock shelving

In difficult conditions of rock hard ground or stony territory, drilling or digging post holes for wooden treated posts can be expensive and time consuming and not necessarily effective.

Provided the anchor braces are solid, large lengths of netting can be supported by steel T-rails or 40 - 50 mm steel pipe, rock drilled into position. (If regulations allow!!) Alternatively tensioned steel cable pulled tight over a double anchor brace assembly can support relatively large netting spans (45 - 60 m in length) with the base wire and underlying ground wire pulled tight towards the ground using a screw or bolt in rock anchors at the normal post spacing of 4 - 6 m. In this situation additional height should be given to the anchor braces so that there is both upwards and downwards tension on the netting and a convenient system of tension to the supporting cable applied to keep it as near horizontal as possible. The fence can also be made more visual and given some additional strength by nailing 50 x 100 mm wooden battens to the fence with the base supported by a simple triangular foot lying flat on the rock surface.

In almost any fence situation ingenuity and adaptation can work well provided the anchors and braces are sound and the netting is tensioned to the manufacturers specifications.

Of course in such areas regular inspection and maintenance are more critical, but difficult terrain need not be an insurmountable problem.

SUMMARY

Common sense, and consultation with existing farmers or experienced contractors will find solutions to most fencing problems.

In general, however, AVOID

- Running water or flood prone gullies and depressions.
- Swamp conditions.
- Erosion risks.
- Clearing topsoil and not reseeding.
- Rapidly changing contour - sharp hump and hollows.
- Traversing across steep terrain where fence height can be compromised upslope.
- Obvious risks of snow drift.
- Obvious risks of tree fall damage.

Note high tensile wire zinc galvanizing can be critically damaged by fire. Perimeter roads can reduce fire risk.

- Short cuts are long term expensive.

And USE

- Prepared and cleared fence lines prior to construction.
- Square or rectangular layout not long narrow fence layout plans as these are up to 30% more expensive per unit area fenced.
- Fence for as large an area as feasible from the start. Continual additions are expensive.
- Fence the most fertile acreage.
- Posts on the outside, netting on the inside on perimeters for maximum strength.
- Quality construction materials and techniques, paying special attention to anchor posts and brace assemblies.
- The skill and knowledge of others.
- Neat and tidy fence finishing.
- Regular routine checking and maintenance.

INTERNAL FENCING, HEIGHTS, MATERIALS

Cheaper netting fencing or other plain wire or combination options are available for internal fencing and are perfectly adequate for stock control and usual farm management needs.

As seen in the catalogues many options are available. Elk can be easily controlled behind 6' (1600 mm) netting, with or without a top up plain wire. Both 6" and 12" stay spacings can be used.

The exception to this might be in weaning pastures which must be calf proof, and in overwintering pens which should be full height and perimeter fence standard because of their heavy use and snow and ice buildup in winter. A 1600 mm high internal fence may also act as a laneway fence. This works well only if the laneway is wide and never under pressure. Narrow laneways should be fenced to full height. It is unwise to try and extend the distance between post spacings too far in the internal fence as fence strength still requires vertical support at 9 m maximum.

Smooth wire options, and electric fence supported on post and fibreglass T-rails and rods have all been used successfully with deer but may not always be adequate for separating calving groups and bulls during the rut. They certainly are not adequate to separate weaned calves and cows, but have their greatest use in cost effective subdivision of pasture for grazing management.

This system can be advanced with time and some patience to establish a temporary electric fence "break feeding" system using plastic high visibility tape made hot with fine stainless steel wire. Tapes are supported on insulated standards and provided the current exceeds 4000V at all times, and the elk are never critically hungry, strong positive control is achieved.

As elk settle into a routine and become established on a new property, rotational grazing using some of these techniques can be usefully employed as group size increases. This is particularly useful for young growing stock.

In all these variations the same basic principles apply - sound well constructed strainers, anchor posts and brace, but appropriately scaled to fence size (i.e. 40% below ground, 60% above) and the fence should always be parallel to the ground, and posts in line with contour. On pasture small depressions and hollows are often better infilled with rock or logs, and in cultivation, care taken not to bury the bottom wires by cultivating too close to the fence.

Many also are quite sadly familiar with the mess machinery can make when accidentally tangled in hi-tensile netting. Implements and trailers can become firmly wedged with huge damage to the netting vertical stays. They don't often break, but will be pulled sideways and downwards often damaging galvanizing and are difficult to repair. It is easier to cut the section out, restrain and splice in a new section, than to patch up damaged areas.

TREE PROTECTION

While it appears that trees, woodlots, or bush areas are not absolutely required within the farm for an adequate elk farm operation, there are many who prefer to include them for reasons of aesthetic appeal or some aspects of welfare.

Often the very trees a farmer may wish to keep are critically damaged before any protection is put in place. Loss occurs through extensive trunk damage from bulls rubbing or cleaning antlers or cut antler buttons, and many species of trees are quite unsuited for heavy livestock presence.

Woodlots have an advantage in creating quiet shade and areas for cow and calf groups, or for bulls after velvetting, but often there is a balance between use of trees and annoyance by insects. The shade of single or groups of trees provides relief from direct sun and still allows the cooling effect of any breeze. Such trees should be protected either by the rather unsightly strapping of slab wood (sawn timber ends), or similar protection directly to the trunk, or by building a relatively expensive post and mesh or post and netting fence around the tree.

The latter is always a high maintenance problem, hard to keep any tension on the fence and subject to a lot of abuse. Trees may remain untouched for some time and can be permanently damaged in just a few days, so the potential should be anticipated and developed as part of the fence strategy.

Corrugated iron has been used but there are several recorded instances of lightening strike passing down the tree and arcing from protection to a nearby fence killing prized breeding bulls sheltering under the tree.

Several tree species (pine, poplar) can be planted outside a perimeter fence in a hedge type density. Elk will keep these trimmed through the fence and this can provide shelter and shield elk from road traffic or other busy farm activities. Species from the cypress family or any toxic plant should be avoided, and there is always the faint risk of fire danger.

In large woodlot areas the susceptible trees and heavy undergrowth are soon eaten out and softwoods, pines, will disappear quickly. However, such areas can be grazed or browsed on a short intensive pattern which allows mature trees that are resistant to damage to grow clear above the browse line and establish, and succulent undergrowth can regenerate for grazing. This may have particular health advantages as these deep rooted plants bring up subsoil micro-nutrients and trace elements that can be deficient in pasture.

Instead of totally grazing these areas which inevitably leads to a slow decline in the tree stand, some managers will fence off smaller areas within a woodlot and allow protection of both full growth and undergrowth areas giving shade and shelter. Such areas can be magnets for calves and their being trapped or hung up in the fence should be anticipated.

In many yard and raceway developments reductions in lane width or sharp corners in yards or lanes may be better fenced out with some sort of wooden barrier. It is ideal to develop these as triangular tree protection areas enclosing existing trees or planting shade trees. However, there are no common rules or "must do", as this decision really comes down to personal choice.

RACEWAYS, PRESSURE AREAS, FEEDLOTS AND WINTERING AREAS

Fence construction in these areas is logically stronger, and more subject to use, abuse and regular maintenance.

Post spacings should be reduced to 4 m in final yarding areas and laneways leading to facilities. As this is an area of heavy traffic the choice should be to have good drainage and use areas that readily dry out during thaw or rain. Fence strength can be enhanced by visual means, using sighter boards and rub rails to both prevent injury and keep stock off the wire.

The most frequent problem is that if fence only covers short distances it can be difficult to keep proper tension under the pressure of heavy usage and short strains. It may pay to consider some type of inline permanent tensioner so that tension is properly maintained. Experience also shows an advantage in keeping the netting a little higher off the ground and using one or two heavy gauge bottom line wires to keep the netting clear of mud created by stock and vehicle traffic. This also moves the impact zone of netting higher so deer are less likely to damage feet in netting when crowding into the final raceway area. About 150 mm is sufficient, and also allows for raceways to be regravelled from time to time without burying netting.

Simple Shelter

Gateways in this area also get heavy use and it is better to hinge gates higher and cover the base gap with a hinged strong mesh or equivalent, and form good foundations with gravel through the entrance into wintering pads.

Again time spent in planning and laying out races, pens, etc. is very well spent and additional time spent in construction will extend the life of these areas considerably. In developing a farm, newcomers should have confidence in their own adaptations and ideas, for if they are simple in concept and fit a routine of handling and systematic approach, elk will learn and adapt very quickly. Basic principles, curves and corners are developed in the next chapter, and there is no substitute for spending the time on other farms and sharing ideas and experience.

GATEWAY PLACEMENT

With a central laneway farm layout, and pastures of roughly equal size and shape being theoretically ideal, there is opportunity to develop gateway systems that complement each other in placement and use.

There has always been strong debate over whether gates should be at the sides/ends of pastures or situated in the middle. A common compromise that works well is to locate gates between pastures in the middle of fence lines, on any rise in the terrain, and gates leading from pastures to laneways at the corners.

If the farm is small it is easier and good stock sense to use the laneway to shift stock and does not require additional expense for little gain.

Gates on the perimeter should be kept to a minimum for security reasons but may be needed in hilly terrain to allow entry from the lane and exit at the bottom of slope for hay cartage or in wet conditions. Consider double gates at all perimeter openings.

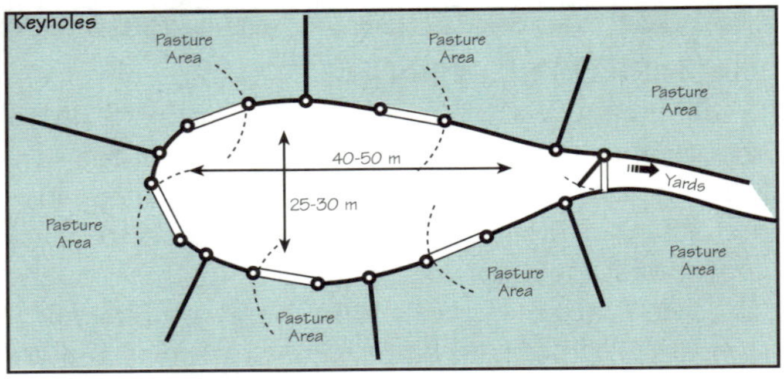

The Keyhole concept is the best choice for moving stock from pasture to pasture. Long straight laneways with gates set in line with the fence are not as effective at allowing stock to exit the pasture as an angled approach. While it might be easy to swing a gate across a narrow raceway and block it for re-entering, it is often difficult to push elk up to this barrier until they see the opening and turn. While more expensive, a wider raceway and a purpose built gate and barrier across the lane in two or three locations is more effective.

Under this construction, if shifting stock from one pasture to another half the animals can break back along the fence as the other half move around a sharp turn and they are then very difficult to regroup and move. Diagonal crossing or straight ahead crossing is generally trouble free, and it is advisable to exit into the lane and move the elk down the lane to a gateway in the other pasture corner. (Plan A)

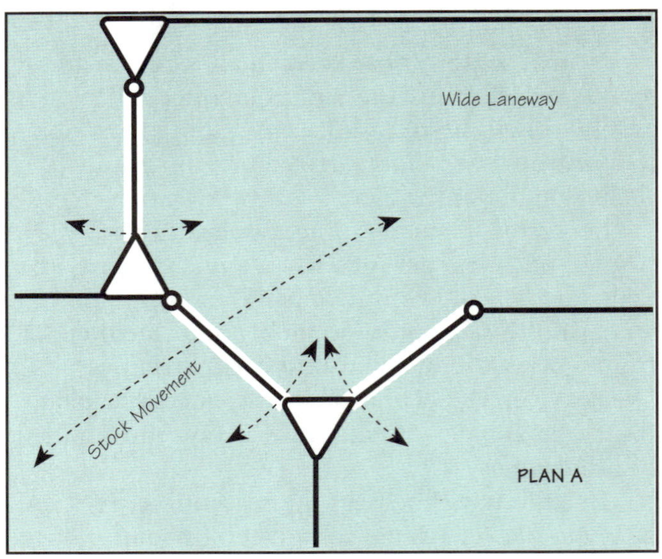

Intermediate between this and a full keyhole layout is the diagonal gate layout (Plan B). The advantage of this is that:

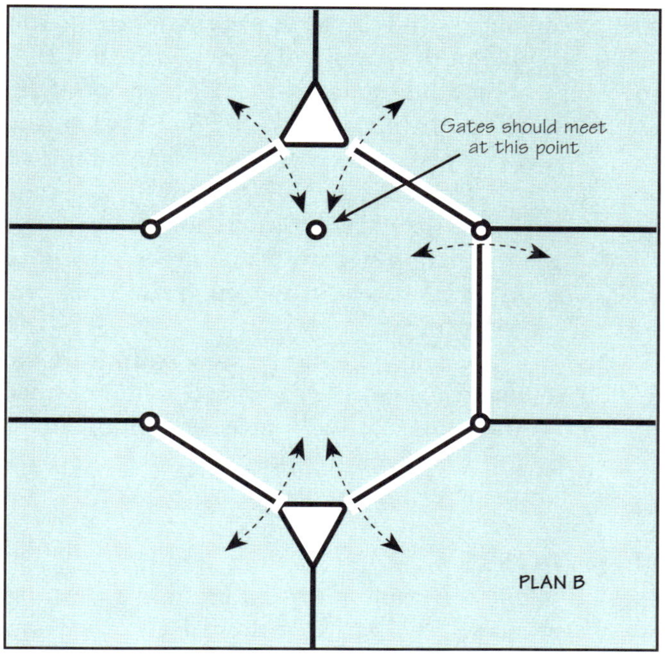

i) Elk travelling up the laneway will always break toward some relief from long straight laneways and naturally curve towards the gateway area even if there is a closed gate ahead of them blocking further movement down the farm length.

ii) From the pasture the open gate is obvious and elk will follow either fence line and exit without fuss, although if the gate swings into the pasture it must close hard against the fence line. Elk will turn at that point from one pasture to the next and the gates can be each swung through 90° and lock. This opening can be left to allow stock to drift through if moving cows and calves or using feed enticement. However, it is still preferable to encourage elk to use the lane and re-enter them at the next entrance.

This gate placement also allows easier vehicle access and is a more effective placement in narrow lanes. In wide lanes a standard size gate can be used and extra lane width covered by building a wooden barrier from the fence to accommodate the extra width. This may be an ideal place for shade tree planting, or a temporary feed storage area. This is the next step up to the full keyhole construction in its variety of forms. As a final but important detail, all gates should be hung and swing freely and be properly and securely fastened with chain and locking clip and supported at the non-hinge end. The investment in gates and a suitable design will save time, frustration and potential animal injury and is part of a sound investment in an elk fence.

The detailed diagrams of fence anchors and brace systems, and netting installation techniques in the following pages are supplied by Bill Daugherty, representing Langley Wire Industries and Ray Nixdorf, of Saskatchewan Agriculture, as a gesture of support for this handbook and sound fencing practices.

DETAILED CONSTRUCTION TECHNIQUES FOR ELK FENCING

Thanks to Ray Nixdorf of Saskatchewan Agriculture and Food for the text and drawings.

The intent of this part is to provide installation tips, techniques and recommendations to assist in the construction of high tensile wire fences for elk farming. Included are guidelines for material required, construction techniques, a detailed, step-by-step outline for building a page wire fence. and selected tips and techniques for working with high tensile wire.

Fence construction techniques for high tensile wire vary greatly from the low tensile barbed wire systems many farmers are familiar with. Many producers have little installation experience or training to construct fences that fully utilize the properties of the high tensile wire.

High tensile wire fences are designed to be elastic. They should act as a long spring that is held up by the end braces. Any force applied to the fence is distributed along the entire fenceline. The high tensile wire can stretch and return to its original length if its yield point (about 80 to 90 per cent of its breaking strength) has not been reached. After reaching this point, it will remain stretched.

High tensile fences that are not constructed properly may have a shortened life expectancy and necessitate costly, time consuming repairs and replacement. The possibility of animal injury and escape increases with inadequate or failed fencing. Although the best fence is the one with the lowest installation cost that will do the required job for the longest time, fence construction is not the place to cut corners in order to save money. Fence height requirements vary from jurisdiction to jurisdiction, and you must make sure that your perimeter fence and gates comply with the regulations in your jurisdiction. The standard for elk farm perimeter is 20 - 96 - 6 High Tensile page wire. That is eight feet tall, with 20 horizontal wires increasing in spacing from bottom to top, and consistent six inch vertical spacing. Gates should be the same height and provide similar containment ability, with less than a six inch space above ground.

Because the perimeter fence is the final barrier to livestock escape, its construction is typically taller than that of interior fences. In some operations, the perimeter fence is page wire netting that is not only taller but also has closer wire spacing than that of the interior cross fences.

MATERIALS SELECTION

Choice of construction material is entirely dependent upon the producer. Important factors to consider when selecting construction materials are longevity, strength, cost, and the producer's ability and skill. Economy and availability of the fencing material are often the primary influences on materials selection. Some resourceful producers have successfully utilized salvage materials (e.g. used wire); however, the strength and longevity of the material cannot be accurately evaluated.

FENCE WIRE

There is a wide variety of page wire available differing in tensile strength, thickness (gauge), corrosion protection (galvanizing) and wire spacings. The most common high tensile wire used for elk fencing is 12-1/2 gauge page wire. Fig. 1 shows a sample of the variety of page wire available.

It should be noted that all types of high tensile wire have a high recoil tendency. When the wire is cut or broken it will recoil in an attempt to re-spool. This characteristic will result in a tangle of wire that may act as a multi-loop tourniquet for any animal body part. If an entangled animal is not freed immediately, it is at high risk for a crippling injury or death.

Page wire fencing is made from high tensile steel and is available in several different heights and wire spacings. It is the most effective fencing method for limiting interaction between immediately adjacent pasture groups.

POSTS

When selecting posts, elk farmers have a variety of choices. Posts can be either metal or wood, with a variety of diameter sizes and lengths available. Both wood and metal posts have worked well on elk farms and it remains a preference choice for the producer. The most important factors in post selection are the intended application, the cost and the availability.

Metal posts are stronger and have a longer life span than wooden posts; however, the fence wire is subject to more corrosion. Metal posts are easier to set in rocky ground. Wooden posts require less specialized equipment and skills to work with than metal posts and should be pressure treated for longevity. Sharpened or pointed wood posts are easier to pound than blunt ended posts. Pointed posts are more likely to wander away from

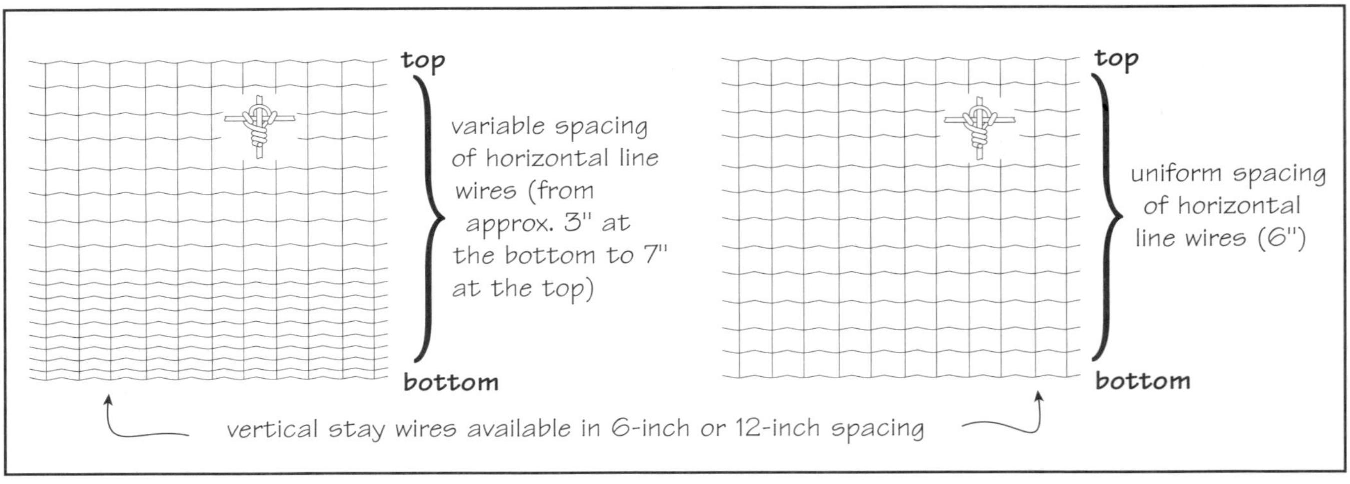

Fig. 1 Examples of high-tensile fencing available for deer farming

vertical during pounding after striking buried rocks. Wood flexes slightly and is more forgiving on animals when used in the construction of handling facilities.

CONSTRUCTION TECHNIQUES

BRACE ASSEMBLIES

The brace assemblies are the foundation of a good fence. As the anchor points of a fence, brace assemblies must withstand many pressures such as the constant static load of the tensioned wires and the dynamic load of animal and environmental pressures. Failure of the brace typically results in the failure of a whole section of fence. The greatest amount of time in fence building must therefore be devoted to proper brace construction.

There are many types and variations of brace assemblies that differ in terms of ease of construction, labour costs and strength of the finished product. The horizontal-rail brace is the most commonly used because it is easy to build and is both effective and strong (Fig. 2). End braces that follow the same construction principles as hydroelectric transmission posts are also effective. These end braces are often called tieback or anchored braces (Fig. 3).

HORIZONTAL-RAIL BRACE CONSTRUCTION

The box brace can be used as a single span brace for most applications. The single span brace consists of two posts (the end and brace posts) connected by a brace-rail and tensioned by a brace tension-wire (often called a twitch-wire). The fence wires are tied off and tensioned at the end-post. The addition of a second span increases the brace strength needed for longer fence pulls or softer ground.

All posts used in the construction of brace assemblies should be a minimum of one diameter size larger than the line posts. Ideally, the end-post should also be larger than the brace-post. It is important to remember that the load bearing capacity of a post increases with the depth that it is set into the ground. All brace assembly posts ideally should be set with an equal amount of post below

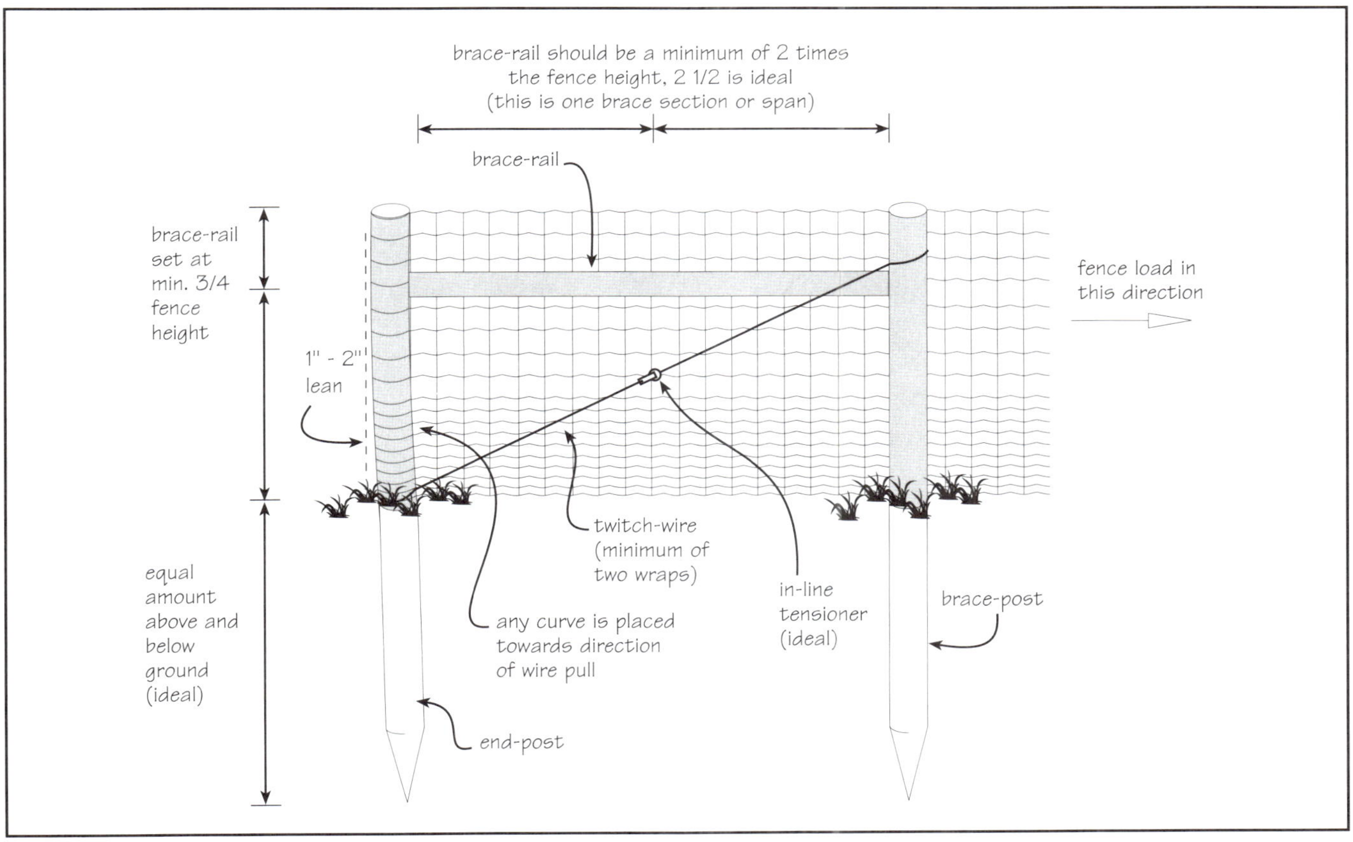

Fig. 2 Horizontal-rail brace construction

the surface as above. This is not always practical or possible, therefore a minimum of half of fence height is recommended (e.g. a six-foot fence would need a nine-foot post driven into the ground three feet).

The end-post should be set with a slight lean (one to two inches) away from the direction of fence wire pull. Any curve in the end-post should be set towards the direction of wire pull. This helps prevent brace failure caused by any twisting of the end post during wire tensioning. The brace post should also be set with a slight lean. After the fence is fully tensioned the brace assembly posts will have very little or no lean remaining.

A general rule is that the holding ability of the brace assembly increases with the length of the brace-rail. The brace-rail length should ideally be two and a half times the height of the fence; however, this rule is often impractical due to materials availability and cost. Therefore, brace-rails that are twice the fence height are often used and are sufficient. The brace-rail should be positioned at least three-quarters of the fence height to function properly. The twitch-wire should be placed about four inches above the ground on the end-post and four inches above the brace-rail on the brace-post to properly strengthen the finished assembly. The twitch-wire must be attached securely to prevent complete brace assembly failure. Ideally an in-line wire tensioner should be installed in the twitch-wire to allow future adjustment of the brace assembly.

TIEBACK BRACE CONSTRUCTION

The end-post of a tieback or anchored brace should be one or two diameter size larger than the line-posts. The end-post should be set with an equal amount below the surface as above. This becomes more important for this bracing technique than for construction of a box brace. The end-post should be set with a slight lean (one to two inches) away from the direction of fence wire pull. Any curve in the end-post should be set towards the direction of wire pull.

Any buried anchor will be suitable as long as it is immobile. The easiest method to create an anchor is to drive a second post so that between two thirds and three quarters is buried. This anchor post should be the same dimension as the end-post. It must be placed directly in-line with the end-post at a distance equal to the height of the finished fence (e.g. if the fence is eight feet high, the

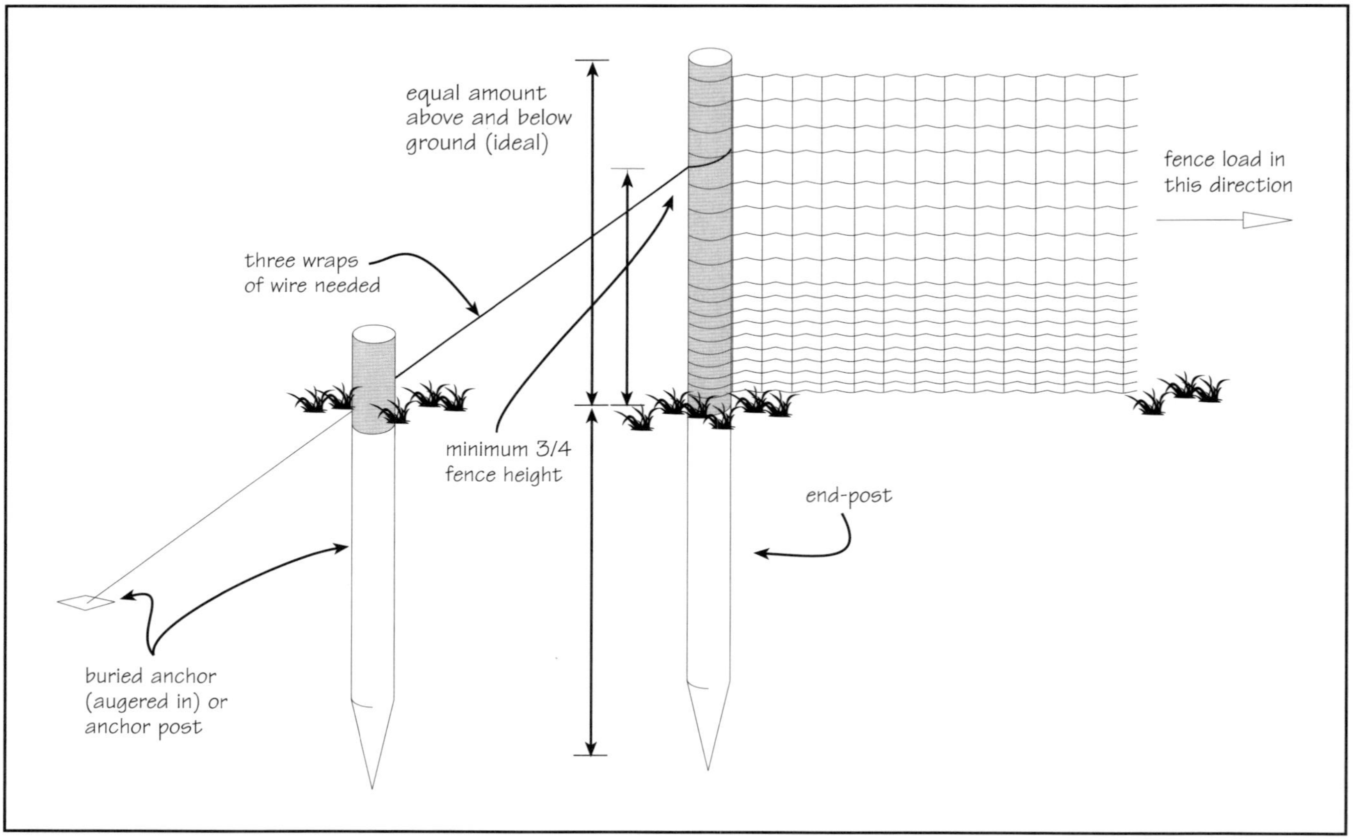

Fig. 3 Tieback anchored end-brace

PIPE LINE-BRACE ASSEMBLY: 8' ELK FENCE

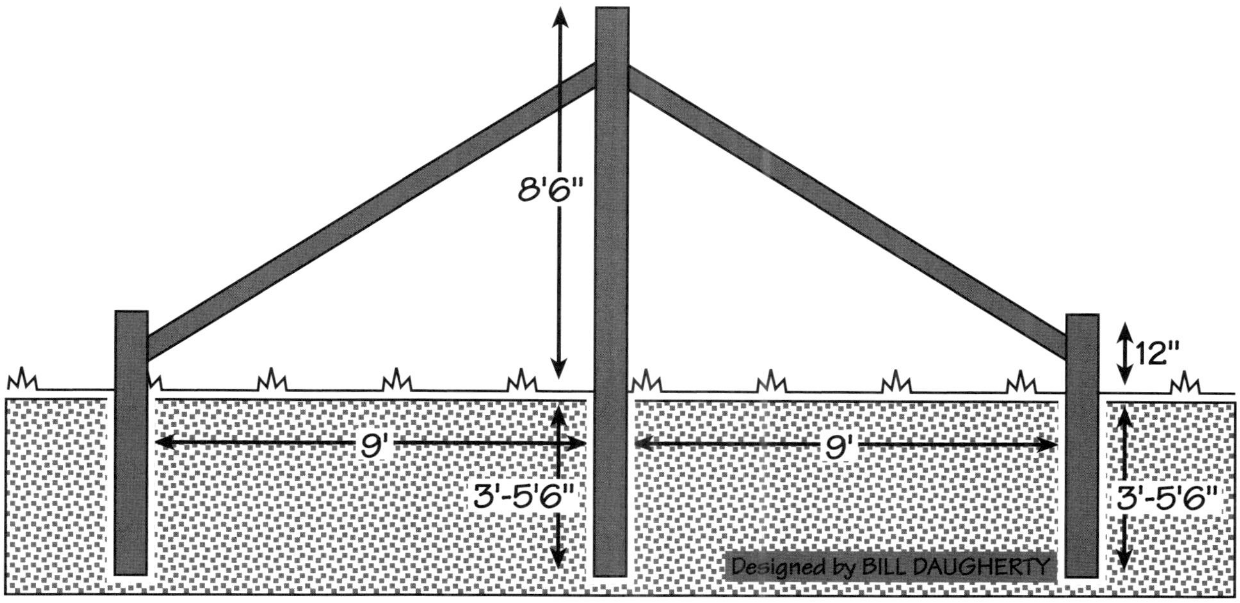

COMPONENTS

Brace Post................14' x 2 7/8" Structural Tubing
Angle Braces.............12' x 2 3/8" Structural Tubing
Angle Braces Feet.........6' x 2 7/8" Structural Tubing

Rust Proof Metal Primer
Oil Base Porch Enamel

INSTALLATION

Note: Pipe Braces set in mixed soils are set 5'6" deep in concrete. The minimum diameter of the hole being 12". Pipe set in solid rock are set 3' deep in concrete with a hole diameter of 4" to 9". Pipe may also be driven in mixed soils to a minimum depth of 6'.

1. Prime pipe with a rust proof metal primer before setting.
2. Set Brace Post.
3. Set Angle Feet 9' from Brace Post on either side and in line with fence line.
4. Notch both Angle Feet on Brace Post side to accept Angle Brace.
5. Cut and saddle Angle Brace so it will meet the Brace Post 7' from the ground. Weld solid, top and bottom.
6. Cap all pipe with cement. If pipe is left open, water will rust the pipe off at ground level.
7. Brush all welds. Treat with rust proof metal primer and paint with enamel.

Caution: Allow two days for cement to cure before pulling on brace.

BRACE ASSEMBLY: 8' ELK FENCE

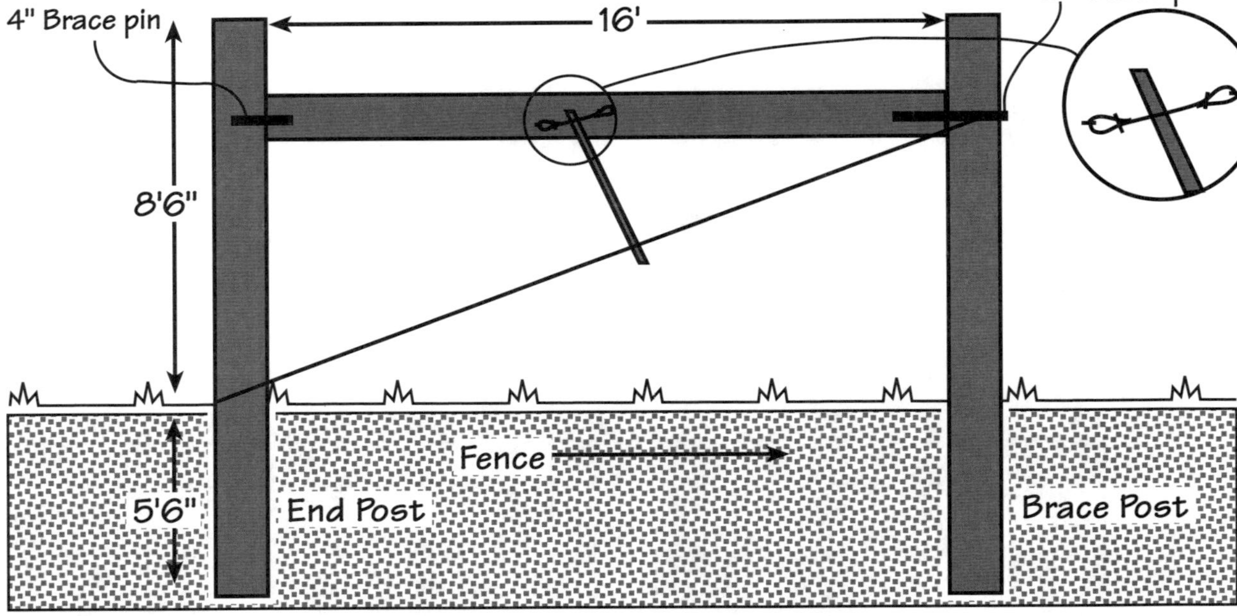

COMPONENTS

Brace Posts..............14' x 7" .40 CCA Treated Pine
Cross Members.........16' x 5" .40 CCA Treated Pine
.........16' x 3" Schedule 40 Pipe
Brace Pins............... 4" #5 Rebar
.........10" #5 Rebar

Twitch Wire.......Double Wrap 9 Ga. Wire
.......Double Wrap 12 1/2 Ga.
.......Hi-Tensile Wire
.......Single Wrap 5/16" Cable
Twitch Stick.......3/4" x 36" Sucker Rod Steel
.......3/4" x 36" Pipe
.........1" x 36" Fiberglass Sucker Rod

INSTALLATION

Note: Braces must be installed in fence line, regardless of the length of pull.
Braces should be placed no more than 1320 ft. apart.
Brace width must be a minimum of 2 times the height of the fence (2.5 times is preferred).

1. Earth-set the End Post.
2. Set the Brace Posts, using the Cross Member for measure.
3. Position the Cross Member between the 2nd and 3rd wire of Langley Wire Industries fence fabric. Drill and set the 4" pin first. Drill only 2" into end post. Drill all the way through the Brace Post. Position the Cross Member and drive the 10" pin into end, leaving 1" exposed for the Twitch Wire.
4. Wrap Twitch Wire, anchoring at the base of the End Post with staples, then double wrap.
5. Twist the Twitch Wire with Twitch Stick until the End Post moves 1/4" in soil. Anchor the twitch Stick to the cross member. Fabric should be on opposite side of the Twitch Stick.

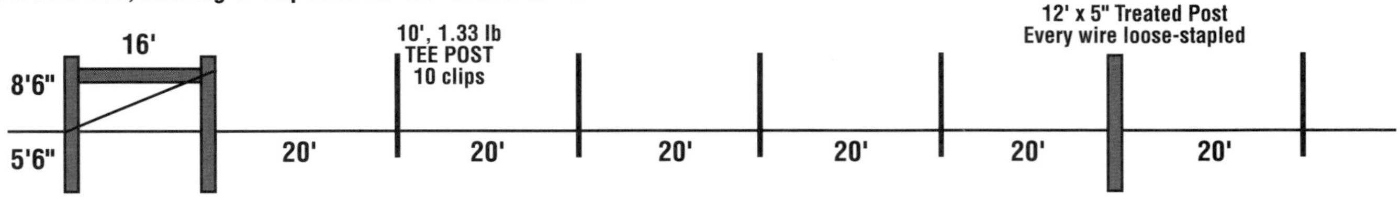

Langley Wire Industries Ltd. Bill Daugherty fax: (210) 620-7019

PIPE BRACE ASSEMBLY: 8' ELK FENCE

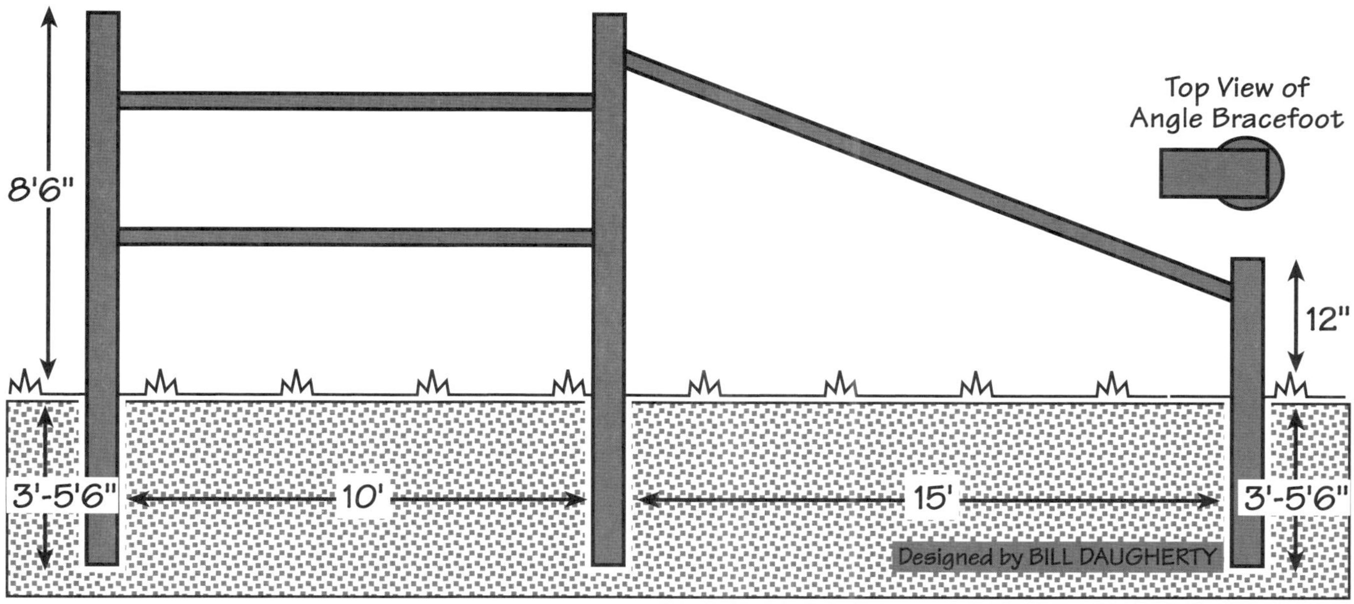

COMPONENTS

Brace Posts.............14' x 2 7/8" Structural Tubing
Cross Members.........10' x 2 3/8" Structural Tubing
Angle Brace............18' x 2 3/8" Structural Tubing
Angle Brace Foot.........6' x 2 7/8" Structural Tubing
Rust Proof Metal Primer
Oil Base Porch Enamel

INSTALLATION

Note: Pipe Braces set in mixed soils are set 5'6" deep in concrete. The minimum diameter of the hole being 12". Pipe set in solid rock are set 3' deep in concrete with a hole diameter of 4" to 9". Pipe may also be driven in mixed soils to a minimum depth of 6'.

1. Prime pipe with a rust proof metal primer before setting.
2. Set End Post and pull guide wire.
3. Set Brace Post 10' from End Post - then Angle Foot 15'.
4. Measure inside brace width, cut and saddle two cross members. Position top cross member between 2nd and 3rd wires of Langley Wire Industries fence fabric and the middle cross member half the distance between the ground and top cross member. Try to set the middle cross member so it will fall between horizontal wires of the fabric. Weld solid.
5. Notch angle brace foot so 2 3/8" angle brace can lay inside foot. Cut and saddle angle brace to meet the top cross member of the brace. Weld solid.
6. Cap all pipe with cement. If pipe is left open, water will rust the pipe off at ground level.
7. Brush all welds. Treat with rust proof metal primer and paint with enamel.

Caution: Allow two days for cement to cure before pulling on brace.

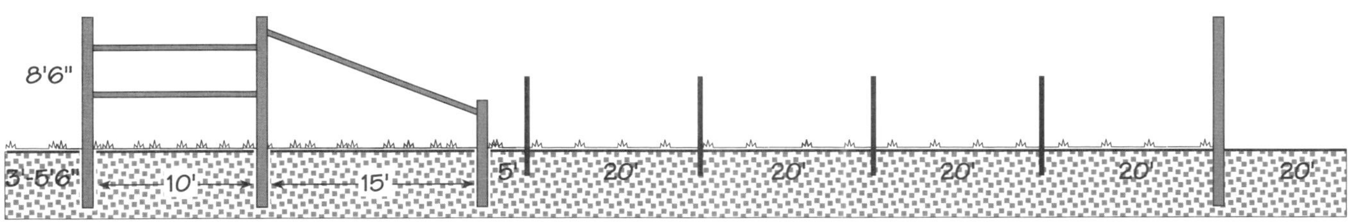

Langley Wire Industries Ltd. Bill Daugherty fax: (210) 620-7019

anchor should be placed eight feet away from the end-post).

A minimum of three wraps of high tensile wire is needed to complete the brace. This brace wire should be attached to the end-post at a minimum of three-quarters of the fence height.

LINE POSTS

Posts and their installation are the most expensive items in fence construction. The purposes of the line-posts are simply to hold up the weight of fence wires and maintain constant wire spacing.

The most common method used for setting posts is a mechanical post pounder. Posts can also be set by augering a hole and then back filling once the post is in place. When setting brace posts using an auger, it is recommended that extra footing be attached to the bottom of the posts (Fig. 4). This will increase the resistance of the post to twisting and pulling out.

Fences constructed using page wire should have line-posts set approximately 15 to 20 feet apart. Wide post spacings can be used with high tensile wire only if the wire is properly tensioned and the terrain allows for it. It is important to note that closely spaced posts are more prone to breaking when severe pressure is applied to the fence wires.

When fencing uneven ground the posts should be set perpendicular (in contour) and not plumb (vertical to gravity). This general rule is most applicable to fencing extremely hilly ground when working with page wire (Fig. 5). If the choice is made to set the posts plumb, they should be set a little higher to allow for constant wire spacing.

Gullies and hills require special attention with respect to line-posts. Line-post spacing needs to be adjusted so that a post is set in the bottoms of gullies and on the tops of hills. Posts on top of hills should be set a little higher than other line-posts. Posts at the bottom of gullies need to be set slightly deeper than other line-posts to resist upward pressure and pulling out. In extreme cases, such as steep angles or soft ground, posts in gullies may need to be one diameter size larger and require additional footing or anchoring to remain in place. Placing a deeply set second post next to the line-post and joining the two with low-tensile wire will provide extra footing (Fig. 6).

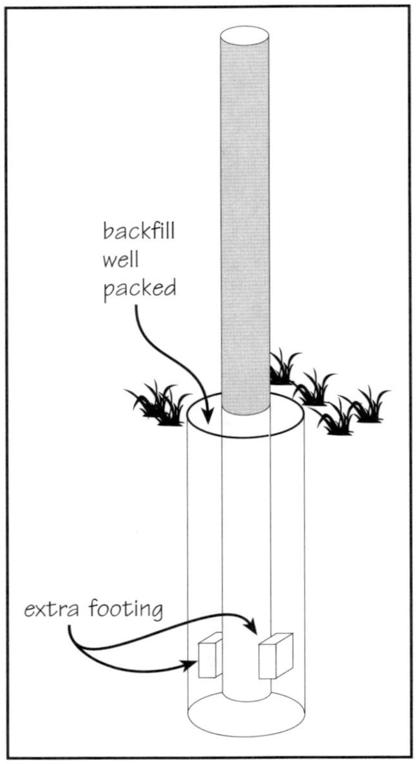

Fig. 4 Extra footing for auger set posts

TYING-OFF AND SPLICING WIRE

Tying-off and splicing high tensile wire requires careful attention due to the high finished fence

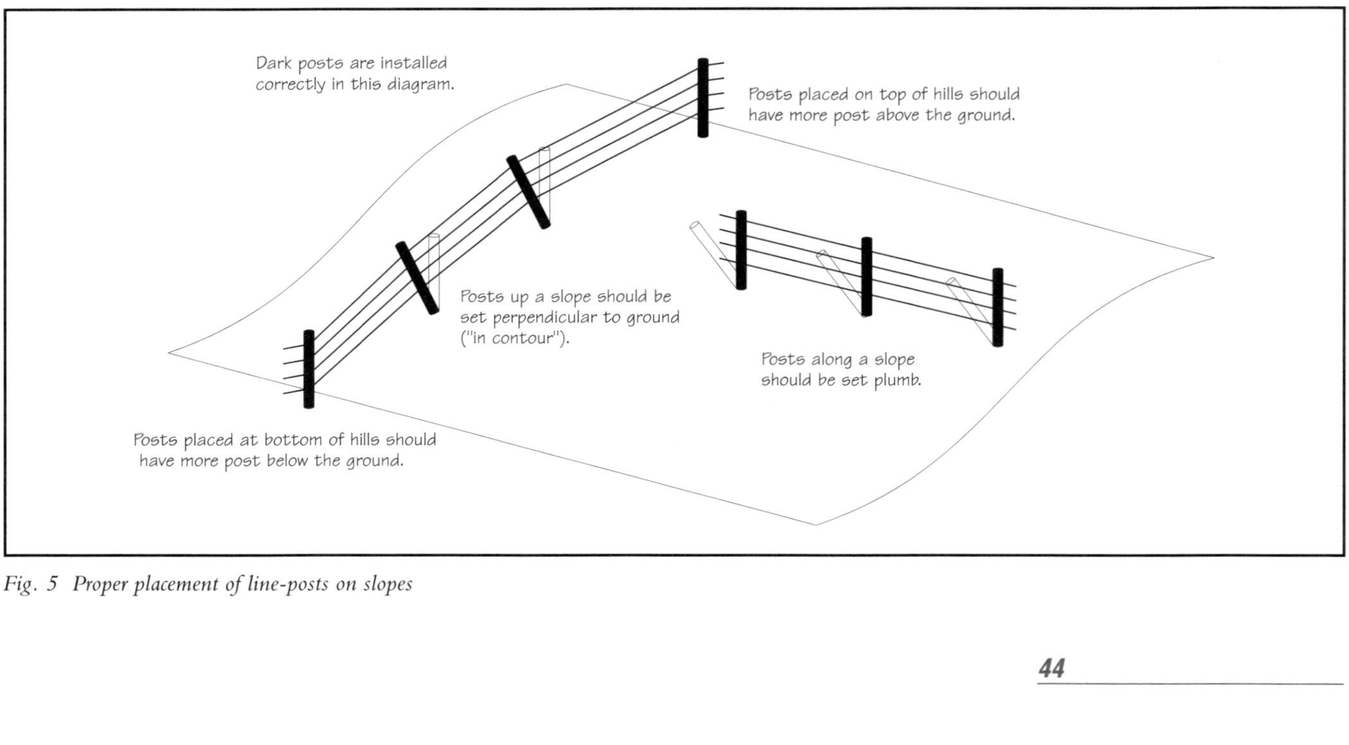

Fig. 5 Proper placement of line-posts on slopes

tensions. Tying off and splicing may be done by either using knots or mechanical splices. Mechanical splices (compression sleeves) are relatively easy to use and offer the advantage of maintaining 100 per cent of wire strength when properly installed. It is important to remember that mechanical splices are costly, may require a special crimping tool and may be manufactured for a specific wire gauge (size). The manufacturer indicates the number of compression sleeves needed for effective splicing.

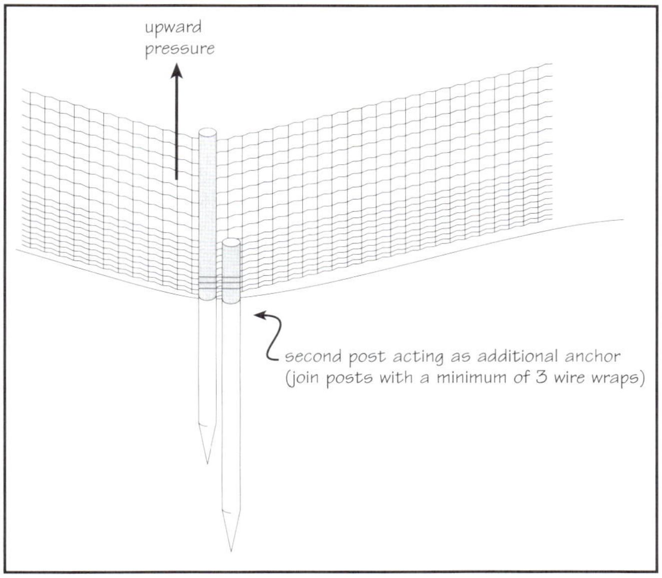

Fig. 6 Extra footing provided by second post

Because of the difficulty in knotting high tensile wire, mechanical splices are generally recommended. However, high tensile wire can be tied off and spliced effectively using the following methods.

SPLICING SINGLE LINES

Single lines of high tensile wire can easily be spliced using a knot or a mechanical compression sleeve (Fig. 7).

The most effective knotted splice for high tensile wire is called the 'Figure 8' splice and is shown in Figure 7A. This knot self-tightens when the wire is tensioned. It is important to leave about six inches of wire on each free end in order to finish the knot properly. After tensioning and finishing, the free ends may be trimmed off as close as possible. This splice is effective to about 89 per cent breaking strength of unspliced wire.

When using compression sleeves (Fig. 7B), it is important to leave about one inch of wire on either side of the splice. The manufacturer will also specify how many should be used to create an effective splice.

SPLICING PAGE WIRE

In order to effectively splice page wire using knots, overlap the netting along a vertical wire. Each free end must be wrapped three times around a horizontal wire, and then may be trimmed off closely. See Fig. 8A.

Splicing page wire with compression sleeves is identical to that of single lines. The only difficulty that may arise is caused by the tension curve present in page wire. This curve can simply be

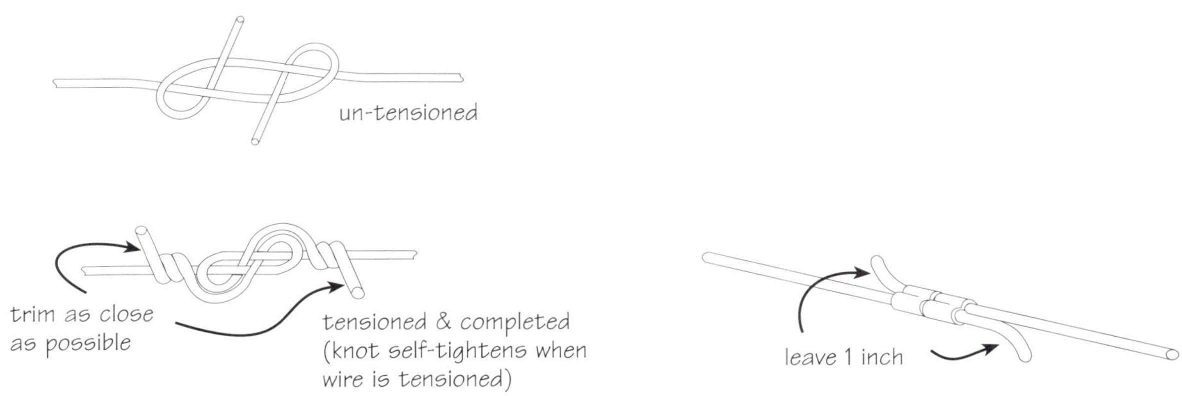

Figure 7A: Splice using 'Figure 8' knot.

Figure 7B: Splice using compression sleeves.

Fig. 7 Methods of splicing single lines of high tensile wire

straightened out with pliers prior to using the compression sleeves (Fig. 8B).

TYING-OFF WIRES TO END-POSTS

The quickest way to tie-off high tensile wire to end-posts is by using the knot shown in Fig. 9A.

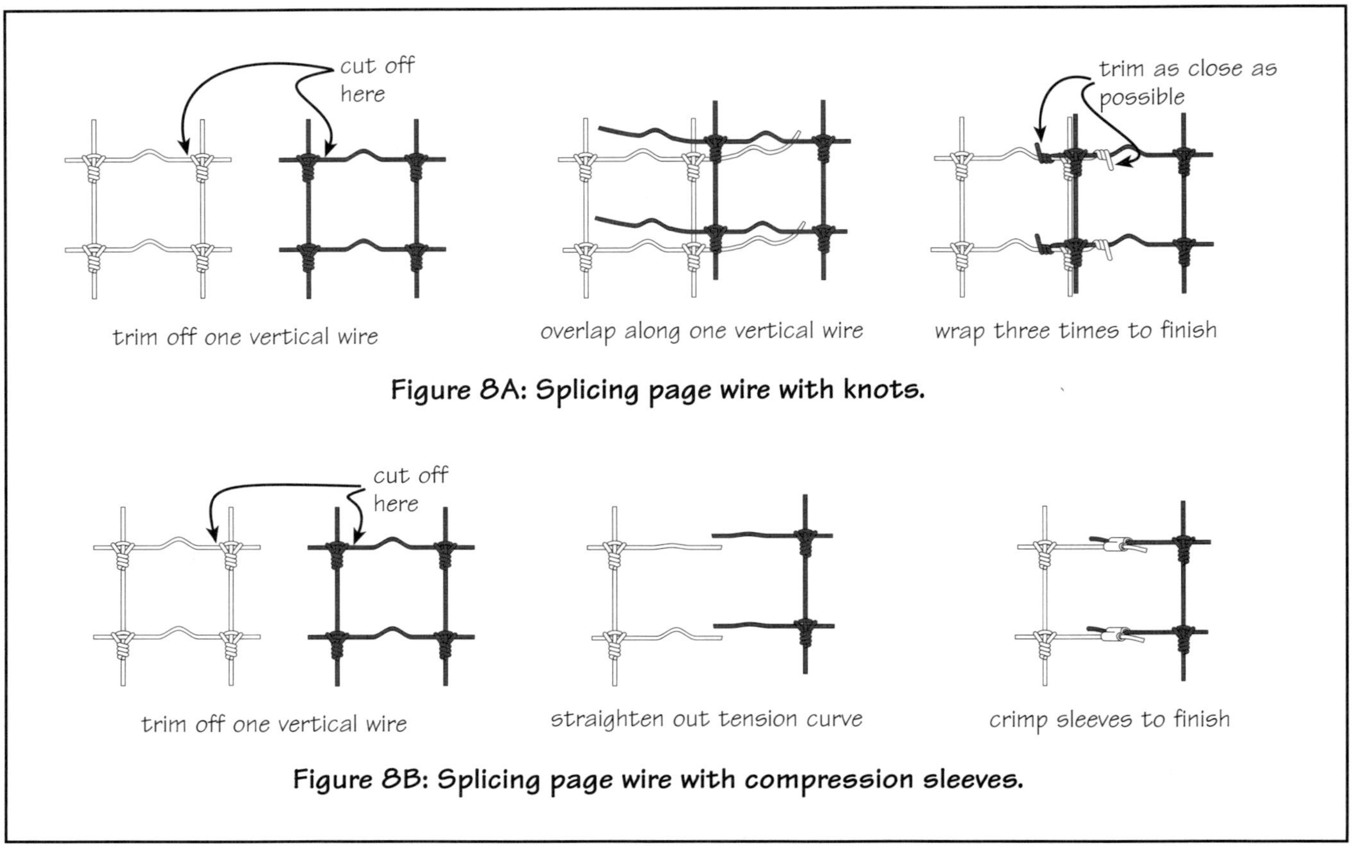

Figure 8A: Splicing page wire with knots.

Figure 8B: Splicing page wire with compression sleeves.

Fig. 8 Methods for splicing page wire

Using mechanical splices may provide a stronger finished product (Fig. 9B); however, stripping off all of the knots and straightening out the tension curve from page wire would prove to be too labour intensive.

WIRE TENSION AND STRAINING

There are a variety of commercially available tensioning tools to choose from that may each require separate techniques to operate.

The finished tension of high tensile fences at 20°C should be approximately 250 pounds. Table 1 shows the recommended wire tensions for a variety of temperatures.

A simple device for measuring wire tension can easily be constructed (Fig. 10). Materials needed include a spring scale that measures in pounds,

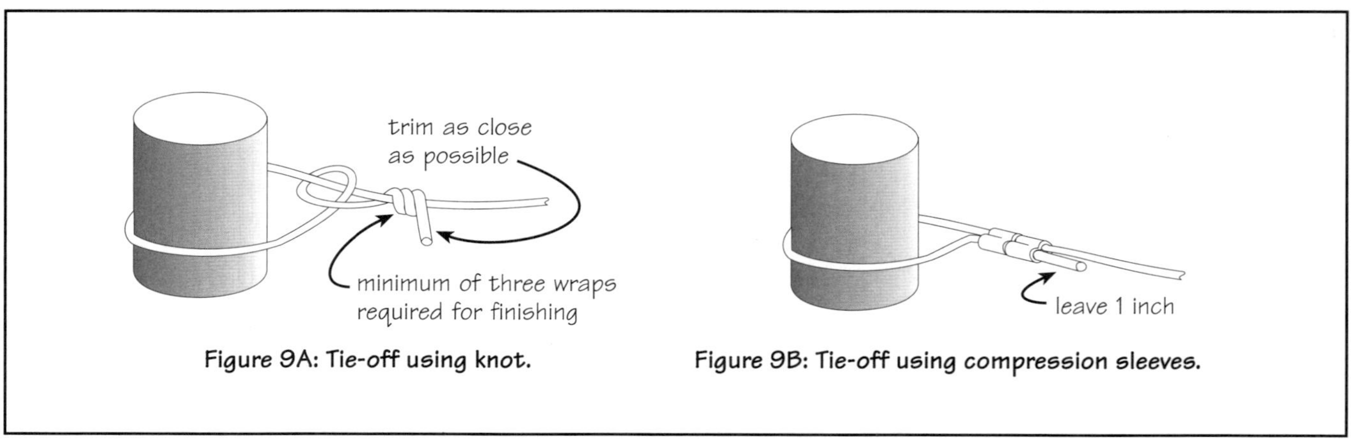

Figure 9A: Tie-off using knot.

Figure 9B: Tie-off using compression sleeves.

Fig. 9 Methods for tying-off wire to posts

three nails, and 1" x 2" wood boards. Drive two nails 40 inches apart on a straight line (Fig. 8, points A & B). Drive the third nail midway between these nails at a point half an inch below a line connecting the first two nails (Fig. 8, point C).

Temperature (ºC)	Tension (lbs.)
30	30
20	20
10	10
0	0
Source: Adapted from the Beef Herd Management Reference Binder & Study Guide, Alberta Agriculture.	

Table 1. Recommended wire tensions

With the nails A and B touching the wire, pull the wire with the scale until it touches the nail at point C. The wire tension is the scale reading multiplied by 20 (e.g. if the scale reads 12.5 lb., the wire tension is 12.5 x 20 = 250 lb.).

Remember that the fence wire will contract and expand as the temperature changes. Air temperature has the effect of increasing tension by 22 pounds for every 10°C drop in temperature. Due to this seasonal variation, fence wires should not be over-tensioned in the summer. An example calculation is shown in Table 2.

Permanent in-line tensioners can be installed during fence construction. These self-ratcheting devices are installed on each strand of high tensile wire and allow the producer to easily and efficiently adjust the fence tension throughout the year (Fig. 11). They also facilitate quick repairs and re-tightening should the fence wire break. A special handle is needed to properly operate permanent tensioners. The cost and the added installation effort often limit the use of permanent in-line tensioners. If used, permanent tensioners should be placed in the middle of the fence to ensure that the end braces bear the wire tension equally.

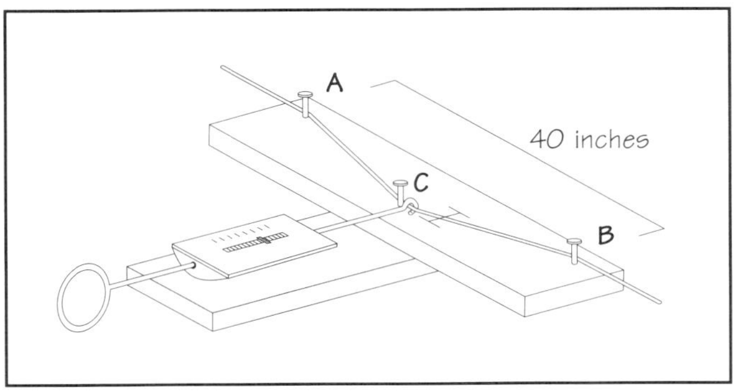

Fig. 10 Simple tension gauge

Changes in Wire Tension

For one strand of 12-1/2 gauge steel wire:

$F = 2.6 \times \Delta T$

where: F = force or tension (lb.) as a result of temperature change

and: ΔT= change in temperature (°C)

Change in wire tension:

$F = 2.6 \times \Delta T$

$\Delta T = -30°C \text{ to } +30°C = 60°C$

$F = 2.6 \times 60°C = 156$ lb. added tension to each strand

New winter tension:

250 lb. original tension + 156 lb. increase due to temperature change = 406 lb. tension per strand

Change in brace forces:

Wire force on end braces when built:
5 strands x 250 lb./strand = 1250 lb.

Increased force due to 60°C temperature change:
5 strands x 156 lb./strand = 780 lb.

Total force at -30°C = 1250 lb.+ 780 lb. = 2030 lb.

Source: Adapted from BC Agricultural Fencing Handbook

Table 2

Tension springs can also be installed on each strand of high tensile wire during fence construction (Fig. 12). These springs allow the wire to expand and contract with the temperature while maintaining the installed tension. The main disadvantage to utilizing these springs is the added installation costs. These tension springs are also primarily intended for use with single wires and not page wire fencing.

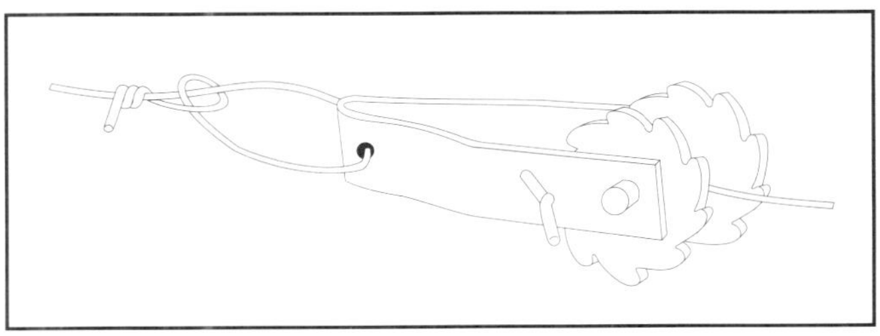

Fig. 11 Permanent in-line wire tensioner.

To ensure that the end brace assemblies are bearing the fence tension equally during strain-up, wire tensioning should be done from as close to the centre of the fenceline as possible. In special situations, where there is more resistance to wire movement such as hills or gullies, tensioning may be done closer to the point of greatest resistance.

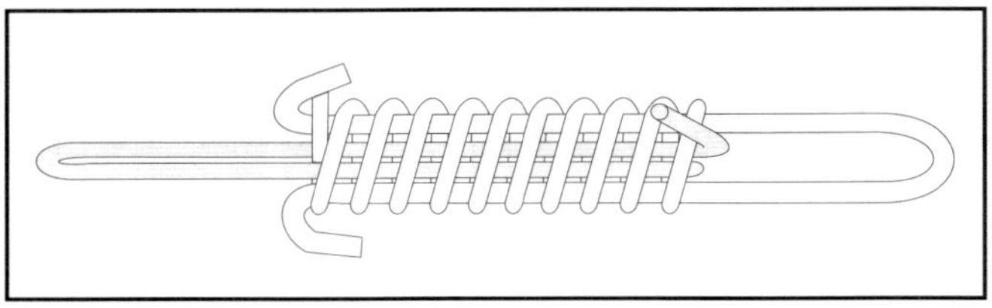

Fig. 12 Tension spring for single lines of high tensile wire

Tensioning the fenceline close to one end increases the load on that end brace assembly during strain-up and increases the chance of brace failure.

Final tensioning is done after the netting is standing fully upright along the entire length of the fenceline. Trimming out (permanently attaching the wire to the posts) is done after the entire fence has been fully tensioned. This technique is used because the vertical stays of the page wire will hang up or catch on staples if the fence is tensioned after the attachment to the posts. The netting should be first attached to the highest points of the fence, followed by the lowest points, and finally to the remainder of the line-posts. Correct tension of page wire netting is achieved when half of the tension curve is removed (Fig. 13). Page wire netting should be installed slightly above the ground (one to two inches) to reduce the possibility of corrosion.

ATTACHING THE WIRE

The method used for attaching wires to posts will depend upon the type of post selected. Wooden posts require stapling, while metal posts require tying and hooks.

TYING WIRE

Attaching wire to metal posts requires the use of ties. Many producers have reported success using

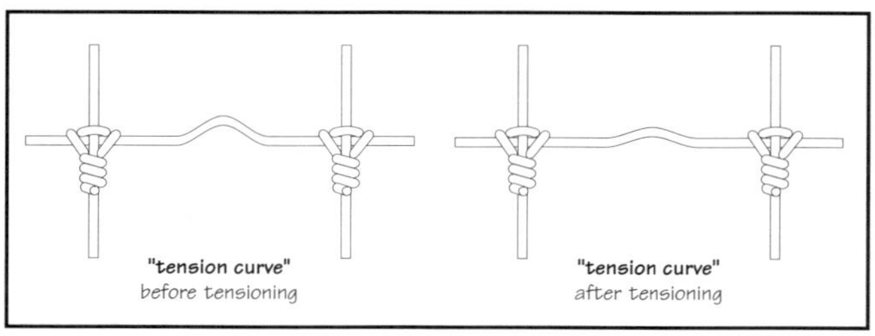

Fig. 13 Tension curve in page wire fencing

the tying method shown in Fig. 14. It is important to remember that the "holding-power" of this method relies on the wire being twisted in opposite directions.

Fig. 9 illustrates the most secure method of tying. High tensile wire should be avoided for use as a tying material because of the difficulty in twisting. However, with a portable power drill and a special attachment high tensile wire can be used.

The wire used for tying should be galvanized to reduce the amount of corrosion.

In order to maintain constant wire spacing, special attention is required on hills and in dips. One solution used by producers is by welding hooks or half links of chain to the tops or bottoms of the metal posts to hold the wire in place. The condition of the ties should be checked frequently and repaired as necessary. There is considerably more wear on fence wires due to friction and corrosion when using metal posts than for wooden posts.

When working with page wire netting, every horizontal line need not be tied. Several ties should be placed at the bottom, top and middle to hold the netting in place. Additional ties may be required on hills, gullies and areas of high animal pressure.

STAPLING WIRE

There is a correct way to place and drive staples. Staples should never be driven in completely, but rather a small space should be left to allow the horizontal movement of the wire (Fig. 14A). Stapling in hills and gullies requires extra attention. In order to prevent staple pullout, the tension load needs to be transferred to the post rather than the staple. This is easily accomplished by driving the staples in at an angle as in Fig. 14B.

To prevent splitting of posts, staples should never be driven along the grain. Rotating the staple to about 30 to 45 degrees from vertical reduces the chances of post splitting. The direction of rotation is determined by the manufacture of the staple (Fig. 14C). Staples have a flat slash-cut surface on opposite sides of the points. These slash cuts act as wedges and force the staple legs to curve away from the flat surface when hammered. Proper placement and driving of staples increases resistance to pullout by 40 per cent.

There are several simple techniques that should be employed when stapling corners, hills and gullies to reduce stress on wires and result in stronger, long lasting fences that require less maintenance. When stretching around corners, the

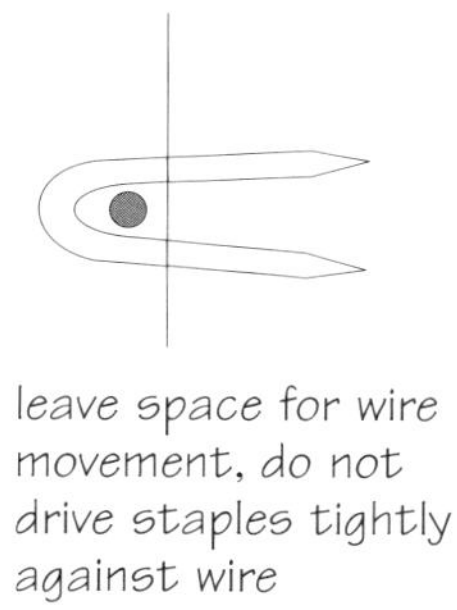

Figure 14A: Space left between staple and post.

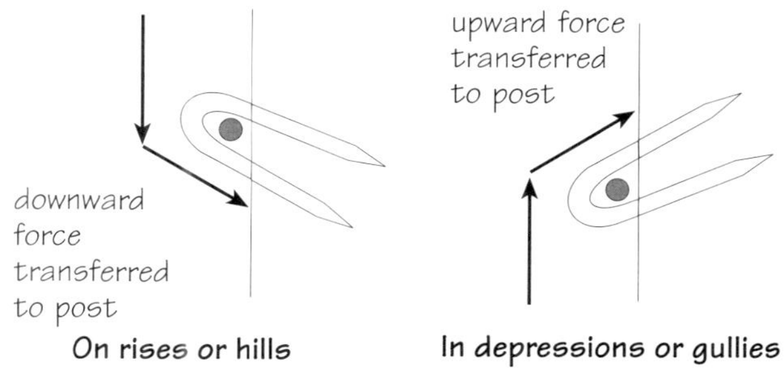

Figure 14B: Drive staples in at an angle.

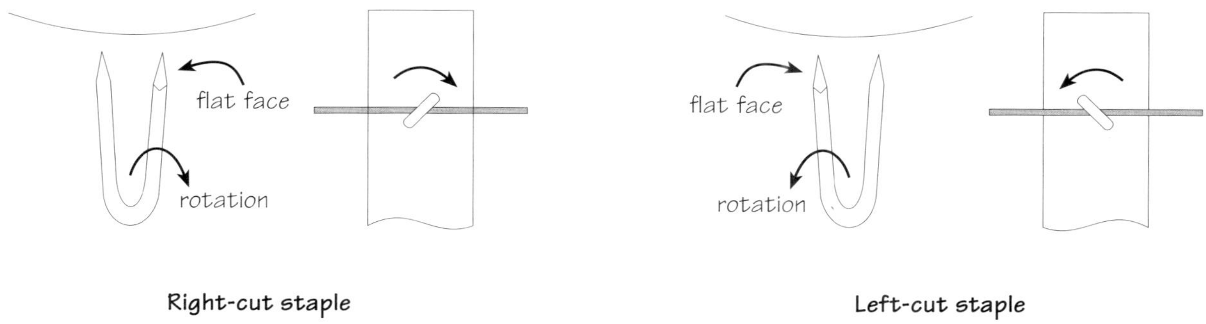

Figure 14C: Correct rotation and placement of staples.

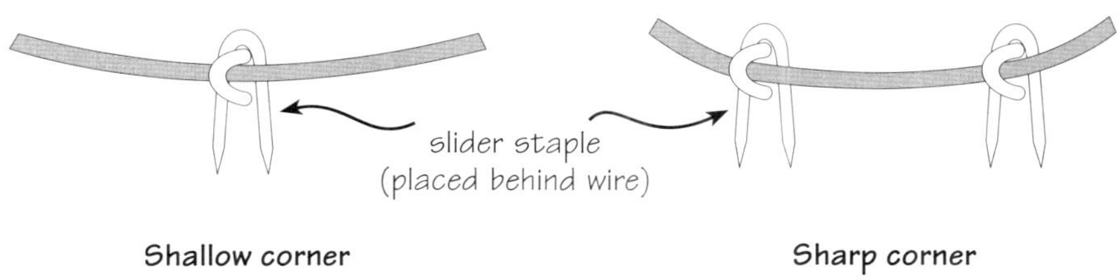

Figure 14D: Use extra staples as "sliders" to reduce friction.

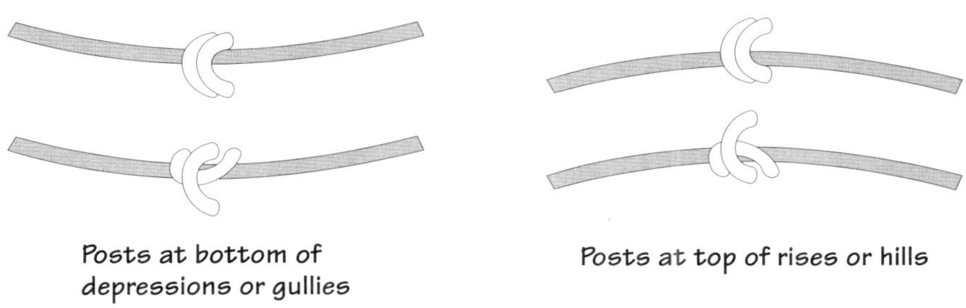

Figure 14E: Double staple posts on extreme hills and gullies to prevent staple failure.

Fig. 14 Stapling Techniques

wires should be strung on the outside of posts to prevent staple pullout. Placing a second "slider" staple between the wire and post to act as a sliding surface can reduce wire friction on posts (Fig. 14D). It is recommended to double staple the wire on extreme hills and gullies to prevent staple failure (Fig. 14E).

When working with page wire netting, every horizontal line need not be stapled. A minimum of two or three staples should be placed at the bottom, top and middle to hold the netting in place. Additional stapling may be required on hills, gullies and areas of high animal pressure.

FENCE DIRECTION CHANGES

The following section outlines some methods of bracing corners to prevent corner failure. Small angle changes in fencelines (up to 60 degrees) are difficult to hold even when in-line braces are used. To quickly calculate the angle of change for a corner, refer to Table 3 and Fig. 17.

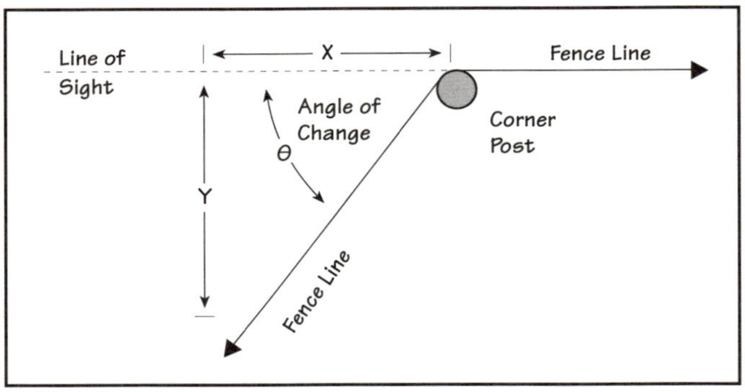

Fig. 17 Calculating angle of change

A quick method for accurately planning a square or 90 degree corner is shown in Figure 18.

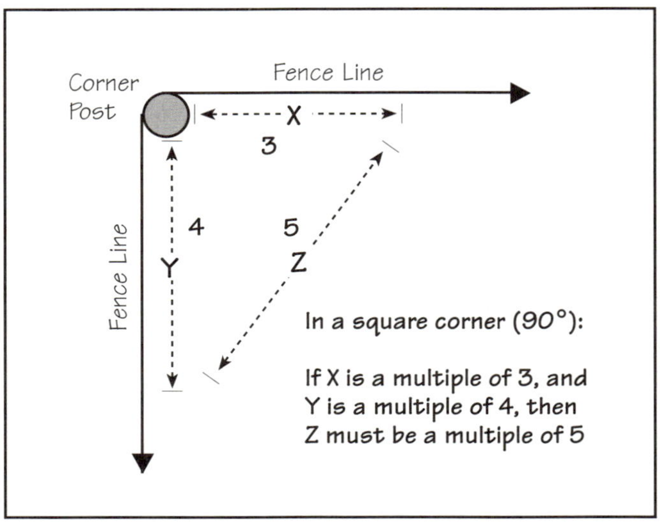

Fig. 18 Technique for planning a square corner

Several measures can be taken to reduce the chance of corner failure. Use larger diameter posts for all corner braces and set them deeper than the line posts. Setting the corner post with a slight lean (one to two inches) away from the angle of direction change will aid in reducing failures. Always run the wire on the outside of the corner brace assemblies.

A shared post corner can be used for changes in fence direction from 60 to 90 degrees (Fig. 19A). Another solution is to create two fencelines and end each fenceline with separate end braces (Fig. 19B). Adding more brace panels can reduce strain on the corner post (Fig. 19C).

The use of buried anchors or tiebacks to the outside of the angle can be effective in eliminating corner failure (Fig. 19D and Fig. 19E). Additional

For X = 10 feet	
Y	**Angle Ø**
1 ft. 9 in.	10°
3 ft. 8 in.	20°
5 ft. 9 in.	30°
8 ft. 5 in.	40°
11 ft. 11 in.	50°
17 ft. 4 in.	60°

Table 3. Calculating angle of change

bracing to the inside of the angle is an alternative method of preventing corner failures (Fig. 19F). Both the internal and external bracing must equally bisect the angle to be effective.

SPECIAL CONSIDERATIONS

WATER SOURCES

It is important that water sources such as ponds or dugouts be fenced. This will allow the producer to control the animals access to the water during certain times of the year. Thin ice can pose a serious drowning hazard.

OVERHEAD AND BURIED UTILITIES

When planning the construction of fences it is important to locate all buried utilities (such as electrical, telephone, natural gas, etc). Call the local service providers at least 48 hours in advance to have the locations of underground facilities properly marked. Check with the utility surveyor about how long the survey is valid. Often the fence construction must be completed within a specific time after marking, otherwise a second survey is required.

The grounding of fences constructed from nonconductive posts (wood) is often neglected. Accidental electrocution can occur from lightning or fallen overhead utility wires. Precautions should be taken to adequately ground fences that run parallel to or cross under overhead powerlines.

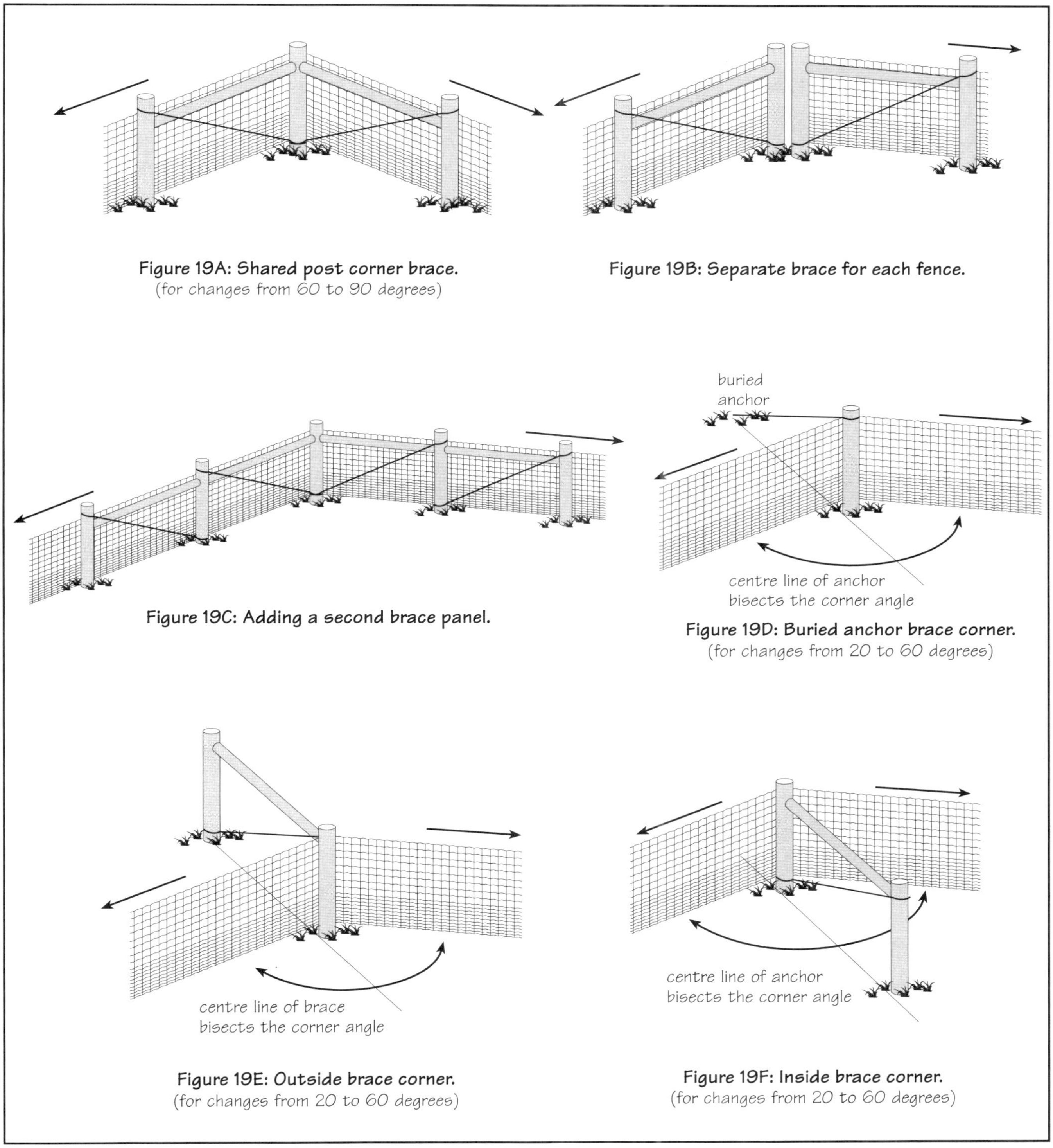

Figure 19A: Shared post corner brace.
(for changes from 60 to 90 degrees)

Figure 19B: Separate brace for each fence.

Figure 19C: Adding a second brace panel.

Figure 19D: Buried anchor brace corner.
(for changes from 20 to 60 degrees)

Figure 19E: Outside brace corner.
(for changes from 20 to 60 degrees)

Figure 19F: Inside brace corner.
(for changes from 20 to 60 degrees)

Fig. 19 Corner bracing techniques

FENCE VISIBILITY

The visibility of the fence to the animals is extremely important in high pressure areas such as alleyways with abrupt direction changes or narrowing into handling facilities. An effective solution devised by some producers is to attach plastic shade netting or snow fence to the fencing in these critical areas. It is important that any plastic netting product being used for visibility be ultraviolet resistant to ensure longevity.

FENCING AGAINST PREDATORS

The most commonly used page wire is manufactured with variable wire spacing where the horizontal lines are closely spaced near the bottom. This design offers more defense against predators than page wire that has uniformly spaced wires. To prevent coyotes and dogs from digging under the perimeter fence, many producers add an 'apron' to the outside of the fence. Typically, this apron consists of a tight mesh wire such as stucco wire or chicken wire that is attached to the bottom two or three feet of the fence and extends out two or three feet (Fig. 20).

The apron is typically held tight to the ground by pegs or by soil. The effectiveness of the apron will increase as grass and other vegetation grows up through it. It is important to install the apron tightly to the ground at the bottom of hills, low spots, and places along the fence that have cover.

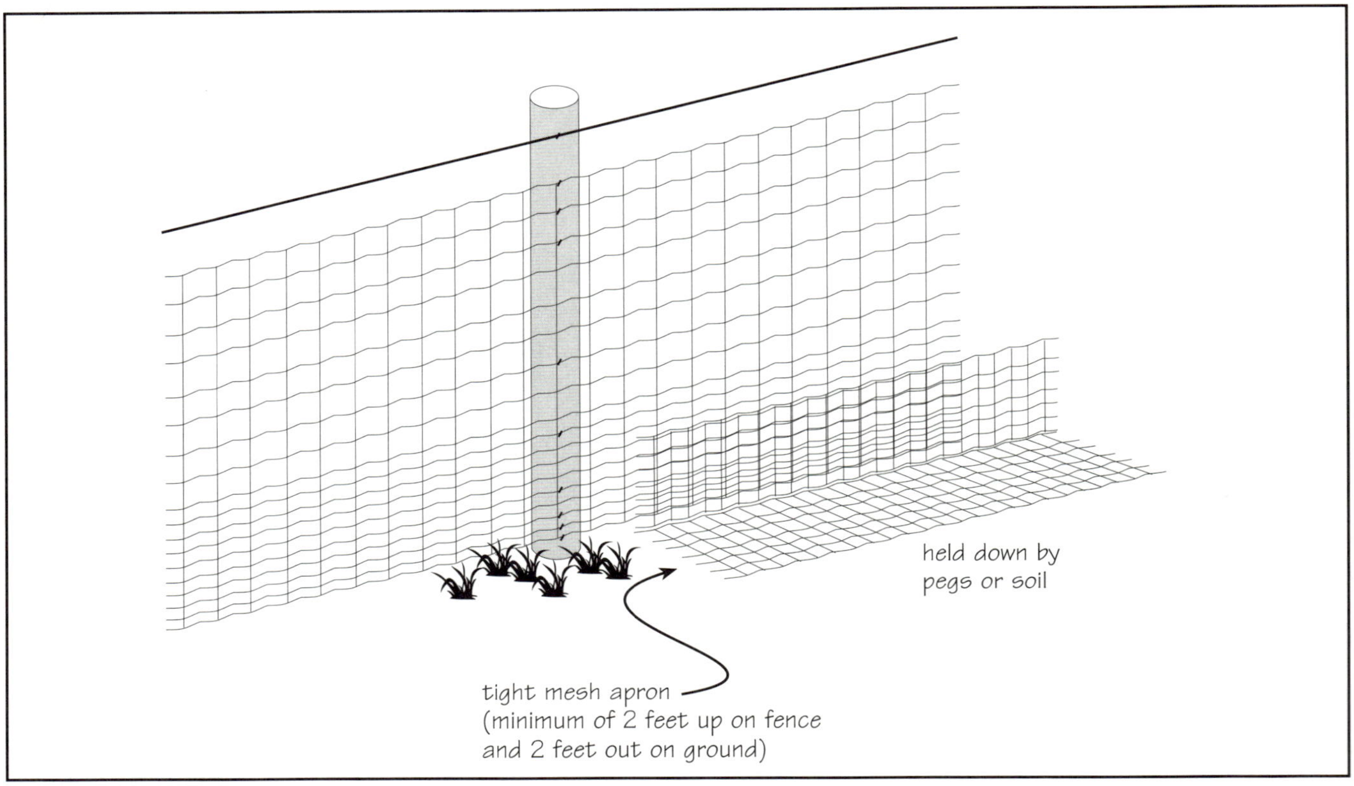

Fig. 20 Fencing against preditors

Page Wire Fence Building Steps

1. GATHER REQUIRED TOOLS

e.g.
- hammer
- crimping tool
- fencing pliers
- compression sleeves
- spinning-jenny
- pliers (lineman & needlenose)
- wire cutters
- 2-1/2 inch barbed staples

2. END-POSTS AND GUIDE WIRE

To begin the construction of the fenceline, locate and set the end-posts of the corner braces. All brace posts should be one diameter size larger than the line-posts (minimum of 6" diameter) and pressure-treated. The depth of setting will depend upon the soil type; however, the ideal is to have an equal amount above and below the ground. In rocky soils, the minimum is 4 feet, and in sandy soils the minimum is 6 feet. Set the end-posts with a slight lean (up to 4 inches) away from the direction of pull. After setting end-posts, a single guide wire of 12 gauge high tensile is pulled tight between the posts. Use a turntable or 'spinning jenny' to dispense the wire. To avoid tangles, start at one fence end and walk at a steady pace gradually slowing to a stop. This guide wire should be set along the ground and can remain there until all line-posts are set.

3. BRACING

The brace assemblies are the foundation of any fence and must be built accordingly. The braces may be single or double span and should ideally be 2x times the fence height (minimum of 2 times). The brace rail should be pipe, tubing or round wood posts; never a 4 x 4 or landscape timbers (these will warp!). The brace-rail should be set at 75% of fence height. Cut the brace-rail so that it fits snug between the end-post and brace-post; do not notch or cut the end or brace-posts. A minimum of 2 wraps (3 is ideal) of high tensile wire is needed for the twitch wire. The twitch wire must be anchored securely or the brace will fail. The inclusion of a permanent in-line wire strainer in the twitch wire will allow for future adjustment of the brace.

4. LINE-POST PLACEMENT

Generally the line-posts are set between 15 to 20 feet apart; however, spacings of up to 30 feet can be used. The topography of the ground should be used as the final determiner of post spacing. Adjust your spacings so that posts are set on the crowns of hills and in the lowest points of gullies. These posts will bear more tension than the rest of the line-posts and may need to be one diameter size larger. Posts on top of steep hills may need to be slightly taller to ensure proper spacing of the wire. Utilize additional footing for posts in gullies where necessary. Be sure to place all line-posts on the same side of the guide wire (this guide does NOT mark post centres, but rather one side of the fence).

5. WIRE

Before rolling out the page wire, move the guide wire up to the top of the posts (you can move it down later). Typically, the wire is put on the animal-side of the fence. Make sure that the closely spaced wires are at the bottom (close to the post). If you are using uniform spaced page wire, the twisted ends of the knots in the wire should point down. Run all wire around the outside of corners. The fenceline is tied off at both ends and tensioned in the middle. This allows the fence to be tied off without being under tension. Use the knot described earlier for both wooden and metal posts. Splicing may be done using compression sleeves or knots. When using sleeves, be sure to use the size designed for the gauge of wire you are using. When knotting, neatly overlap one vertical stay and wrap the free end around the horizontal line wire a minimum of 3 times.

6. TENSIONING

Tensioning should be done from the middle of the fenceline (or frictional centre) to ensure that equal strain is applied to each end. Begin tensioning the fence while the page wire is lying down; when the tension becomes firm, stand the wire up and continue. The tension curve will be half the size when the fence is correctly tensioned. In order to keep the tension on the end-posts uniform for each line wire, it is important to keep the vertical wires straight. When tensioning, be sure to account for hills or dips; do not tension fully, the tension will increase as you pull the wire up hills or down into gullies.

7. TRIMMING OUT

Attach the wire to all high points along the fence first, followed by the low points, and finally the remainder of the fence. The page wire should be positioned about one inch off the ground. If stapling, use barbed staples. Be sure to rotate the staples to prevent post splitting and to leave a gap to allow wire movement.

Selected Tips & Techniques

TIP 1: STEPS FOR REMOVING KNOT FROM HIGH TENSILE PAGE WIRE

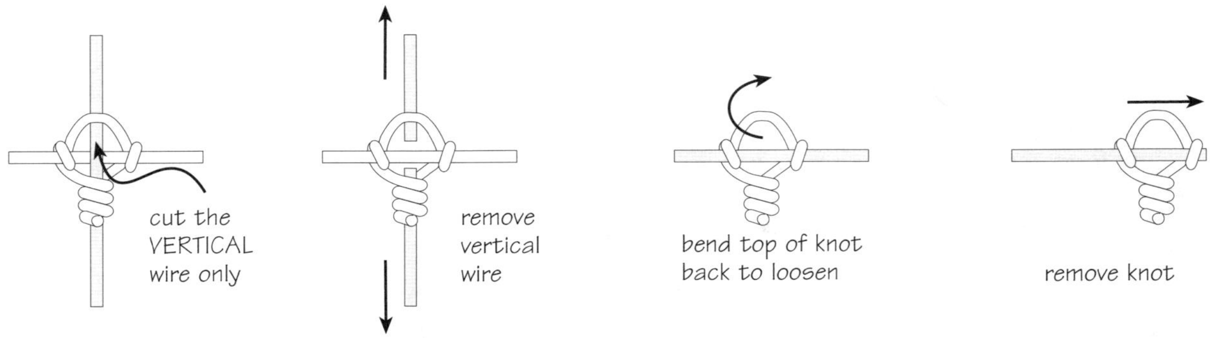

TIP 2: STEPS FOR BREAKING HIGH TENSILE WIRE BY HAND

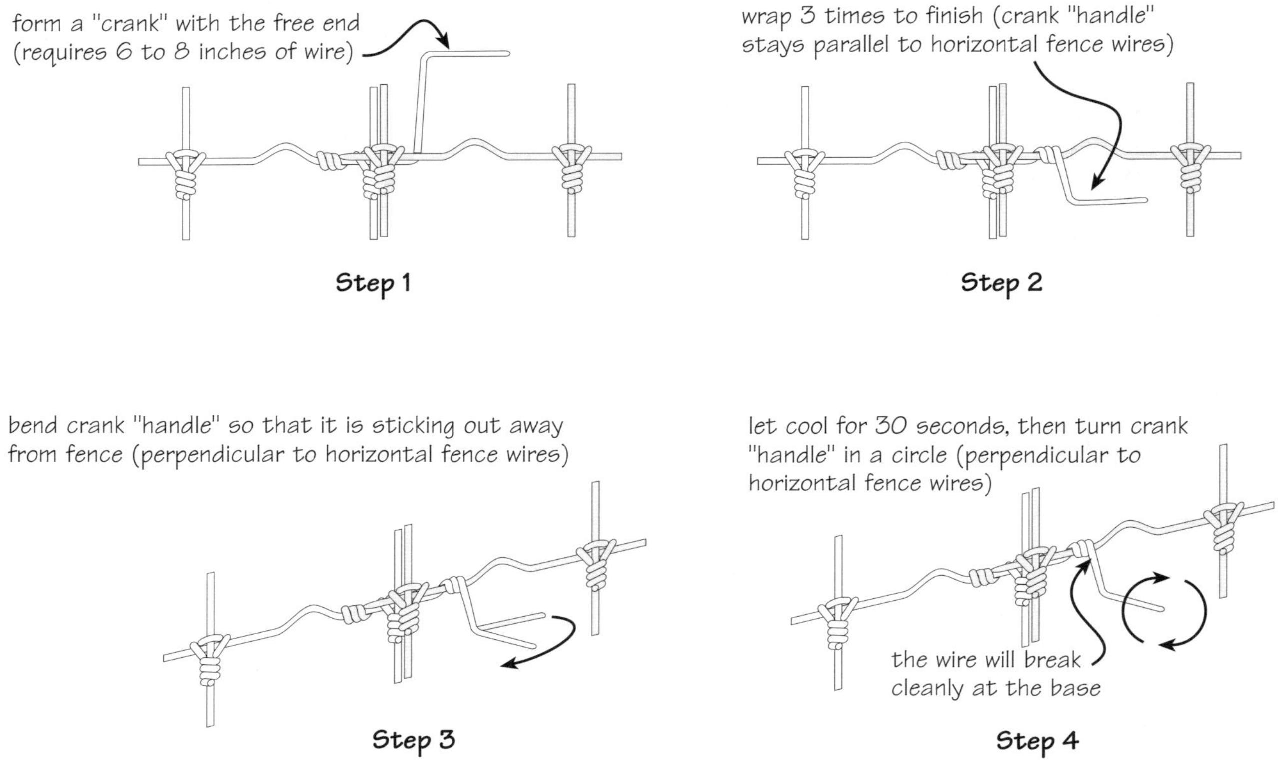

Chapter 1 Yards/Handling Facilities (curves & corners)

BASIC PRINCIPLES

Successful management of elk requires that the animals can be collected from pastures or winter feed lots at any time for any reason without difficulty or creation of stress in the elk or the farmer. Once mustered they need to be sorted into appropriate size groups for handling in a purpose-built facility.

Stockmanship is often more important than a set pattern of yard design, but there are standard basic principles required to meet the combination of the elk's natural response to this somewhat artificial situation, and the necessary management task.

Yards and facilities can be as varied as the farm setting, existing buildings and myriad of construction materials available. However, there are basic principles that once understood, and incorporated, enhance the natural instincts of elk to move or flow away from pressure, and thereby add to the farmer's management objectives, without stress or risk to safety.

Farmed elk are creatures of routine, and will over time quickly reinforce both positive and negative experiences.

The key points in design and handling are

- Simplicity & Routine
- Control & Pressure
- Letting the elk react naturally
- Curves & Corners
- Safety and avoidance of over crowding
- Efficiency and patience

RACEWAYS AND ENTRANCES TO YARDS

Once elk are moving in the central raceway toward yards they are generally being pressured from behind and will move easily as a tight group. However, if they are confronted with oncoming areas of resistance their first instinct is to turn and rush back the way they came, or worse, panic, charge fences, etc. Areas of resistance include abrupt changes in race direction where the next area of flow is not visible or obvious i.e. contrast in light and dark areas, or distractions like road traffic or unfamiliar activity, new or altered construction to a familiar system.

Therefore, if raceways curve or turn several times prior to entry to a holding or milling yard, the animals view that as an "escape" or relief avenue and will comfortably take that option.

In design AVOID

- Long straight approaches directly into buildings
- Sudden transitions of construction materials (e.g. fence - solid plywood)
- Movement into direct sunlight or sharp shadow/light - contrast
- Natural hazards in pressure areas
- Pooled water, uneven ground

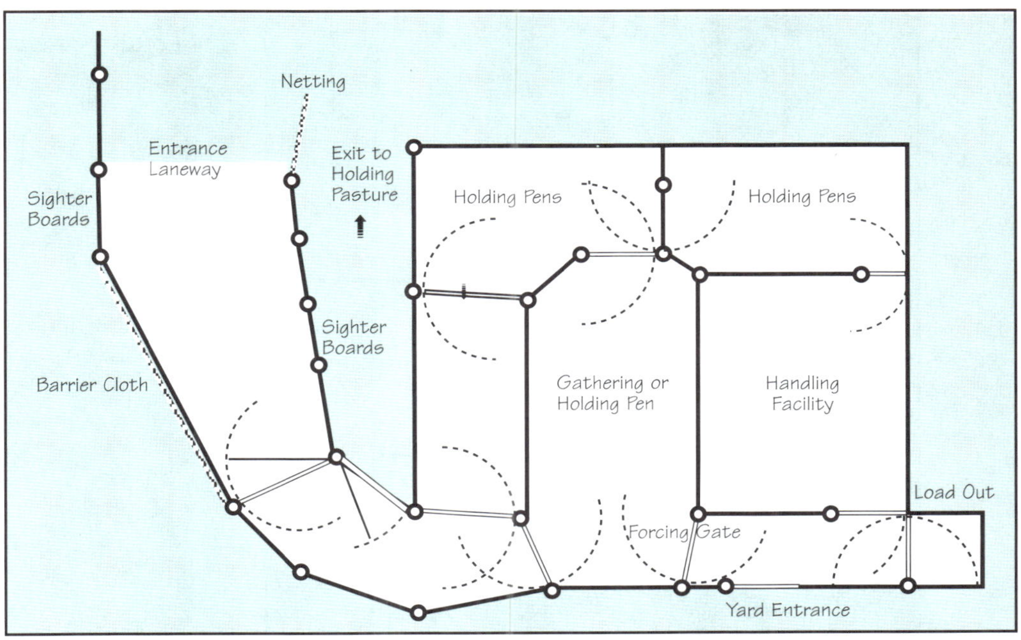

Fig. 1 Terminology and layout for approaches and holding pens for a standard deer or elk facility

USE

- Curves, turns and contours
- Approach yards obliquely
- Gradually strengthen fence and merge introduction into the pressure areas using sighter boards on top of netting, snow battens, snow cloth and open boarding prior to the handling yard
- Yarding raceway along with other farm features - overwintering pens near the barn
- Gradual reduction in raceway width
- Race width, receiving pens at an appropriate size for the elk farm numbers

For small units where there may be only 2 or 3 pastures a central laneway may not be practical. Elk can be efficiently yarded from a small holding paddock, using a wing fence that can be converted to a raceway for yarding. When not in use the elk can have access in and around it, and the area can also be used to establish feed stations.

Small elk units quickly produce quiet, extremely responsive animals with good relationship to the farmer and therefore, elaborate yarding/handling facilities are not required.

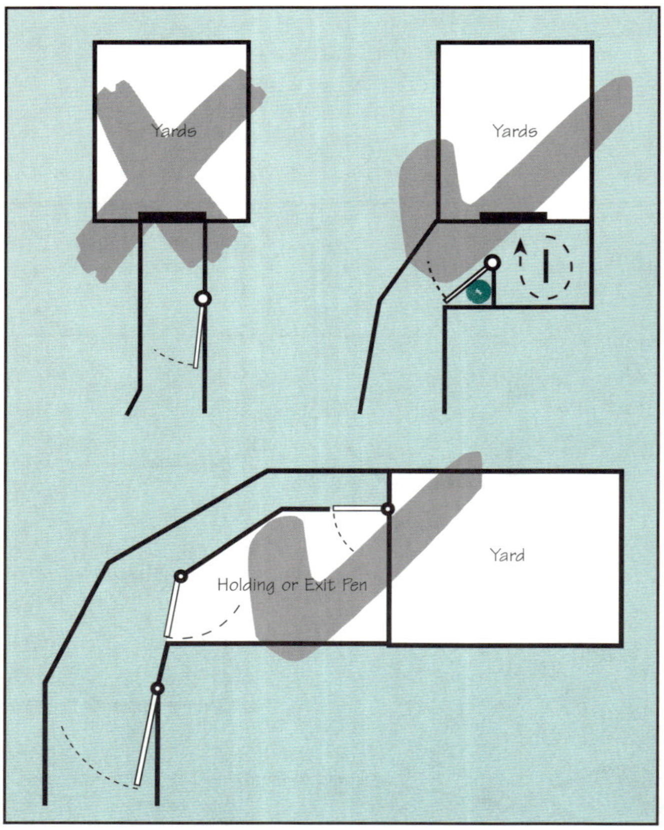

Fig. 2 Variations on laneway approaches to handling facilities

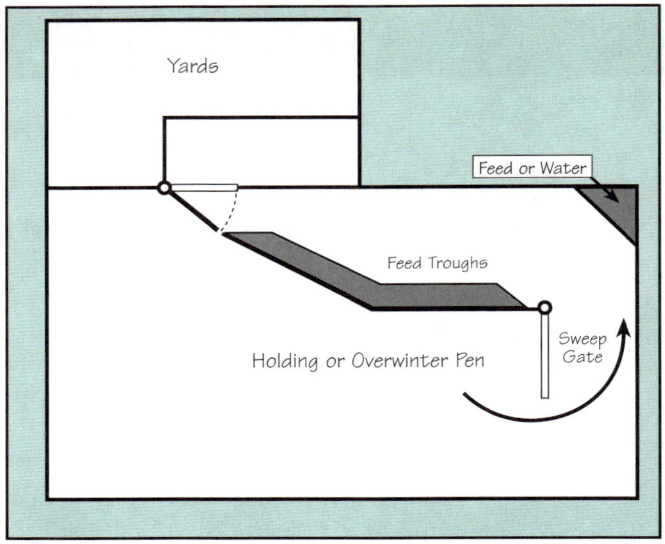

Fig. 3

CONSTRUCTION IN RACEWAYS, YARD ENTRANCES, HOLDING YARDS

Width There is a difference of opinion over ideal race widths, which can also vary according to stock numbers farmed and breed. Some favour narrow width (10 - 12') for the full length for stock control. The raceway can easily be closed using the gates from the pastures. This is useful for shifting animals from pasture to pasture or to ensure that elk returning from handling have no other options.

Others favour a wide race that gives flexibility in a number of areas, but may increase costs. On the whole, narrow raceways are becoming less widely used as farms expand in size. However, regardless of the size of the central raceway in the

Raceway Width Design

Advantages	Disadvantages
1) Narrow: (10'-12') (3.2-3.8 m)	
Stock control Flexibility in design - no need for double gates or barriers More effective use of pasture land Suits small stock numbers	Stock more likely to hit fence if panicked Can be visually confusing Subject to heavy stock pressure and surface damage Not suitable for larger herds Restrictive vehicle and farm machinery access
2) Medium (16-20') (4.8-6.4 m)	
Stock flow is better Less pressure on elk Vehicle access improved	Less control Extra expense in gateways and stock control Land lost to grazing
3) Wide (24-100') (7.0-30 m)	
Relaxed stock flow Suits development of keyhole holding areas Can be maintained in part in pasture and used for grazing small groups	Difficult to control small groups/individuals May need additional labour for yarding Extra expense in gates for subdivision within the lane

Table 1

area preceding the yards, raceways should narrow to between 10 - 14' to maintain the yarding pressure and allow strong and manageable gates to be used. A gradual width reduction, either curved or decreasing over distance is recommended. Table 1 summarizes the options:

Raceway features Gates in laneways and holding yards should close behind the direction of elk movement and towards the yards to maintain the pressure. During yarding all such open gates/ doors must fit flat against the fence or wall and be secure against wind or disturbance to avoid stock fright or the possibility of accident. If any gaps are left between an open gate and a fence there is a strong chance one or more animals will get trapped here with risk of panic or injury. Gates at the final pressure points of yards/races should be of more solid construction than pasture gates as these can be subject to considerable stock force and over use.

It pays to cover gate netting with snow/shade cloth or fill in the panels with plywood. If solid wooden gates are used there is additional strain on hinges and posts. All such gates are subject to greater wind stress and should be well supported at the locking end when closed, and positively locked with a steel pin into a supporting post, or chained tight. It is recommended that a chain goes right around a supporting post as there are many reports of animal pressure bursting open gates when only locked with a staple hook on the post. In a holding yard this area will come under the greatest pressure. Elk seeking escape once finally yarded regard gates as the weak point and will challenge any gaps.

Fig. 4 Illustrates the variety of common successful techniques and materials in laneway strengthening construction

For the final 20 - 30 m of laneway approach and in the holding yard itself, it is advisable to increase fence height, decrease the spacing between posts (approx 3 - 4 m) and improve construction strength as this is the area where livestock will charge and test fences.

Timber, plywood, and road construction cloth are all used successfully. Current trends include solid boarding or plywood up to 1.5 m (4') above ground and open boarding above that up to 2.4 - 3.0 m (8 - 10') depending on individual preference. Commonly, as a consequence of farm layout, the yards and holding areas are at the perimeters of the elk farm. The requirements of security and risk of potential escape in this area are at a maximum, and building style and height should reflect that.

HANDLING YARDS

Receiving Yards, Holding Yards, or Gathering Yards

The use of one or more receiving and holding yards gives a variety of advantages. In new yard construction it minimizes expensive shed building as the elk can be happily held outside the building.

They permit a multi-use role, particularly on small operations as shelter, protected winter feeding areas and water stations. In larger situations they are often useful in displaying stock for sale, for weaning procedures, and most importantly as a relief/relaxation area for elk after round-up and yarding prior to intensive handling.

However, if the yard size is too large the gains made in control during yarding can be lost as elk will bunch in corners or move rapidly around the yard making movement into the shed, or division into smaller groups difficult. A division capable of splitting the yard into two, once the herd has entered is useful and multifunctional (Fig. 5). When gates are open they encourage the habit of circular movement and provide a barrier for easy splitting of large herds, or smaller groups of, for example velvetting bulls or breeding bulls in rut. A milling yard should be square or broadly rectangular. Avoid building long narrow pens - width should be no less than 1/3 of the total length, with an area from 4 - 6 m^2 per elk in receiving area.

Similarly, a receiving yard that is too small creates as many problems and stress factors as an oversized pen. Most injuries and damage to animals, velvet antler or quality venison are created by overcrowding - particularly when combined with poor stockmanship. Overcrowded elk will pile up in corners and climb on each other. Cows can inadvertently crush young stock and antlers are easily broken especially in 1 and 2 year old animals. Overcrowding and hard antler are lethal combinations.

The solution lies in stockmanship and in only handling sufficient numbers for the facility design rather than trying to build for every possibility.

Attention to detail is necessary in construction. Projections like nails, door catches, water troughs or feeders should be eliminated or protected. Gaps greater than 150 mm under walls or gates are an opportunity to test escape. Often this area has to contend with heavy traffic and mud build up so good drainage must be considered. Consideration might be given to a suitable flooring - coarse sand, washed smooth river gravels or wood chips all work well, but concrete use in these holding areas is not advised.

If possible such holding yards may incorporate large trees, well protected and used to seal off right angled corners in the pen perimeter as well as providing shade and a comfort zone. If elk are held in such pens for a long time in hot conditions an artificial shade barrier is necessary (cloth or snow battens forming a partial roof are advised). Water troughs can be installed behind solid panels in the yard corners. Well constructed holding yards can be the most useful part of any facility with multiple uses. Positive associations with food, water, etc. can make the whole task of round-up and yarding an easy and enjoyable task, a proverbial "stroll in the park". Once the system is established, experienced animals should be yarded with newcomers as a calming and educational influence.

Handling Facilities - Design & Function

Elk handling facilities are a series of pens and gates that allow groups of animals to be received, broken down into smaller groups before entering a processing area that can handle one or two individuals or groups of three to five.

The aim of any design is to reduce stress,

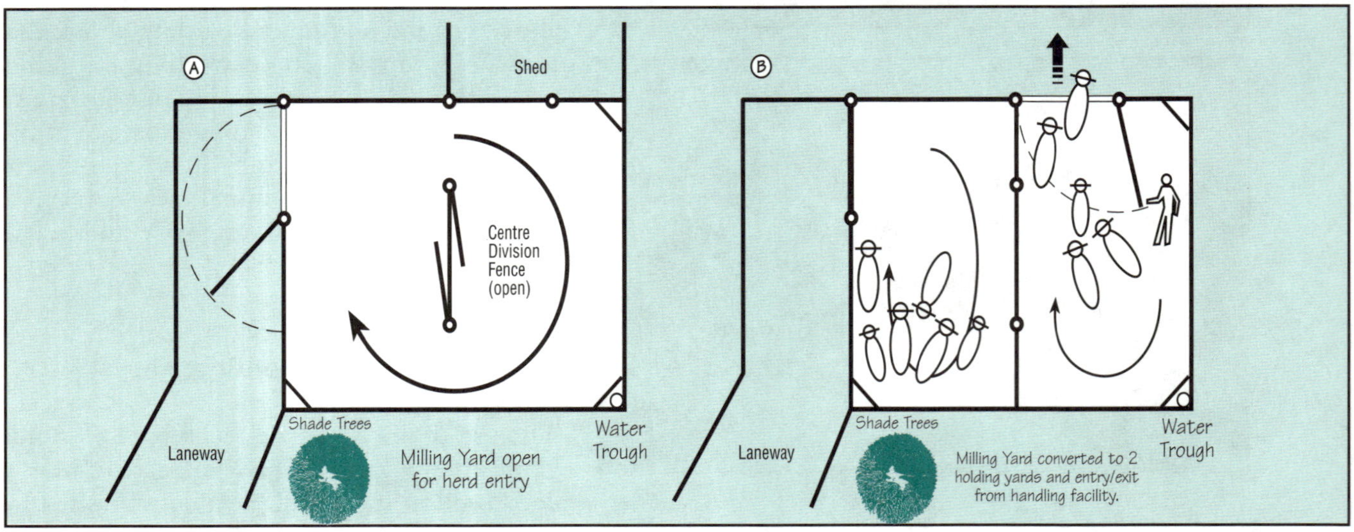

Fig. 5 Milling or Receiving Yard Operation

aggression, or any potential injury to elk or handler, and to achieve the management task efficiently and effectively. It is not necessary to consider great speed of operation as a priority as elk are more comfortable when handled under less pressure, but an efficient routine without interruption.

Good working conditions are essential - and planning involves consideration of many features summarized below.

Handling Facility Design - Keypoints to Consider
Access to essential services (power, water, vehicles)
Size: Adjust animal numbers, don't over build
Simplicity and Efficiency
Welfare and Safety
Light and Ventillation
Quality Materials / Sound Construction
Curves / Corners - Flow Through Pattern
Specialized Equipment Meets Animal Needs
Routine and Familiarization

KEY FEATURES TO FACILITY DESIGN

Light - natural or artificial Elk respond well to enclosure and subdued light on first entering a barn. It is not necessary to have darkness as many operations require tag reading, veterinary examination and animal health treatments that are specific in placement. Good lighting in a work area is essential and it is often practical to have lighting controlled by a dimmer switch.

User friendly materials Elk have a habit of staying in tight groups and trying to move around pens and through gateways as a group. Wooden boards must be thick enough to absorb this pressure (1" - 1 1/4" x 6" - 8" planks supported at 1.2 m intervals). Plywood (5/8" - 3/4") is an ideal construction material as it can be easily framed and supported, and used to form curves or to construct chutes. Timber is quieter and more forgiving than steel/sheet metal construction, although care must be taken in repair and maintenance to replace damaged areas as they happen. Young elk in particular can easily break legs or damage faces and ribs on any broken boards or projections.

Doors carry a lot of pressure at the hinges and locks. Most accidents happen in this area.

Common faults include:

- Doors bursting open under stock pressure, or being buckled or broken.
- Hinges sagging - doors fail to latch or close easily.
- Gaps between hinges/door frames - creating foot traps.
- Door latches projecting into stock path, risking bruising/eye injuries.

As doors are one of the single biggest investments in yard construction they must be well framed and hung with locks and catches that are easy to operate and shut securely.

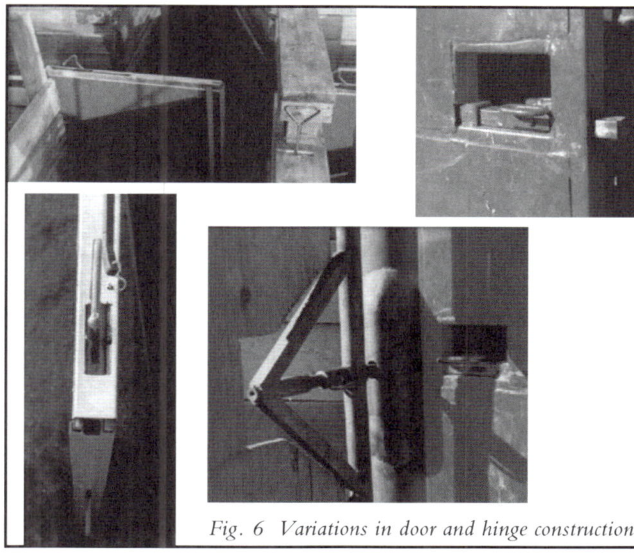

Fig. 6 Variations in door and hinge construction

Flooring/Surface Elk farmers are faced with a difficult choice balancing cleanliness and ease of maintenance with suitability for their animals. There are a number of choices:

(a) Concrete is an ideal medium for the farmer and is effective in elk yards with some limitations.

- Too smooth and elk will slip and fall with potential serious injury through tendon or ligament damage and loss of grip - worsened by heavy use and fouling.
- Too rough - coarsely grooved, etc. Concrete is difficult to clean and is an uncomfortable surface.
- Frequent yarding on concrete can create damage to hooves through excessive wear and presents the risk of infection.

Cleanliness and function can be combined with a spreading of coarse sawdust or wood chips over a concrete base with routine cleaning 2 - 3 times a year.

Concrete need not be used throughout elk yards but is ideal for heavy work areas and special facilities such as velvet antler removal areas and work chutes and weighing platforms. Any area used for surgery (AI reproduction, etc.) should be smooth concrete.

If the surface is roughened or raked just prior to setting an acceptable compromise can be reached.

b) Sand is favoured by many farmers - it is quiet, dust free and absorbs dung and urine well, but like other loose surface material, it migrates to the sides of pens after a time.

c) Sawdust and wood chips are quiet, absorbent and cheap, but often prove to be dusty unless replaced frequently.

d) Earth flooring and coarse gravels should be avoided. Elk in confinement urinate frequently and quickly create wet, slippery, uncomfortable conditions that pose risks to hygiene and animal health and become unpleasant to work in. Gravels "migrate", jamming doors and weigh scales and are noisy and unsettling. Make provision for drainage.

Walls Elk yard walls are similar to doors and safety and strength are essential in construction. Modern facilities use open boarded arrangements from 1.2 - 1.5 m above floor height and only solid boarded in the working and full pressure areas (Fig. 7). An open plan appears to allow groups to maintain usual contact, and causes less disturbance when handlers enter the pens. Such construction adds to openness and airiness and is conducive to stock well-being and the general cleanliness and appearance of yards. This construction can also save costs appreciably.

Care must be taken, however, to avoid significant gaps under walls, or spaces between boarding that become potential foot, or head traps. Gaps under doors and walls should not exceed 100 mm, or 100 mm at eye level, but can be increased to 150 mm towards the top of the yards. Pen height is recommended at at least 2.5 m, and preferably 3 m, as occasionally elk will jump when confined, or in a single animal/pen situation, as is often the case in handling situations.

Fig. 7 Handling facility wall construction

When building pens inside existing barns it's useful to reflect that elk move quickly against walls. Any projections, (stone walls, roof beams, etc.) can be very damaging to the animals and must be avoided. Feet, antlers, eyes and rib cages, and indeed the whole animal are vulnerable to damage through poor construction. Elk move in tight groups at speed in pens and rarely move in single file of their own accord, therefore, there is always the chance of injury through the speed and pressure created.

HANDLING FACILITY DESIGN

The common theme in elk yard design is individual variation, but the common features are SIMPLICITY and FLOW. Elk become quickly used to a routine and will learn a working pattern very quickly, and there are many adaptations of existing buildings and plans, modified to suit, that work extremely well. Before building it is advisable to see as many facilities as possible and plan well using consultation and comment as much as possible. In all good yards:

- The principle of elk moving around curves or corners applies.
- Pressure areas can be relieved by building forcing gates.
- Doors should open into pens against the traffic flow so that control is maintained and the doors shut following the group.
- Ideally the flow should circle around the barn so that after the elk are released from individual handling (work room, scales etc.) they exit adjacent to where they have come from.
- Yards have two functions,
  1) Break down groups into smaller manageable units (4 - 6 head).
  2) From these small groups animals can be handled individually in specially designed equipment or construction for the management task.

Remember: elk as individuals react differently to elk in groups. Once animals are alone their safety and handler safety must be accommodated with care and caution.

COMMON HANDLING FEATURES
Octagon, Tub, Rotunda Circle

There are many variations on this central feature of many elk yards. Such design allows animals to be collected or drafted to/from holding pens, or used as a marshalling area for elk to be lead into chutes, squeezes, scales, etc.

Doors, except for the main entry and group exit door, should open into the circle and shut positively against the door frame. This allows great flexibility in drafting and control on a single animal basis. An internal diameter of 2.4 m allows gates of 1 m width. The shape need not be exact. Often the entry and exit areas are widened to 1.2 m for ease of access and the smallest width becomes a blank - useful for building an equipment cupboard. The most common variation is to extend the width and install swinging squeeze or squeeze doors, centrally pivoted, to allow options of individual animal handling, remote handling or forcing. Widths range from 3.7 m to 5 m. Doors can be solid, with inbuilt sliding panels for access, combined solid at the base with horizontal spaced (150 mm) piping above 1.2 m for access, and often contain a small inbuilt door (0.8 - 1 m wide) for people and animals to pass from one side to the other.

The central pivot pipe must be well supported at base and top to avoid gates dropping and binding on floor and walls. If this design is used as a squeeze it must be of strong construction.

The same principle can be applied in square pens, using just one squeeze gate as a forcing door, or for an animal squeeze against an existing wall. Such a door should be offset from the corner to give adequate head room and in effect form a work race.

Fig. 8 "Rub Strips" prevent hair loss

THE "BRINGANS" CHUTE

This versatile handling race has been developed from the realization that two or three animals together are more comfortable than one and it is ideal for handling yearlings, cows and non-antlered bulls for most of the routine animal health treatments. Weigh scales are easily incorporated, and the chute also serves as a holding area for animals prior to entering a squeeze.

Fig. 9 Bringans Chute. Note: Raised work platform, Adjustable side height sliding door control - Acknowledgement, Dr. Mike Bringans MVSc. Orr Lake Elk, Ontario.

Fig. 10 Bringans Chute, variations

SPECIALIZED HANDLING EQUIPMENT

For the larger units, specialist velvetting operations, or for the greatest ease in handling of adult elk, a number of purpose built individual handlers are available. Such squeezes are an expensive investment but a boon to veterinarians, farmers and animals, although such investment should reflect the scale and size of the operation.

A squeeze must restrain the animal quickly, comfortably and positively, with particular

Fig. 11 Hydraulic Elk Squeeze

Fig. 12 Hydraulic Elk Squeeze, another variation

attention to restraint of the head. If the squeeze is noisy, uncomfortable or ineffective, aversion will build. Any squeeze is only as good as its positioning in the yards and the ease of animal entry. Good movement can be aided by the use of short curving raceways - wide enough to let antlered animals comfortably move. Full pen width forcing doors are also valuable tools, but either way with any new facility or alteration it is important to run the elk in small groups through scales, squeezes, circles etc. 3 - 4 times to develop a routine before holding or working on the animals.

It can also be argued, for highly responsive easily worked elk, the contact between the farmer and his animals - the mysterious "stockmanship" - should be encouraged from an early age. If it's not necessary to use squeezes, etc. for a task - then don't.

LOADOUT CHUTES

One of the most commonly overlooked features of facilities is the loadout chute. This final handling area is often the last area where the farmer has control over the quality and presentation of his product. Usually there is an increased stress level associated with loading into vehicles.

Loadouts work best when they are as wide as the receiving vehicles access doors (1 - 1.2 m), as elk try and move together. Some sort of turn prior to entry works exceptionally well. If a ramp is required it must fit exactly with the floor of the loading vehicle to avoid leg or foot damage and may need to be adjustable. This area can often be the only link with the great outdoors, so security-roofed or a solid funnel between the yards and trailer prevent escape. This area should also be

Fig. 13 Access doors on squeeze for semen collection, AI, Calving

surrounded by a secure fence. See Transportation section for loadout design.

PEOPLE SPACE

Any handling facility must also accommodate handlers, visitors, and particularly veterinarians needs. No working in yards should proceed without some form of record keeping, and there is always the need for storage of a variety of animal remedies, veterinary drugs or accommodation for stock testing and identification procedures.

Within the facility a work area for elk handlers is critical and is often associated with a chute, scales or squeeze working area.

Consider the need for the following:

- Secure (lockable) cabinets for drug and animal remedies.
- Safe disposal for needles, glassware, used containers.
- Adequate adjustable light sources and power supply.
- Water supply for washing facilities.
- Clean surroundings.
- Notice board/whiteboard for shed records.
- Data recording facilities.
- Antler and meat cold storage.
- Storage for elk handling equipment.

Trouble free elk handling can't be done with equipment lying in corners or on top of walls. Basic consideration to the task ahead and checking and maintaining equipment will ensure that elk handling is trouble free and, most of all, enjoyable for both the elk and handlers.

ELK IN HANDLING FACILITIES

In handling areas in yards elk tend to behave less as a group animal and are treated as individuals during intense handling, mainly because of their size and the speed and force with which they move.

Calmness, patience and confidence are important considerations with elk, and rushing or tension and nervousness are readily recognized.

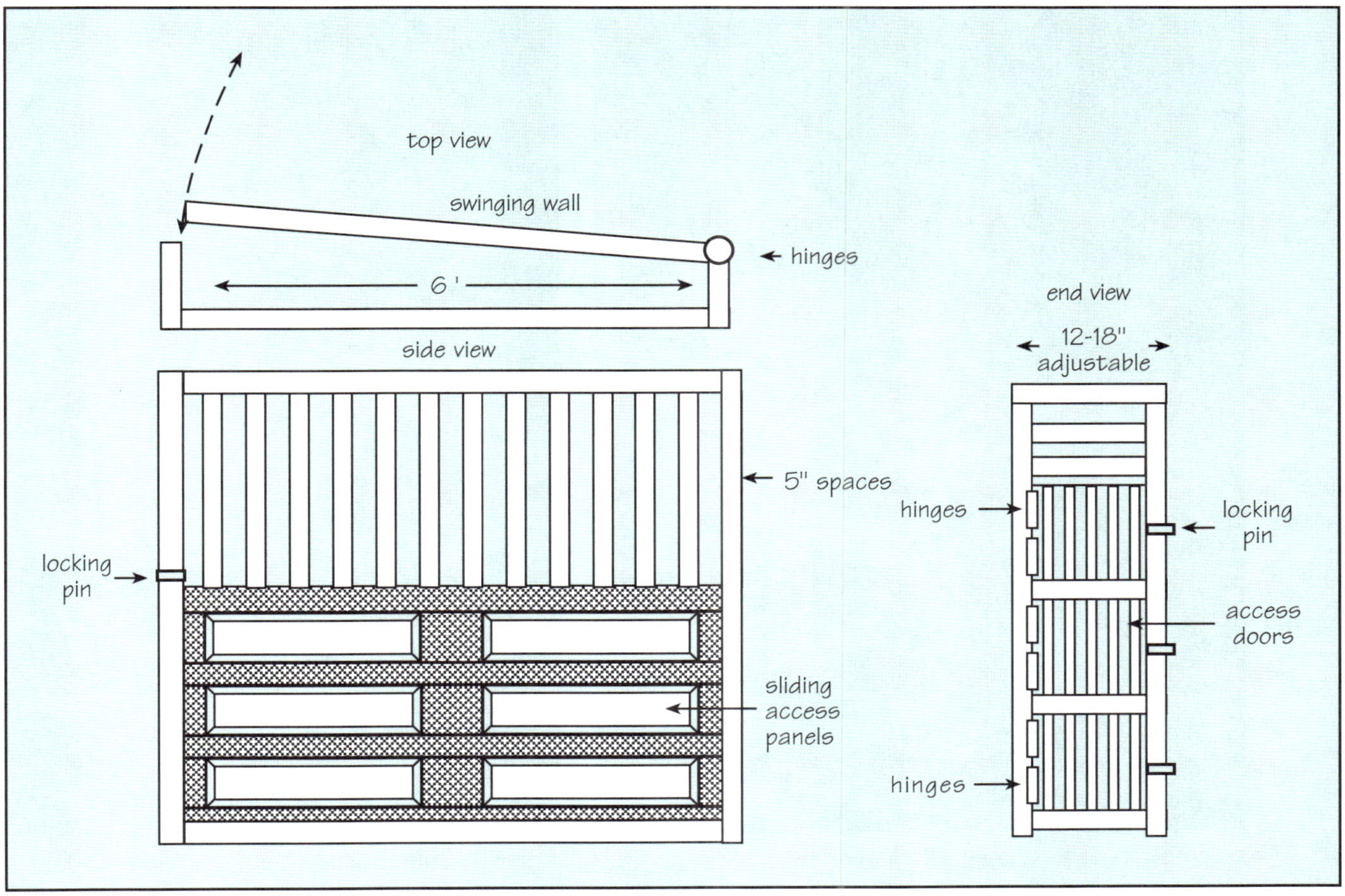

Fig. 13 Swinging Wall Squeeze, (Cow Box)

Elk do not appreciate being held in confinement for long periods and it is much better to use milling pens or keyhole areas to hold the animals, yarding only when the task can be done simply and efficiently before release.

Elk require reasonable space when entering yards to spread out and settle first. Modern systems use a wide (3 - 4 m) corridor approach with full width doors swung opposingly to divide animals into smaller groups, either from ground level, or above from cat walks. Working pen size ranges from 3 - 4 m^2, and the main work area generally involves a 4 - 5 m diameter circular swing gate arrangement, useful for forcing elk into other work areas (i.e. chutes or squeezes) or as an individual animal squeeze. Protection with some sort of shield or the use of doors to push and sweep animals creates confidence for both.

Elk work well in raceways 1.2 - 2.0 m in width. Solid bases for walls and gates up to 1.5 m in height then an open construction of pipe/steel bars, etc. give more comfort than solid walls to the necessary 2.5 m. If catwalks are built across the top of pens or doorways they must be high enough to allow movement of antlered animals. Again simplicity of design and a recognizable flow are critical. Elk are difficult to move from one blind pen to the next and prefer space and some movement first. They run extremely well around corners.

WESTERN DESIGN - SWEEPS & SLIDES

Farmers in Western Canada have continued to improve their handling systems since elk farming began - build one, tear it down, build another, and another improvement. The "State of the Art", primarily designed for elk, but also useable for bison and red deer, is illustrated on the pages following. The "all doors" design keeps a door between the farmer and the livestock at all times. The addition of a swinging wall squeeze, plus a hydraulic handling squeeze makes this the most efficient and safe system yet.

Fig. 14 "Cow box" or swinging wall squeeze

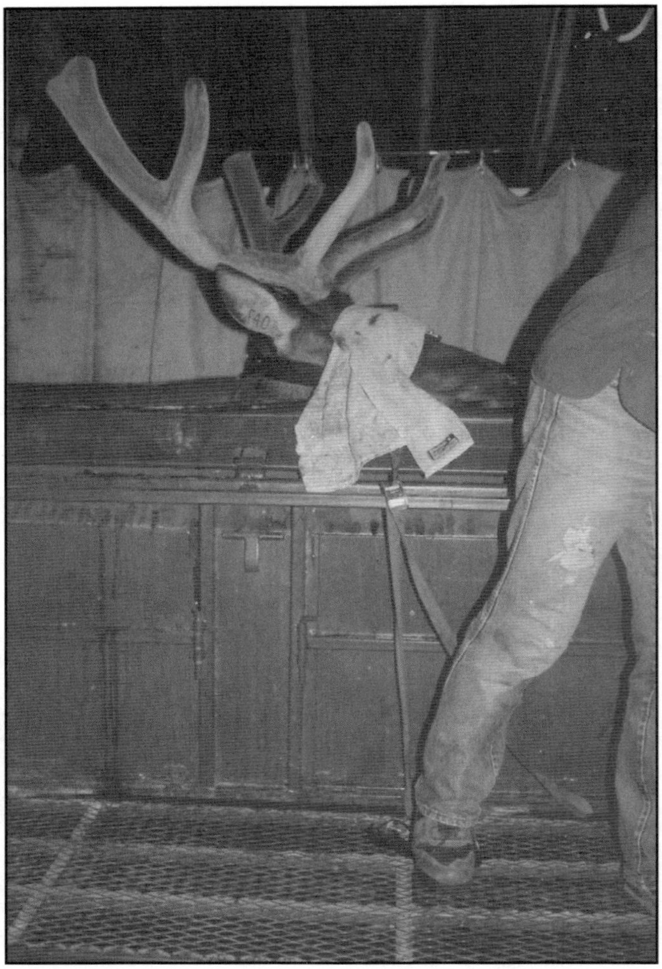

Fig. 15 Properly restrained bull in hydraulic squeeze

SLICK HANDLING FACILITY FOR ELK

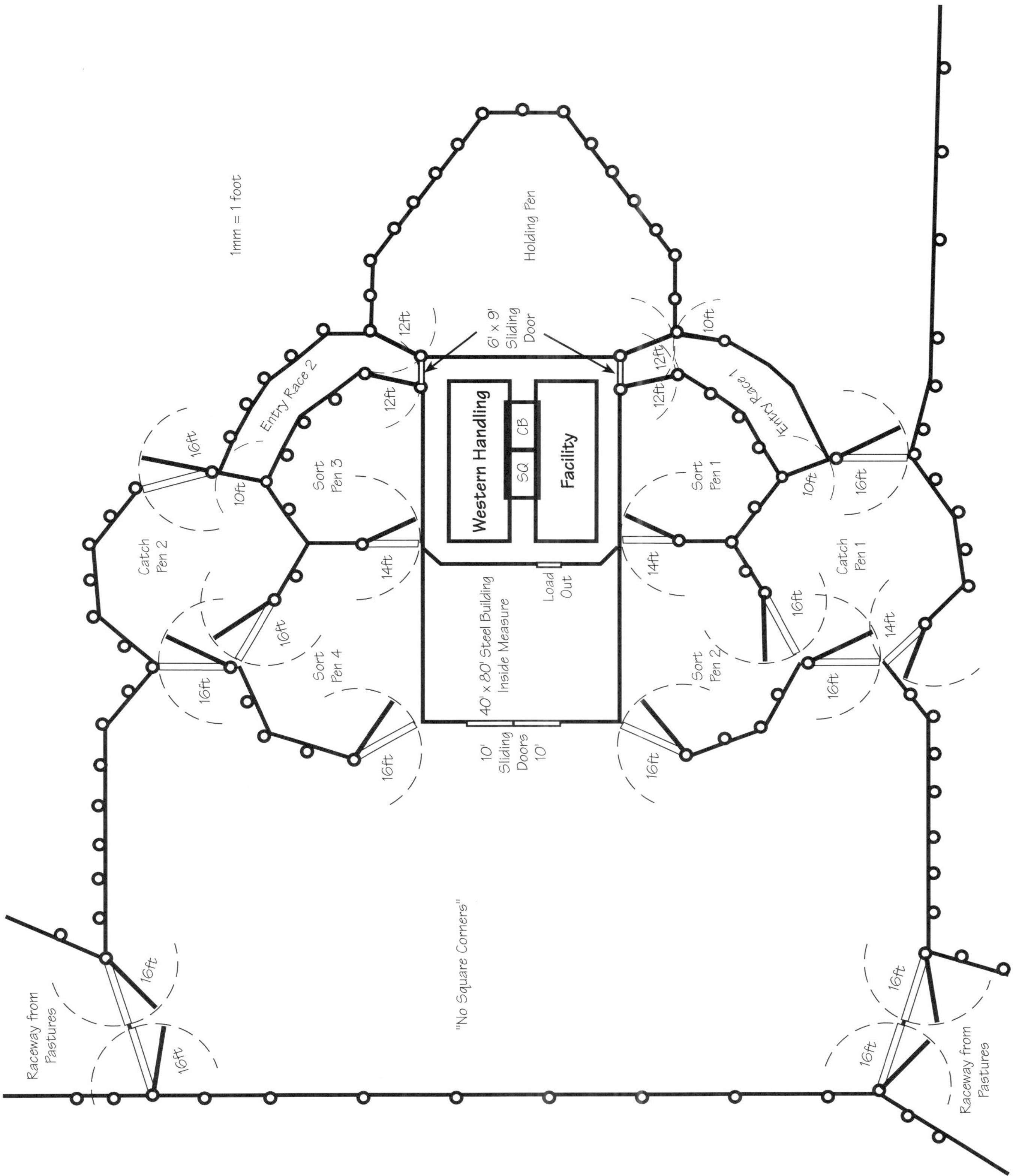

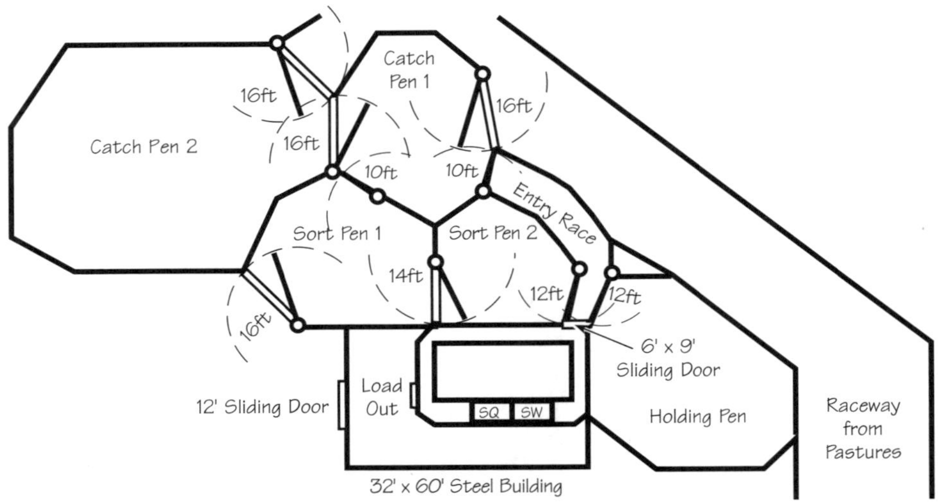

MORE ECONOMICAL VARIATIONS ON SLICK HANDLING FACILITY

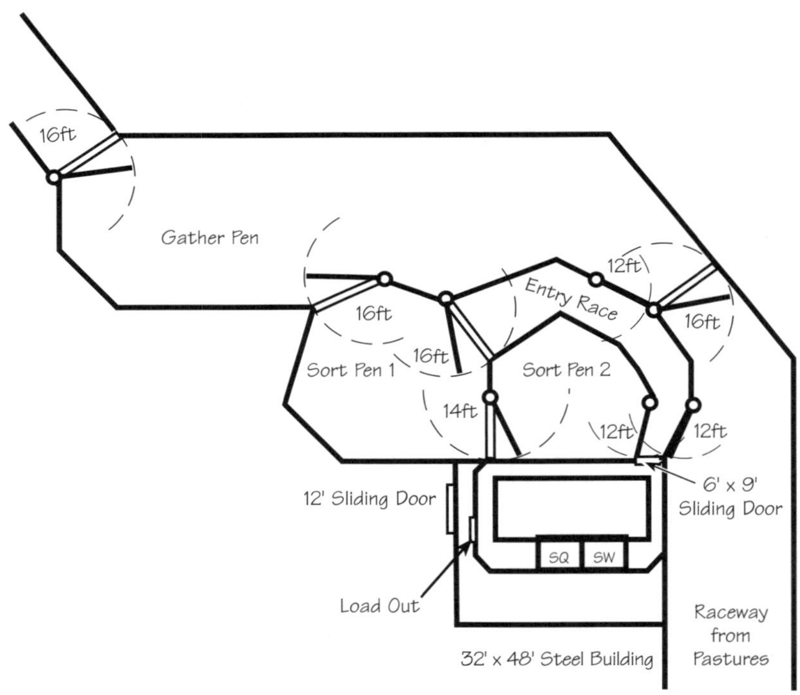

WESTERN ELK HANDLING FACILITY STEPPED EXPANSION

Note: *Measurements may vary due to material selection and space limitations*

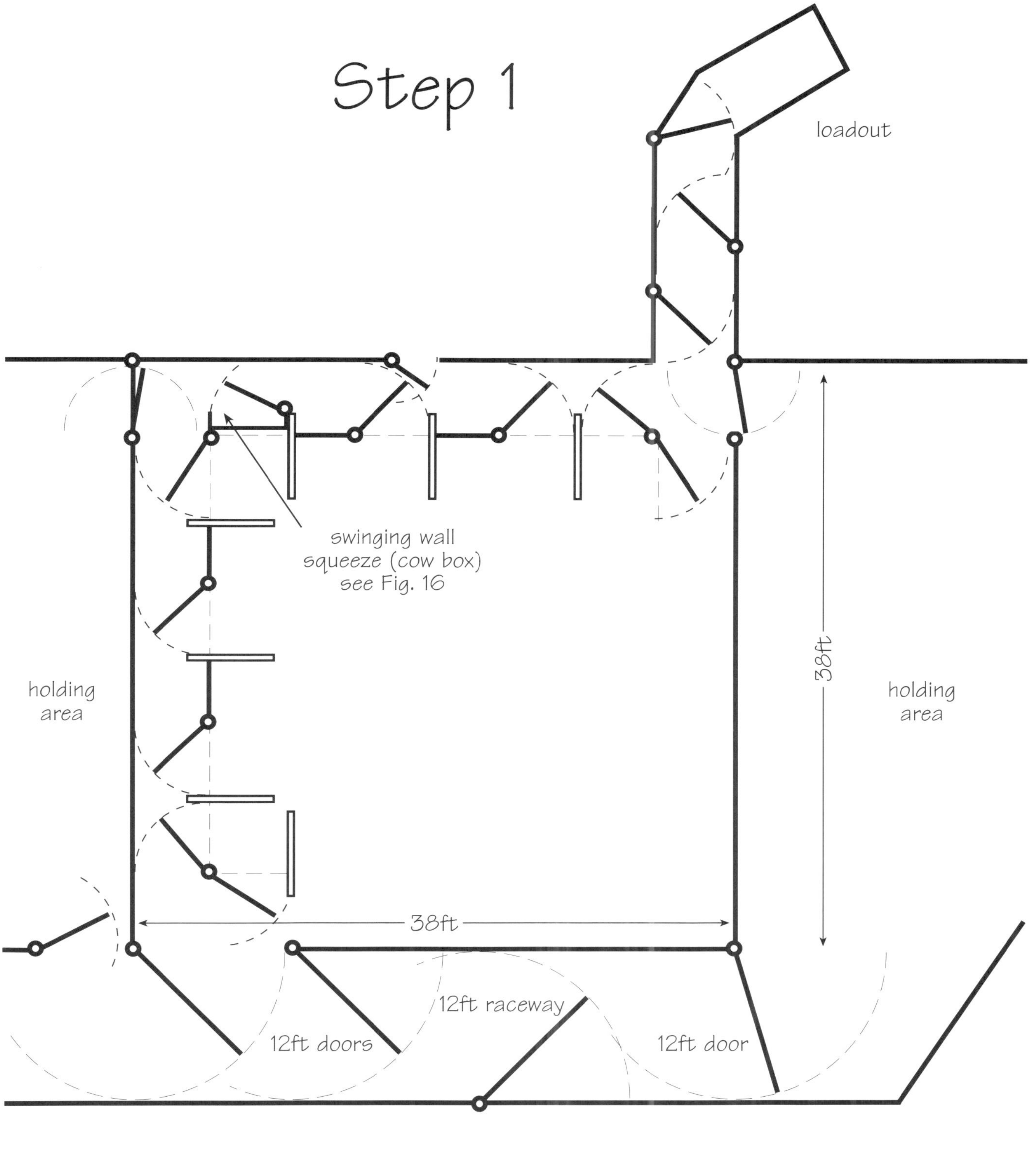

Modify this layout to suit the building you may have available

WESTERN ELK HANDLING FACILITY STEPPED EXPANSION

***Note:** Measurements may vary due to material selection and space limitations*

Step 2

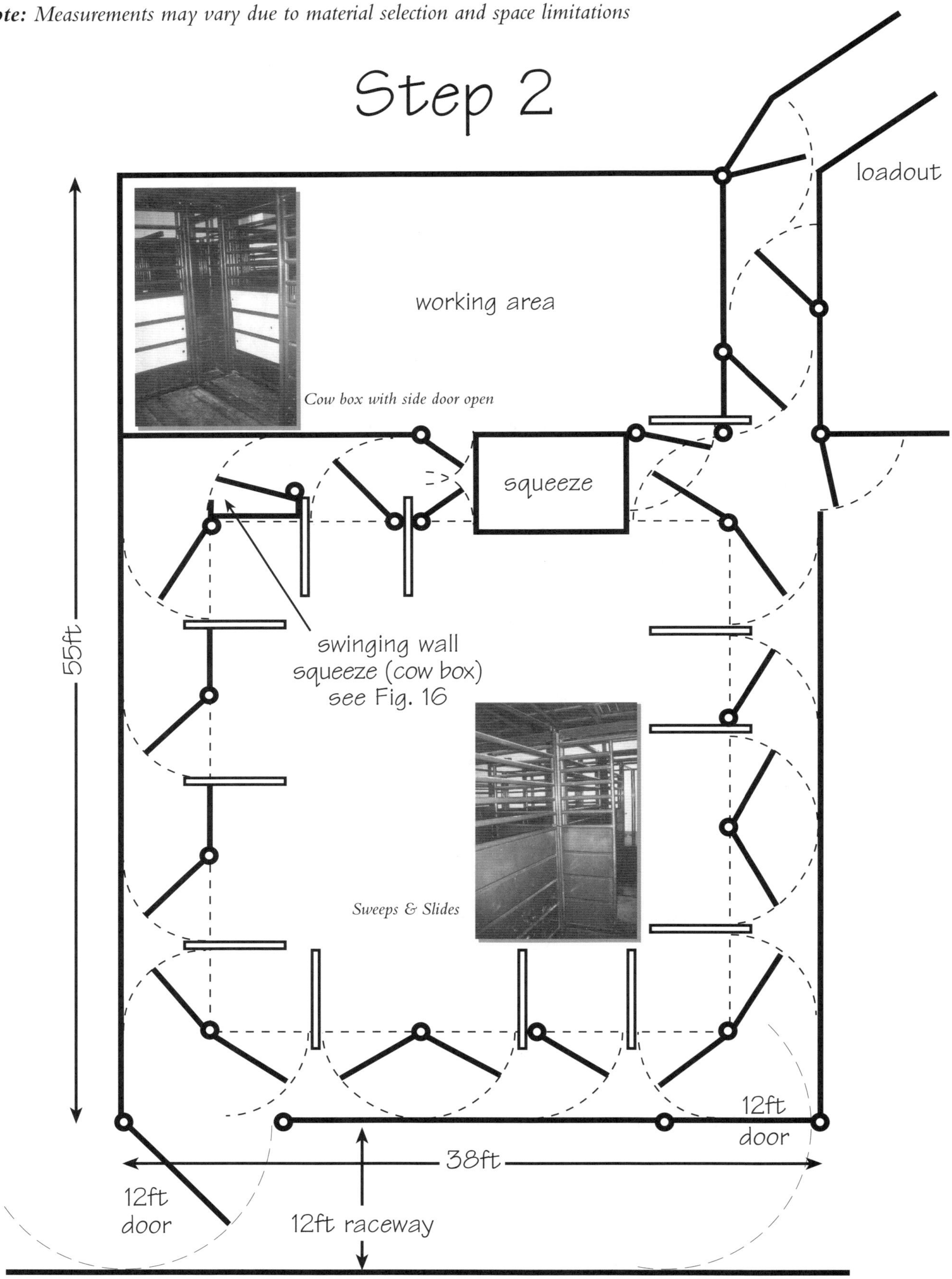

WESTERN ELK HANDLING FACILITY STEPPED EXPANSION

Note: *Measurements may vary due to material selection and space limitations*

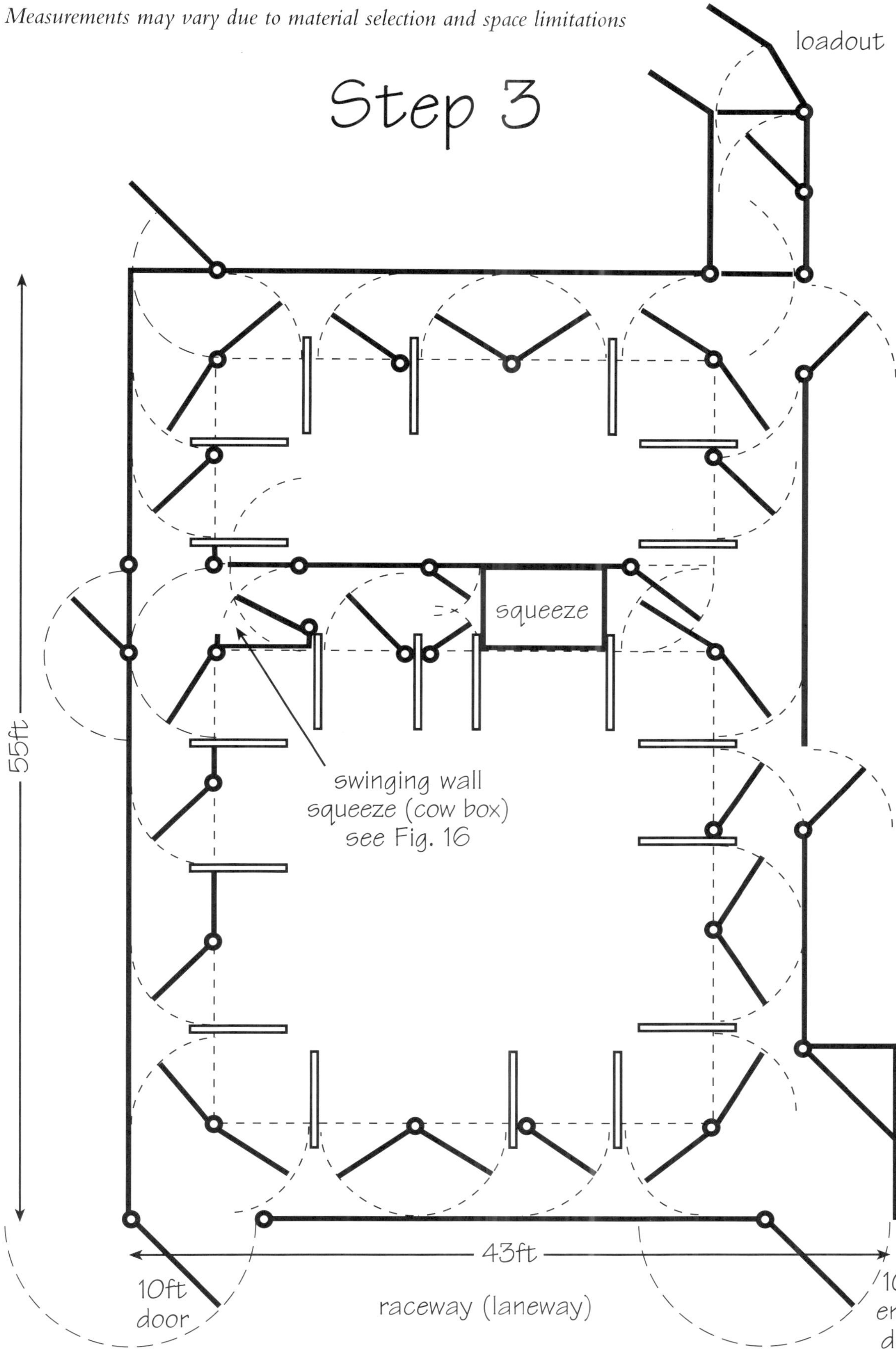

Step 3

Top View, with squeeze in outer row

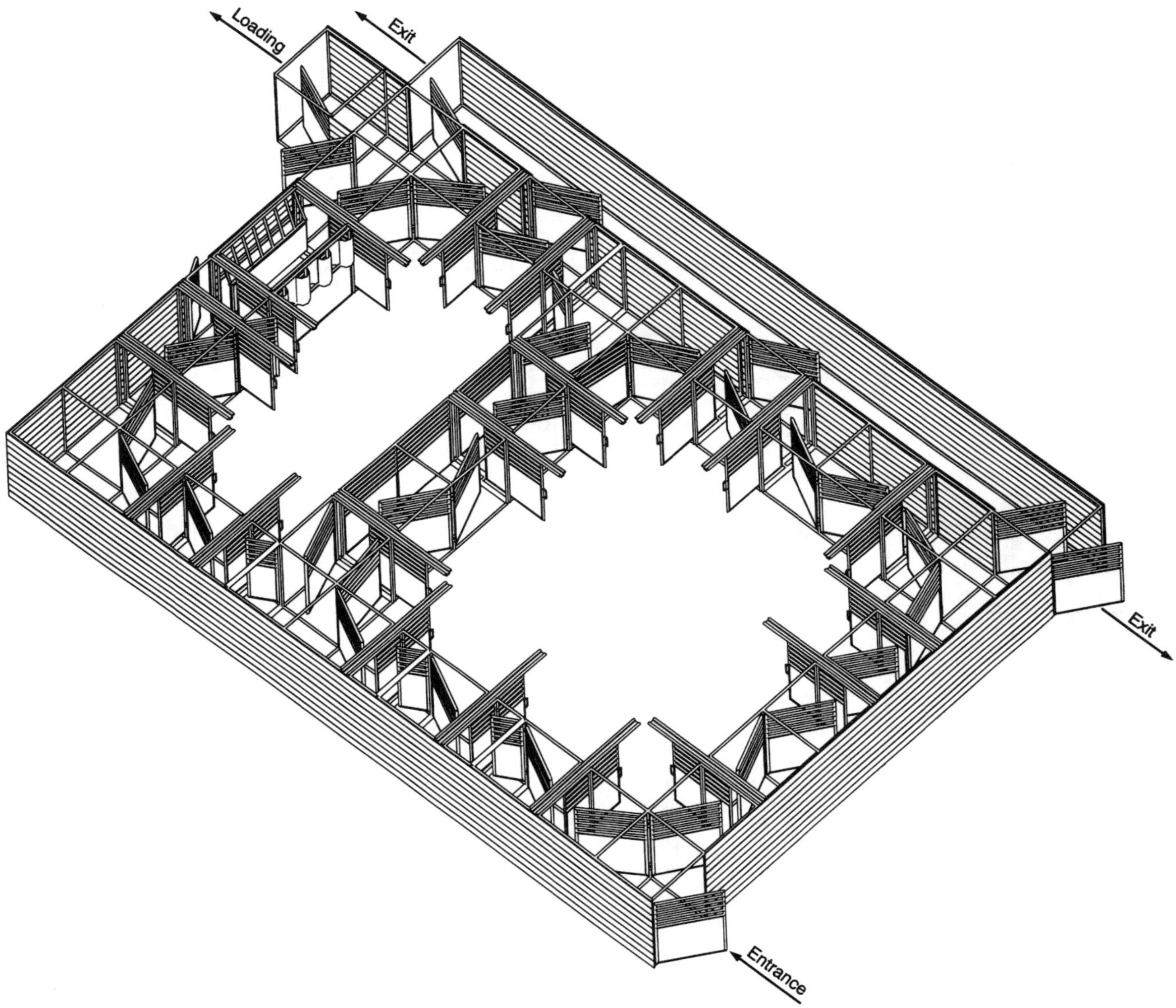

CONSTRUCTION GUIDELINES FOR WESTERN DESIGN

This system is a series of swinging gate and sliding gate modules arranged in a rectangular shape. The operator works at ground level in the centre of the rectangle. Each combination of swinging gate and sliding gate can be used to hold, separate and move the elk. The swinging gates are used to "sweep" the animals out of the module while maintaining operator safety behind the gate.

The bottom 4' of all swing and sliding gates is constructed of solid material. Spaced pipe is used for the upper section. The material used for the solid section can be plank or panel. It must be durable to withstand kicking and smooth to prevent injury.

The next two pages give suggested construction details of the sliding door and swinging gates. Note that the gate can be locked in both the open and closed position. The latch is a 1/2" U-shaped rod which straddles the pipe frame uprights. A self locking tab prevents the latch from being accidentally opened. The construction shown maintains a 5' width of clear access for the elk between all steel posts.

For the sliding gate, the track is conventional heavy duty door track fastened to the gate frame. Guide bars (one on each side) are attached rigidly to the frame at the top of the gate to prevent the gate from twisting when partially open.

The swinging and sliding gates, with an outer wall make complete stand alone modules. Each module can be bolted, clamped or welded to any adjacent module. With less modular off-site construction and more direct on-site fabrication, the amount of pipe framing can be substantially reduced as adjacent gate units share common framing.

It is recommended that elk should have about 8' of unobstructed height inside the chute. However, some producers may want to construct these units inside existing buildings with only 8' ceilings. The drawings all show 8' overall height of the steel structure. Clear height inside the chute is therefore 8' minus the diameter of the top crossover pipe.

Variations of the swinging gate module are used within the system. These variations include the corners, double gate units and the squeeze.

The system can be used either inside a building or outside. In either case, the perimeter walls of the rectangle should be relatively solid and a minimum of 8' in height. Use narrow spacing to prevent hooves and antlers from being caught and injured. These perimeter walls can be free standing post and plank placed adjacent to the steel modules. The walls of a building can be incorporated into the design.

When a building is being constructed specifically for a low level handling system, recommended ceiling height is 12'. This height allows for adequate ventilation and proper lighting.

The corners of the gates and frames are shown with 45° mitred connections using a cutoff wheel. If crimped butt joints are used, arrange the joints so the pipes will drain water from rain, snow and condensation.

If producers start with only a few animals, they may not want to initially commit the money or the time to build a complete system. Steps 1 to 3 offer staged construction alternatives. The slide and swing gates are the heart of the system. When reducing the number of gates for staged construction, consider the associated reduction in handling effectiveness that will be encountered.

Although systems like Step 3 are currently in operation, the direction of gate swing and the number of swinging gates shown is variable and dependent upon the handling characteristics required by the individual elk producer. Producers must design accordingly to meet their requirements and accommodate any restrictions at the site.

SLIDING GATE FOR WESTERN ELK HANDLING FACILITY

Note: *Measurements may vary due to material selection and space limitations*

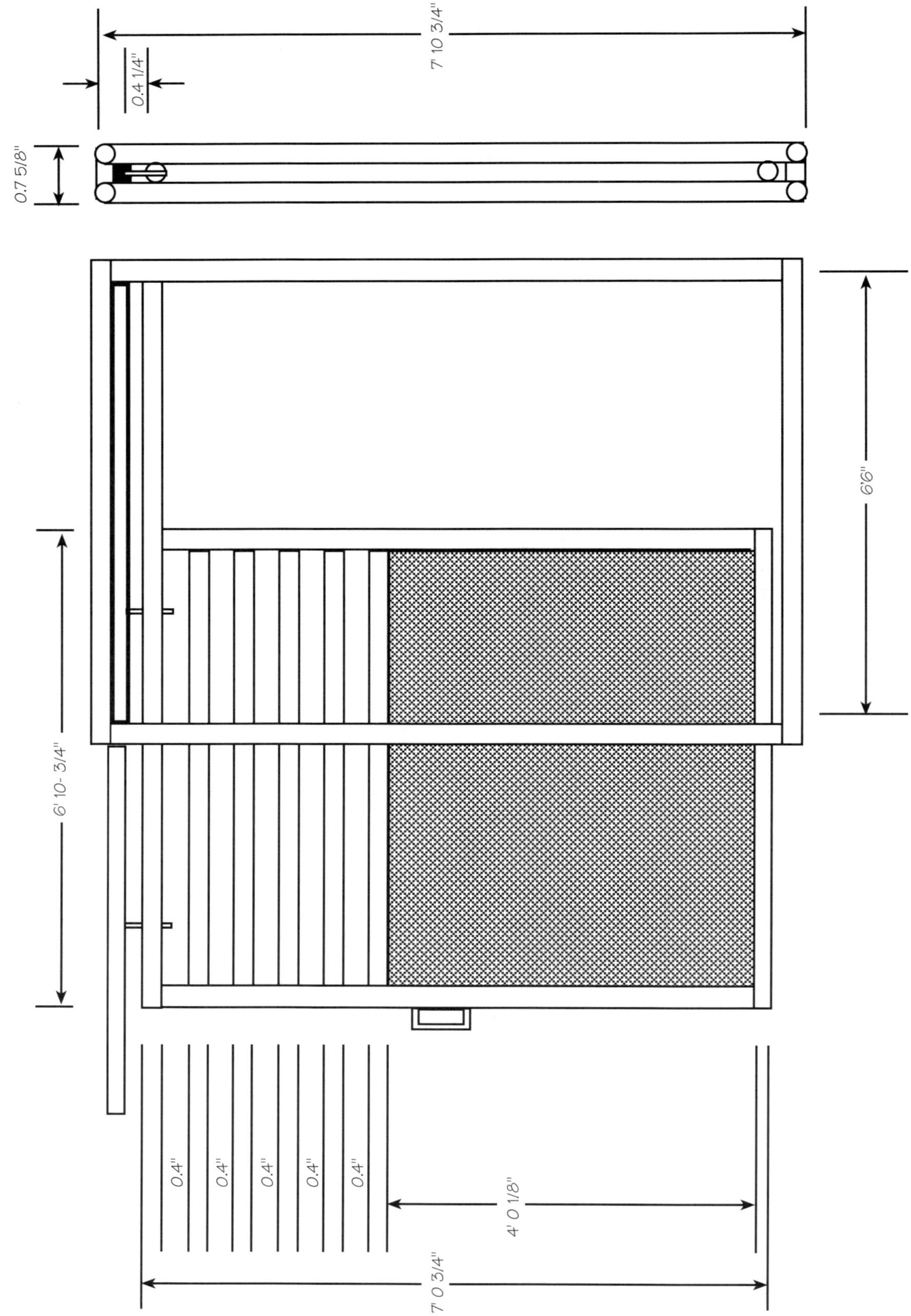

SWEEP GATE FOR WESTERN ELK HANDLING FACILITY

Note: *Measurements may vary due to material selection and space limitations*

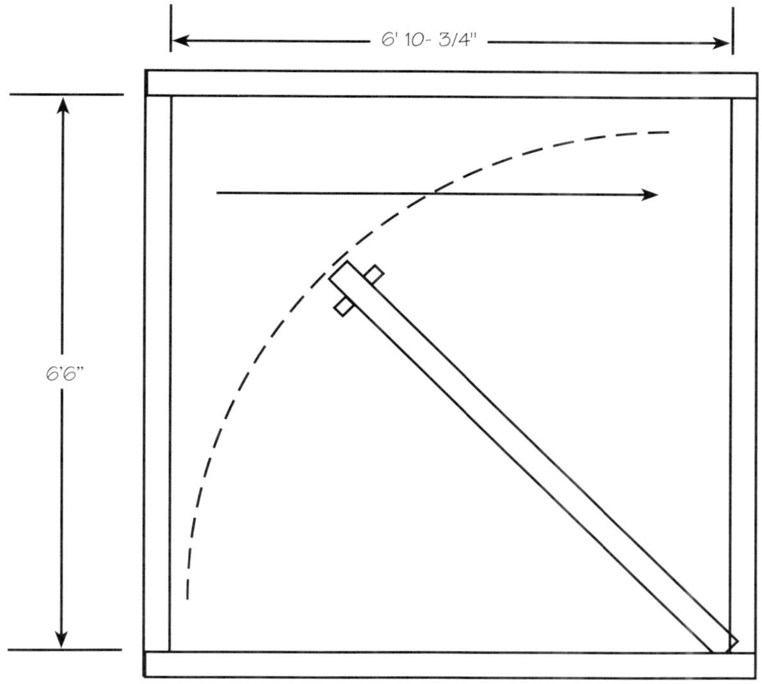

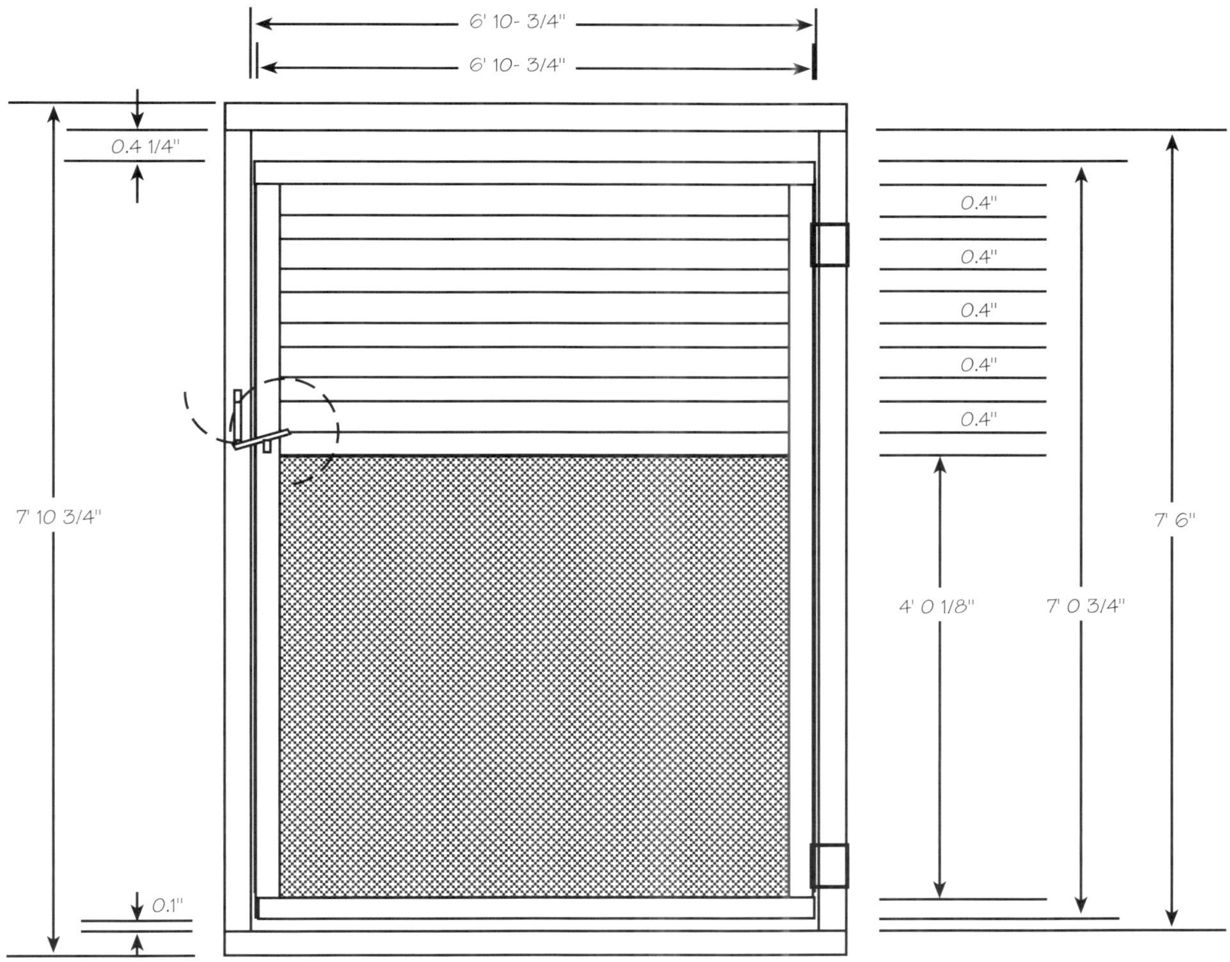

Elk travel very well, given proper conditions. Safe, efficient and humane transportation requires thoughtful preparation, careful handling and proper design.

Here are some key points:

- prepare the groups for shipment well in advance. Sort them into same size/age/gender groups, and allow them to establish social relationships before loading, preferably for a week or more.
- provide sufficient floor space to allow all the elk to lie down at one time. A useful calculation method is to multiply the actual length of each animal by its actual width, then double that area. This calculation applies to antlerless elk. Adults in velvet need triple the area, rather than double. Adults in hard antler should ideally be individually penned.
- provide bedding material, preferably sand plus sawdust, or wood chips, straw or plenty of hay.
- trip length will govern how the elk are fed and watered.

On trips of twelve hours or less, the elk should be well fed and watered before loading, then further feed and water is unnecessary during transport. For the first several hours, the elk are unlikely to utilize any feed or water provided. For trips longer than twelve hours, fresh hay plus water should be provided every four hours. The easiest way to accomplish this, if the outside temperature is near or above the freezing point, is to push "flakes" of small square bales into their compartments, then hose down the hay and the elk each time you stop. This will provide all the moisture they need for trips up to forty-eight hours in length, and will cool and calm the elk as well.

For trips longer than forty-eight hours, provision should be made to stop and unload the elk for a twelve hour or more rest, feed and water break. Moist feed, such as silage, may also be used as a water source.

Transport compartments should always be well ventilated, particularly in warm weather. Air intakes should be positioned to avoid exhaust fumes, and to avoid direct drafts at eye level, particularly in cold weather.

LOADING AND UNLOADING

Elk should be loaded and unloaded as smoothly and calmly as possible. Always load and unload in small groups of 4 to 8 animals. The ideal loading chute for elk is four feet wide, eight feet tall, with a solid roof and sides. It is two eight foot long sections, with an angle of 30° to 45° between, as in the figure below:

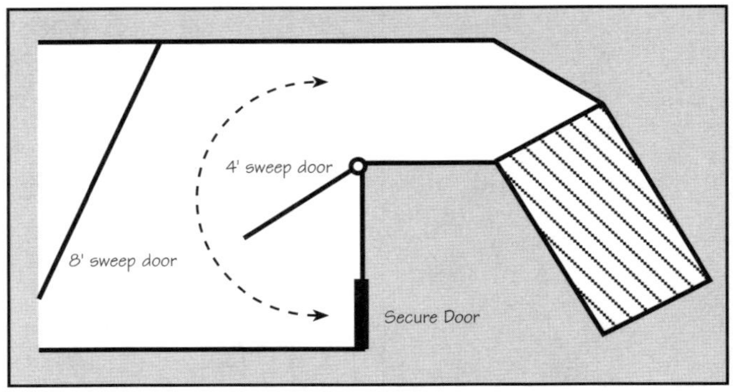

Fig. 15 Adjustable-height Loading Ramp - top view

It also includes an 8 ft. ramp, height adjustable at truck end with a moveable pin on each side, as in the next figure below:

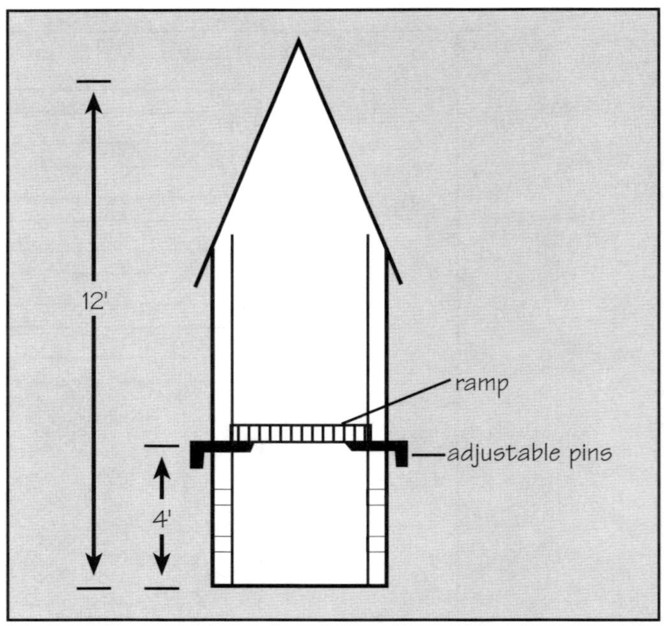

Fig. 16 Adjustable-height Loading Ramp - front view

The adjustable-height loading ramp allows any size of truck or compartment to be safely loaded. The pins fit into grooves on the underside of the ramp, allowing the back end of the ramp to slide as the height is adjusted. If large tractor-trailers with four foot high loading doors are to be used regularly, extend the ramp to twelve feet long to lessen the slope of the ramp. Keep a supply of plywood pieces near the loading chute, and nail them on the inside of the door opening to cover any spaces after the truck is backed in tight. Truck doors may be slightly different shapes.

The sweep doors in the crowding pens before the loading chute should move most animals on to the truck. If they stop or balk in the loading chute, try one or more of these techniques:

- rattle a large sheet of plastic behind them. Various noisemakers will work as well, as long as the elk are not accustomed to the noise;
- walk at the group with a large shield in front of you, and push them on with that;
- install two or three sweep doors in the loading chute, and use these to push them on.

Never beat elk with any tool, and use any type of prod with great caution.

Unloading is best accomplished with patience. Let the elk unload themselves. This may be particularly necessary when handling older males, who occasionally decide they are happy to stay put in the trailer. A sweep gate in the trailer will work well in such cases. Elk will unload easily from the side door of a "pot", given a slow hand and an easy touch.

TRAILER / COMPARTMENT DESIGN

Elk may be transported in almost any type of conveyance built for larger livestock, but a few details can improve handling quite markedly, and reduce injuries.

Walls should be smooth and free of any sharp protrusions. Floors should provide adequate footing, padding and absorption. A layer of sand overlain with wood chips or sawdust works well.

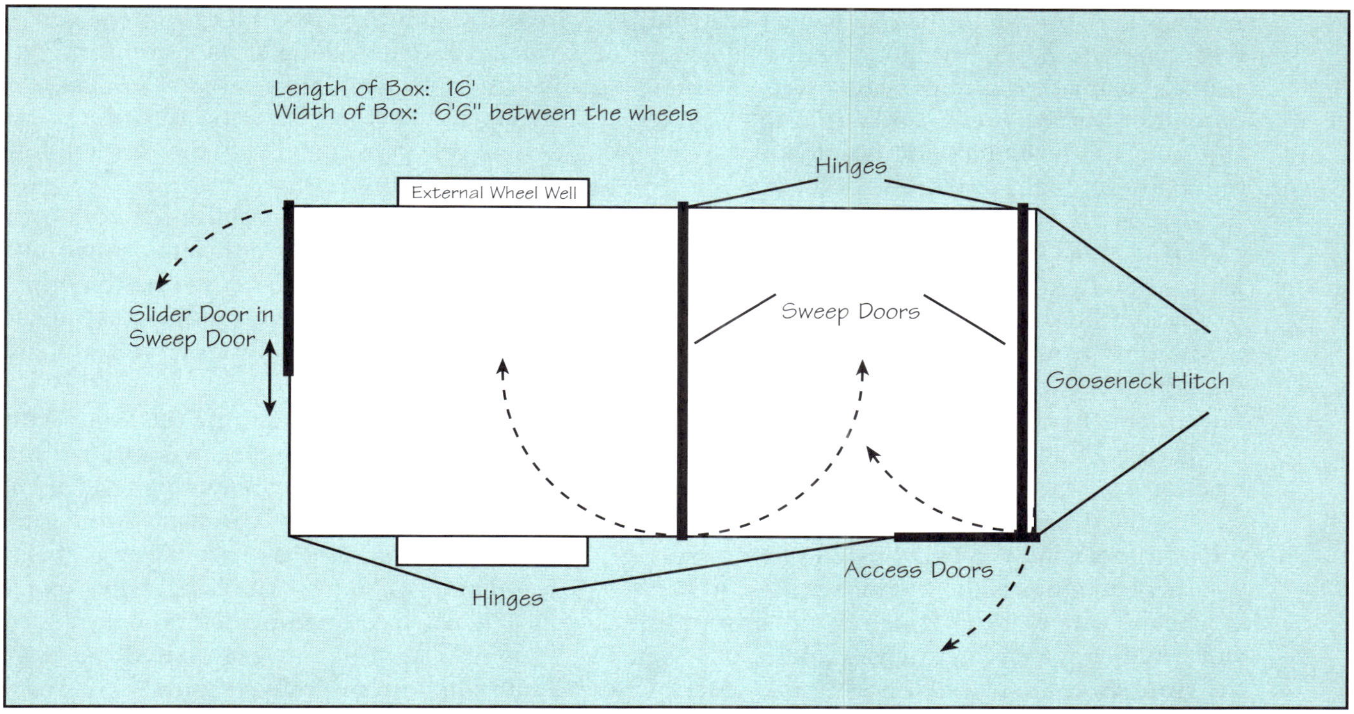

Fig. 17 Ideal Small Trailer for Deer or Elk

The task of handling elk is not difficult, but certainly different than for many other animals. Elk respond to activity dramatically and particularly to things we do badly. Elk are remarkably intuitive and sensitive animals who seem to sense when things are different or about to change. They sense differences in farmer approach. The reaction to a casual and admiring walk or inspection tour around the pasture is totally different than a more business like approach taken if shifting the herd to a new pasture or when yarding and handling is on the agenda. There is often some nervousness or tension associated with moving stock, particularly in the early phases and elk seem to sense that and can become unsettled.

Handling Elk in Pasture Situations

For ease of shifting, moving, mustering, or yarding - good layout, gate siting and access to laneways and keyhole areas are important. A good stockman who knows and understands elk and has developed communication and familiarization with the animals is confident and relaxed and not burdened by haste. He recognizes that the different classes of animals (weaners, yearlings, cows, bulls) need different approaches, but are all strong herding animals. i.e. you cannot handle half a herd, and what happens in one group affects the behaviour of animals in adjacent groups. However, the most important feature of elk handling in the pastures and laneways is the recognition of, and the reduction of, flight distance.

FLIGHT DISTANCE

This is best defined as the comfort zone between farmer and herd where enough pressure is felt by the herd to group together and pay attention, and the farmer is exerting some control on group movement. At this distance the herd will respond predictably as a unit without stress or over reaction.

In newly arrived elk, or groups of young stock, this comfort zone is 60 - 70 m. In older or more experienced animals the distance is 30 - 40 m.

At the fringe of this zone elk will group, watch and may begin a circling motion, more or less in the same spot. Enter this zone and movement will start. It is the control of this movement that is the heart of herd control. Flight distance can be reduced progressively by familiarization with stock, through routine feeding and quiet confident conversation with the animals.

Its quite normal to hear elk farmers talking to, and encouraging, their animals without self consciousness. Often the comments focus on the weather, how much he cares for his animals and how wonderful they are, and do not appear to be all that sensible, but the contact has been made and works well. A comfortable working distance is 25 - 30 metres.

If you attempt to charge within the comfort zone in a burst of activity then the elk will simply react by splitting into two or more groups, racing off and re-establishing at their comfort level. Worse is if one or two nervous animals leave the group and charge off irrespective of fences and where one goes, inevitably the rest will follow and the agitation increases.

The basic rule is: establish the comfort zone, and apply quiet pressure through slow movement towards the animals and conversation with them. When the elk move, you move at a similar or slower speed. When they stop, you stop and repeat the process.

The reality is that in time this will become so automatic that neither party will know it's happening and eventually, with good feeding and understanding, you too can stand at the gateway and call the animals. At the other extreme, if elk become so quiet through over familiarization - (pet and tame elk) then a comfort zone of zero meters is a problem in itself. Elk won't move as desired and can become quite stubborn or pushy in some situations.

Often in the early stages elk are more comfortable if the farmer is on a machine like a 4 wheel bike. The bike is solid and less threatening than a person on foot can seem, and moves predictably.

There are a few tricks and do's and don'ts to learn.

PRACTICALITIES

GENERAL

• Elk are herd animals - use that feature positively.
• Never handle elk as an individual when gathering from pastures. Groups of five minimum are the ideal. The only exceptions may be cows that need calving assistance, or a single older breeding or velvetting animal.
• Never yard groups in numbers greater than the facilities can comfortably accommodate for all the operations planned. Overcrowding in yards and facilities is the biggest cause of injury and stress.
• Always plan your operation from the facility back to the animals. Do not leave any surprises in yards or races - i.e. unexpected doors/gates shut; litter or uncontrolled noises.
• Do not ever be under a time constraint to finish the task.
• Do not mix animal classes (bulls, cows, etc.) when yarding. Reduce group size by half if handling females and their young together at weaning.
• Do not be in a hurry. Move and pressure the animals when they move, stop and maintain contact when the animals stop and increase the pressure slowly.

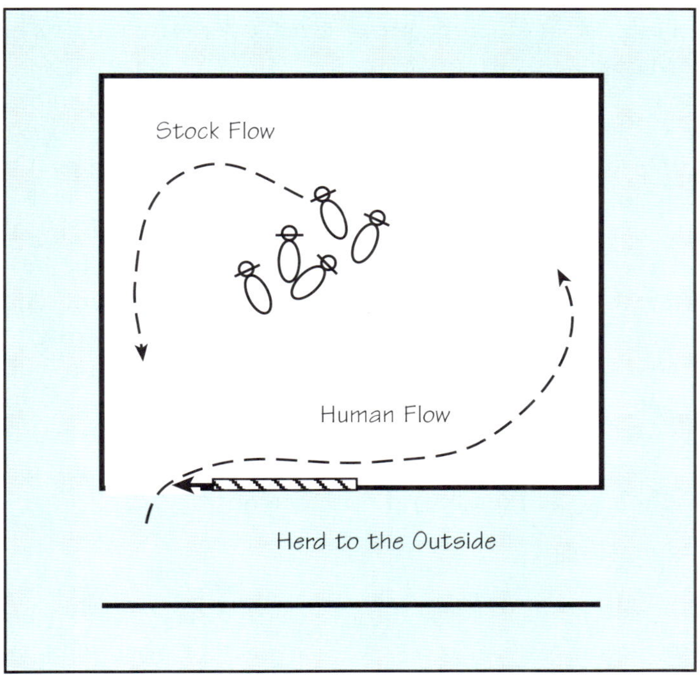

Herding Movement

• Realize it might not always happen as planned on the first try, especially with newly acquired animals. Often a hard sprint three times around the pasture is the norm. Calm perseverance will win in the end, but know when to give up - it's probably the wrong occasion anyway.
• Handling elk is not a spectator sport. Keep strangers, dogs, and small children in a vehicle, or at home.

HINTS

• When entering a pasture or pen to shift animals make it obvious that the gate is open.
• Ensure that gates are open and secure and they cannot blow shut or move as the stock approach, and if the gate opens into the pasture that there is no way animals can get trapped behind it.
• Most problems occur at gateways when there is an abrupt change in surface - grass to mud or gravel, or the opening is not really obvious. If elk have to come out of a gate and turn back along a lane most will go through but some will run parallel with the herd still inside the pasture. If this occurs often, consider adding a gate in the direction most often used. The extra expense is well justified.
• On entering a pasture for round-up, moving on the fence perimeter is more effective than heading straight through the middle and directly at the animals. Elk will have favoured fence lines and will often move to fences adjacent to where other animals are. Use these features positively and plan an approach accordingly.
• For movement of elk from one pasture to another, passive movement, i.e. where the elk shift themselves is a great learning tool. Gates should be securely tied open and feed can be used as an enticement. The danger occurs if one or two animals are left behind and may run fences and be very agitated if separated. Rather than concentrating on the individual, in all cases of separation it is wiser and safer to return the herd, or part of it, to the pasture and begin again.
• The most common fault in handling is when a single elk breaks from the group and charges

down a fence line towards the handler. The common reaction is to use the fence as a barrier and to close the distance to turn the animal. The animal's reaction is often to charge the fence. Injury or worse is inevitable. Once that occurs the rest of the herd is going to disperse anyway. It is better to concentrate on holding the herd and letting the single animal rejoin the herd after its flight. Running, shouting and waving only exaggerates the distraction. Pressure is better maintained through quiet concentration.

DIFFICULTIES

• Weather: All livestock are more difficult to handle in strong winds, or when a weather change (thunderstorms, etc.) is threatening. For repeat operations (like velvetting) yarding at a similar time each day works well as a routine is formed. In extremely hot conditions elk should be given plenty of time and shade to cool off and rest after round-up or handling workouts, otherwise, postpone until weather is cooler.

• Farmers should be aware that elk release a lot of body heat after yarding. If facilities are poorly ventilated heat stress can rapidly build up.

• The single animal: If introducing new stock to the farm (e.g. breeding bulls) it is better to bring the group to the yards and release the newcomer with them, or deliver the animal directly to the pasture and only release it when the group is clearly visible and standing calmly.

• Unplanned events: Elk are very perceptive to changes in their environment. Routine road traffic is no problem, but the truck in the yards filling the silo while you want to yard is. If you have unexpected visitors or unusual activity in the immediate area, it is much wiser, particularly with young stock, to postpone the operation. Resist the temptation to let groups of newcomers help "for the experience". Elk are acutely aware of strangers and will often react quite differently to the usual routine. If you have assistance it is worthwhile to discuss the procedures in advance and have clear communication in the pasture during handling.

SEASONALITY

The seasonality of elk behaviour has implications for stock movement and handling ease.

Autumn is the most complex time of the year. Extra care must be taken when first handling females and their young because of the protection the mothers show and the inexperience of the young.

Adult bulls, under the full influence of their raging hormones, want to fight each other for dominance, protect harems if they are of the chosen few and will walk fence lines, wallow and bugle and generally be disgusting. This occurs even in settled groups that have been together since velvetting.

In moving rutting bulls, passive movement is better than active. There is little need to yard them at this time of year as they are unsuitable for meat production and by placing them in close confinement fighting will be encouraged and these bulls will have scant regard for the farmer or fences. Let the group shift itself and provide as much space as possible for animals two years and older so that territory and comfort zones can be maintained.

In mating groups bulls will defend and challenge farmer entry at gateways or when moving groups. Great patience and care must be shown and you are more likely to succeed in a vehicle or on horseback rather than on foot. Some bulls will even confront through fences and seem to find children and dogs particularly challenging.

During velvet growth the opposite applies and males are at their most placid. However, velvet antler is easily damaged and care must be taken to avoid large group sizes, over crowding or haste in yarding.

Cows in late pregnancy are much more likely to be irritable and it pays to reduce group size in yard handling and avoid pushing groups through narrow gateways or doors.

Time spent with weaners familiarizing them with voice and association to restrict flight distance and response is training time well spent, and after a winter of intensive feeding in wintering pens, most of the agitation and rush response will have disappeared.

In summary, stockmanship is reflected in familiarization, routine, patience and appreciation of the elk and their response.

HANDLING ELK IN LANEWAYS OR RACEWAYS

Centrally located or perimeter raceways are vital to any successful elk farm irrespective of the farm size. Elk flow down through races but it is important not to pressure them until they are in the final curves into the receiving yard. Running behind elk only increases their agitation as they come under the pressure of moving into artificial surroundings. The group seems to need to stop now and again to take stock of their surroundings. A good handler will use that time to regain the comfort contact and wait before moving the group onward.

Occasionally, especially with young animals, one or two will turn back rubbing hard along fence lines often with the head twisted and looking for a way out through the netting. It is not easy to turn such an animal safely, but you can use voice and outstretched arms as a distraction. The importance of rub rails, sighter boards and fence reinforcement nearer the yards will be self evident. However, discretion is often better than heroics and it may be less stressful to stand aside and let the animal pass, return the herd to collect the animal, and start again. It is in this area that the single animal is at its most vulnerable. In such situations people can get frustrated and it pays to have a series of gates behind the group in long raceways. The problem certainly increases in long straight and narrow raceways that lead directly into barns or pressure areas.

For first use of new facilities, two people can move behind the elk with a length of snow cloth or similar material, full race height and width and exert slow and positive pressure.

Yelling, waving arms and general agitation does not help. Often in the early phases it's better to let the elk enter a receiving yard, circulate and return a couple of times before finally following and containing them. Feeding enticement in lanes and yards is not always a good incentive as it is a distraction, unless the raceway is a common feeding ground on small units.

Elk in raceways are very sensitive to the unexpected. Ensure that there are no unusual things i.e. coats or hats left hanging on fences, noise, banging doors, etc. The pathway must be clear to where they are travelling and they will be influenced by the presence and activities of other elk in the immediate vicinity. If disruptions occur, patience and waiting until the group settles is more effective than haste, noise and rush to overcome the problem. Equally, seemingly harmless changes (a new water trough, alterations to yards, etc.) will cause the same response.

KEYHOLES / GATHERING YARDS

Both these features, which are relief areas within a lane system that serve as a collecting site for a number of lanes or pastures on larger farms, or a receiving yard for elk prior to entry to a handling area, are important to relieve stress and serve as auxiliary holding areas. These allow the facilities themselves to be built for smaller more manageable groups and give the elk ample time to relax after herding or before intensive handling, and have their greatest use in reducing group size.

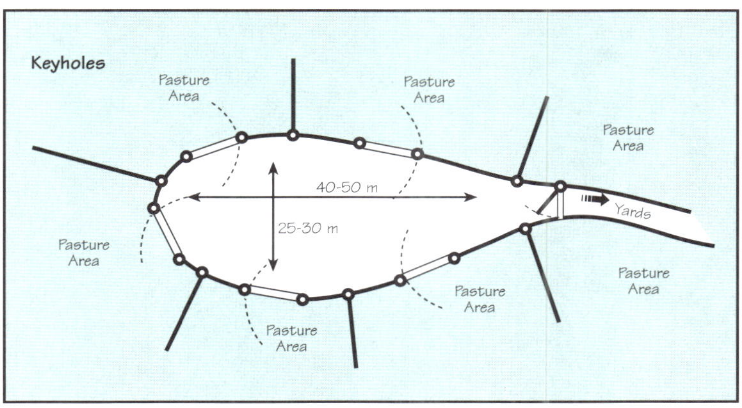

Handling in Yards

The secret of yard handling is to balance manageable group sizes with the facility's ability to cope with numbers, and to have a clear idea of what is to be done so that a regular routine can be established. If you are unsure of where elk are to go after being sorted, then the elk will be too. Again, maintaining a smooth flow is critical. Inside the yards the animals are constantly alert, moving frequently, and will urinate and defecate constantly. To hold animals for prolonged periods only adds to this mild stress and conditions under foot, air temperature and stuffiness can rapidly combine to create additional negative working conditions.

PEN WORKING

If handlers make sudden entry into the pens the animals will bunch and climb on each other, particularly if they are already over crowded. See-through walls and doors relieve this, but all movements should be slow and easy particularly the opening of doors. Many handlers talk their animals through this and make them aware of their position before entering the pens.

You should always move along walls to start the elk flowing through doorways and use the door as a control/drafting measure. Approaches to animals through the middle of the pens is not effective. If two people are working, each should have a clear idea of what is required and will be more effective working together. Elk do not run easily between people.

Visitors are frequently part of handling procedures and can be a disturbing influence unless there is clear communication on what is expected and what is to happen. It is sensible to leave signs requesting quiet and notifying that the yards are in use on a notice board at the yard entry, as there have been accidents to both farmer and livestock because of thoughtless interruption.

A great way to cross the fence - drive your quad right over!

INDIVIDUAL I.D. AND EAR TAGS

Good management of elk farming requires some form of individual stock identification - a unique number that records the animal against a central registry. Tags often include a lot of information - herd number, import number, etc. and may not be suitable for on-farm management systems. Successful elk farming requires some knowledge of what individuals are - age, gender (at a young age), breed type or breeding origin and a unique management number that can be easily read in yards and in the pasture or winter pen situation.

It is recommended that farmers use a double tagging system for permanent lifetime I.D. The management tag should be a large plastic type, double sided with management number on both sides. This tag can display a farm code, including a jurisdiction code (eg. A for Alberta, REF for Roughrider Elk Farm, and a unique individual I.D. number. The I.D. number should indicate year of birth, either as a straight numeric 3 or 4 digit sequence with the first one or two digits indicating year of birth, or as an alpha-numeric sequence with a year letter as the last character. Alternatively, the year of birth may be stamped at the base of the tag stem, or the farm code and an I.D. number may show on the front tag, with year of birth and I.D. number on the back tag. In addition to this management tag, a smaller metal or plastic tag should be affixed permanently in the opposite ear. The number on that tag should be totally unique, applied to that animal alone, and this tag must be as loss-proof as possible.

Fig. 1 Female identification - left ear, double sided tag

For example, a herd size of 200 breeding females would still allow a 3 digit number to be used on a colour base reflecting a single breeding program, and allowing the first digit to represent the year of birth and record 99 male and 99 female calves annually.

Three digit numbers on large tags have proven to be the most effective balance between information storage, and use in pasture and pen situation, with year of birth as a valuable management and sales tool for relatively large herds.

Sex differences in time become obvious, but a standard convention of (with the animal facing away from you) left ear for females, right ear for males is common.

Ear marks denoting age are ineffective as ears can become torn, and the actual placement of ear marks is both traumatic and variable in position.

A separate colour for an ear tag can be used to highlight year of birth, but this somewhat limits the amount of information to be carried. Experience suggests that colour has a more applied use in defining breeding types and different strains of stock as the result of on-farm genetic improvement programs. These can include crosses between red deer and elk, or programs recording individual sire bulls or improved genetics.

Convention in record keeping is to record the colour and then the numbers (G123 is Green 123). In mixed colours or breeding programs the sire tag colour is usually first.

Care must be taken with placement of the larger management tags to avoid the large blood vessels supplying the ear and to tag at the large piece of cartilage tissue found centrally at the divergence of the major vessels. (Fig. 2)

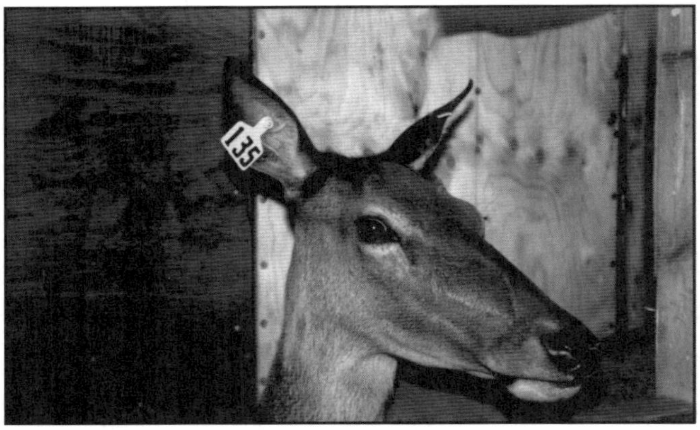

Fig. 2 Front view male. Note permanent metal ID tag and management tag

Note: Elk generally react sharply to ear tagging. The head must be firmly held and the handler ready to let go of the applicator if any sudden movement occurs as the ear tears very easily.

Several Canadian provinces have mandatory tamper-proof tagging systems and an official individual register required to ensure ownership and breed purity. All systems combine to form the basis of animal health monitoring schemes such as TB and brucellosis testing. With accurate and meaningful tagging, the true value of record keeping can be appreciated for both on-farm management tasks and for maintaining accurate data from slaughter animals and for live sale opportunities.

Alternative identification systems i.e. freeze or hot branding, tattooing, or microchip I.D. have not been entirely successful in elk for a variety of reasons, as most are impracticable or unreliable. It is hoped that an ear tag with an electronic chip will be commercially available shortly, and for positive I.D. of elite animals small intradermal electronic tags can be used inserted behind the ear, but a transistor scanner for reading will be required.

The most permanent and possibly most accurate I.D. is a DNA profile, established in an accredited laboratory using hair, blood, semen, antlers or most any body part, and recorded on a registration certificate along with parentage and performance records. DNA is Nature's I.D., the genetic base which defines what an animal is. Many jurisdictions are now moving to mandatory DNA registration of all, or at least all breeding livestock, to allow permanent I.D., and to facilitate matching of products such as antler and meat, and offspring, to the parents and individual producers. The challenge to industry is to participate fully in this development, and to support research which will make this "ultimate I.D." technology more useful, accurate, and reasonably priced.

TAGGING CALVES

Any system of Identification relies on good management to be effective. DNA matching in elk can be an effective tool in confirming what the good manager already suspects or has recorded. But, because the currently - used techniques depend on a very limited number of markers (measurements of genetically based characteristics that differentiate one elk from another), DNA has limitations in its power to sort through an entire herd and accurately identify the sires and dams of each calf. This difficulty increases as a herd

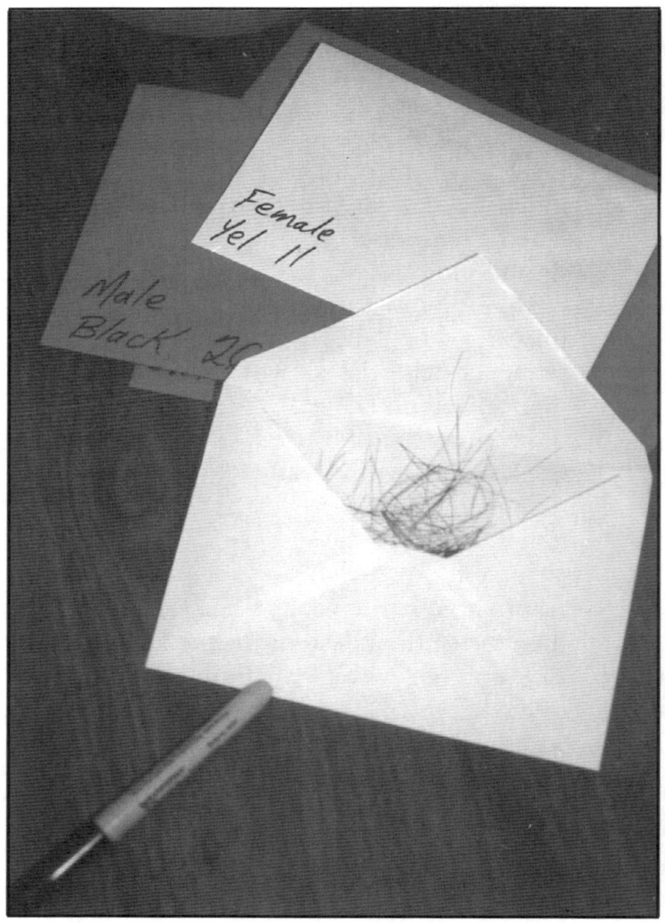

Fig. 3 Hair Samples for DNA analysis - be sure a bit of the skin follicle is attached to the hair roots

becomes more closely related or linebred, and increases with the number of possible sires and dams that could be the parents of each calf. The most effective management techniques to avoid these problems include single sire mating and tagging calves at an early enough age to match them to their dams.

There are two windows of opportunity for tagging calves and matching them to their mothers. One is within a day or two after birth, and the other is when the calves are about 100 days old and still closely attached to their mothers, but old enough to be safely handled with the herd. If you choose to tag them at birth, your first consideration must be safety for both you and the calf. Particularly with young mothers, you must not interfere too early with the calf, as you could interrupt the bonding, or "mothering - up" process. This usually takes about four to twelve hours after birth, or until the calf is up, dried off, nursed at least a couple of times and is hidden away and the mother is grazing at a distance from the calf. The calves will use the "hider" strategy for the first couple of weeks - laying flat and immobile in some cover. For the first two or three days after birth, they will rarely move from that position even if disturbed. That gives you the window of opportunity to find and tag them.

Fig. 4 Elk farmer on quad motorcycle - looking for calves

Finding calves is a learned skill. You will soon learn that calves typically seek cover of some kind - tall grass or weeds, bushes, or shelters you provide, and they are often found along the edges of the pasture - along the fenceline. You can also watch their mothers. When you approach, the cows will often give away their hiding calf's position with a purposeful look in that direction. If you are driving a machine of some kind, you must be extremely careful to avoid driving over the calves, as the little ones just will not move. Follow the same pathways each day as you check. Once you've found a calf, then it's *your* safety that must be you primary concern. Elk mothers can be very defensive of those calves, even if it is not their own.

Fig. 5 Tagging a day-old calf

Producers have come up with a variety of strategies for calf tagging - driving the truck up beside the calf, crawling out the passenger door and under the truck to the calf, building a manually or hydraulically operated cage that fits on the tractor or quad and drops over the calf, with a person door to allow you to enter and tag safely, or scooping up the calf and lifting it into the cab of the truck for tagging. There are merits to each of these, and you should work with an experienced producer before you decide which is the best for you. The most important part of whichever strategy you use is that it is best not to work alone if you can manage that. One person can watch the cow and keep her away - a long pole held up in the air over the cow's head is surprisingly effective - or can otherwise help get the job done.

Once the calf is tagged, you need to sit quietly and watch the behaviour to be confident that you identify the correct dam. This also takes experience, and you won't always be right. Elk are notorious for cross - mothering, and some producers have watched aggressive cows steal calves from a new mother, then the new mother runs over and "adopts" the aggressive cow's calf! That cross - mothering trait is not totally undesirable - if a cow's lactation is somehow interrupted, her calf can survive by nursing many of the other cows in the herd. In fact, an orphaned calf that is bottle fed for three weeks can usually do very well if you just put it back out with the herd, especially if that herd is made up of young mothers.

The next "window of opportunity" for matching calves to dams is in August or so, when you can bring in the whole herd, separate and tag the calves, keep them away from their mothers for a couple of hours, then release them and match them by watching with binoculars. Nursing, nuzzling and other behaviours indicate the relationships, but this is more prone to errors than tagging at birth, as there have been more opportunities for cross - mothering to develop.

With your calves tagged and matched, and only one bull used to breed a group of cows at one time, your records of parentage and performance will be many times more accurate, and the DNA lab people will be pleased to work with you, confirming what you already know!

Age Estimation of Elk by Tooth Eruptions

The eruption, or growth of the incisor teeth at the front of an elk's lower jaw is quite predictable, and may be used to roughly estimate the age of young animals, as shown in the following table, adapted from Haigh & Hudson in ***Farming Wapiti and Red Deer***.

Tooth wear may also be used to estimate age of older animals, but this technique, and many other estimation keys, are highly variable between individuals. General greying of the hair around the muzzle may indicate advanced age, as may the progressive slackening of the external aspects of the vulva in older cows.

TEETH	AGE (MONTHS)
All deciduous	Birth - 17
1 pair permanent incisors	16 - 23
2 pairs permanent incisors	18 - 24
3 pairs permanent incisors	25 - 29
4 pairs permanent incisors	27 - 33

Sample Individual Animal Record Sheet (copy and use)

Tag No. ________ Tag Colour ________ Herd Name ________

Sex ________ Year of Birth ________ Herd Registration # ________

Secondary Tag ________

Registration # ________

Breed Type ________

Pedigree

g.sire	________	g.sire	________
sire	________	dam	________
g.dam	________	g.dam	________

History of Dam & Sire

Dam	________	Sire	________
Breed	________	Breed	________
ID	________	ID	________
Date of Birth	________	Date of Birth	________
Origin	________	Origin	________

Breeding Method

AI ________

ET ________

Natural ________

Details

Date	Weight	Comments

Sold

Purpose ________ Date ________ $ ________

Breeding / Calving Records (copy and use)

Date Exposed	Bull	Method	Date Calved	Weight of Calf & Date	Weight of Calf & Date

Antler Production

Age/Year	Button Drop	Date Cut	Weight	Comments

HERD RECORDS

Herd records can be as simple or as complex as the owner desires, but in general terms the most important data can be recorded as simple entries relating to individual tag number, group, date and a specific event.

Records can be classified as those important for assessing stock genetics and breeding and those involved in a response to management. For example it is possible and often desirable to record growth rates of young animals from weaning to 15 months of age on a monthly or six week basis. For selection criteria only, weaning and 15 month weight and the gain between those two points is genetically significant, but the intervening weights give an accurate check on seasonality, growth and management.

Many computer programs are available to aid data collection and record keeping. It pays to consult widely in the industry before selecting the one that best suits your objectives or, develop your own with a spreadsheet and data base program.

Field note books should be the first line of work in any recording i.e. to record the management events and data e.g. antler casting and birth dates as they happen, not as recollections. However, field note books should always be backed up on computer, record cards or simply recorded in a farm management diary - depending on the scale of operation and the involvement of the individual.

To be useful, records must ultimately relate to the individual animal identification and the relationship of it to other animals in the immediate cohort, i.e. to age group, herd and breed type. An individual's record chart, related to the important aspects of genetics, should include most or all of the information on the Sample Individual Animal Record Sheet.

Such a record should also include animal health treatment records, notes on temperament, TB status, etc. that also relate to significant management events.

Velvet antler, breeding records, calving performances and individual growth rates should be kept as separate files and have more relativity to age and group than an individual pedigree from a management view.

RECORD KEEPING

The significant records to be kept where practical, and where experience allows are:

Note: * Particularly significant for genetics program

YOUNG STOCK

- Dam/Sire Pedigree*
- Birth date
- Weaning weight/date.* Gives some information on genetic potential, (especially if corrected for date of birth) plus information on the breeding and lactation efficiency of the dam (especially if not corrected).
- 15 month weight.* (Say 1st of Sept. as an average) Peak puberty weight. Important selection criteria for future heavy weight animals. Average 15 month weights indicate likely breeding success for females and define management and feeding for meat production.
- Periodic weights from weaning - 15 months e.g. end of autumn, mid and end winter, end of spring, end of summer (15 months). These give checks on growth rates against target figures for that breed type and check on management and feeding. Growth rate from weaning - 15 months can indicate fast growing early maturing stock for meat production.
- Pedicle initiation and spike growth and hardening for breeding and velvet emphasis.
- Temperament.* Record any aggression or confrontational response to handling - 3 T's at a young age and out! Nervous or very flighty animals may need special attention.

STOCK 15 - 27 MONTHS

a) Males

- Weight at end of winter, or at casting. A management check on winter feeding. A couple of weights during winter will reflect management if weight loss is observed.
- Antler casting date.
- Velvet weight* - adjust to 65 days growth for analysis. This requires note of harvest date and velvet grade and style.
- 27 month old peak weight. (1st Sept.) This weight is the best predictor of potential size

and ranking on weight alone, or the best predictor of future size with antler performance when combined with corrected 2 year old velvet production.

- Temperament*

b) Females

- Calving performance - dry, did not calve
  Wet/dry, lost calf
  Wet, reared calf
- 27 month old weight
  * Growth rate 15 - 27 months

ADULT STOCK

a) Males

- Annual lean body weight* - mid winter
- Peak weight, prior to rut, if desired
- Velvet data*, casting, harvest, weight/grade*

b) Females

- Annual mid winter weight* - measure of efficiency - weight of calves born: weight of cow, calving performance

In addition, records can be kept on slaughter data for future use and to have some reflection of sire suitability of purpose.

For slaughter record: Liveweight / off-pasture and Hot Carcass Weight are important to determine dressing % or yield and deciding on the most appropriate liveweight for slaughter targets.

- Hot Carcass Weight (HCW)
- Dressing % $\frac{\text{Hot Carcass Weight}}{\text{Liveweight off-pasture}} \times \frac{100}{1}$
- Fat content (GRD measure)
- Bruising, historical damage etc.

In fact, for any weighing, a routine should involve yarding and weighing at a consistent time interval. With all animals, yarding promotes a loss of body fluids and wastes. 4 - 5 kg can be lost in as little as 6 hours in yarding and transportation. When selling stock by weight be fair to yourself and the buyer. Farmers should not hold stock in yards in readiness for prolonged periods and buyers should be encouraged to hit time schedules. With the great distances to travel and the prevalence of auction sales, both parties should be aware of weight changes from the off-pasture and fasted weight - especially if stock is being bought on liveweight at a slaughter plant.

Buying young stock on a weight basis also can involve some areas of difficulty between yarding, selection and transportation.

Note: Any weight records are only as good as the scales and their operation. Scales should be checked for accuracy and tuned periodically. Electronic scales, both in leads and load bars, can be affected by moisture buildup. Before any weighing program it pays to check scale accuracy.

When recording weights, it also pays to have a printed record of the last weight for that group on hand - to signal scale inaccuracies and as a double check for I.D. and to notice, at that time, any dramatic weight differences that may indicate a management or health problem.

Subjective data - (like body condition scoring) often use a 1 - 5 scale. This should include notes regarding temperament and physical faults or injuries - e.g. hoof overgrowth, undershot jaws, conformation.

Recording and record keeping for their own sake is meaningless unless the farmer has clear objectives in breeding and management in mind. These can evolve as the base data is collected, but at the end of the day when selection i.e. the choice of superiority that is genetically based is made, or a culling decision, i.e. the removal of animals not suited for breeding, the breeding objectives shape the choice. The records simply indicate the soundness of the objective and are the tools to achieve that end.

PROPERTY RECORDS

Just as in any farming operation, an accurate knowledge of the farm geography, contour and layout is important in making good management choices.

Planning layout is greatly aided with aerial photography and the development of a farm map that shows numbering and organization of individual pastures, the potential for expansion and the relationship of woods, water and roadways to development plans.

Many farmers have a detailed farm map permanently mounted on a white board in the office. Here pastures are numbered or named and an accurate diary of stock movements, herd numbers and their on-farm location are kept. Imposed on these or held as separate records are details such as layout of farm water supplies, electric fence breaker and check points, special fence attention areas (e.g. for predator security or risk from water damage), which can be highlighted and can provide very effective communication to farmer and staff.

It takes little effort to keep a file referenced by pasture or pen number to record the following:

- area
- fertilizer, soil testing, pH
- cropping history - pasture species
- stocking rates
- hay or silage production,

and to use these for management advances in future years. It has been shown that, for example tracking stock movements and classes of stock can be absolutely critical in tracing disease concerns and risk assessment.

Again, record keeping for its own sake ultimately becomes a frustration. Records only form part of an overall objective. It is as important to have that vision focused first as it is to collect the tools that can aid that development. Good property records will certainly allow effective use of the expertise that is available to farmers through Department of Agriculture specialists, consultants, veterinarians, and feed analysis laboratories. Information is relatively easy to collect and store, elk farmers tend to be an enthusiastic group that have a need to know and there is a huge array of information and interpretations available.

Fig. 6 Cow Herd

JUDGING THE CONFORMATION OF ELK LIVESTOCK - WITH AN EMPHASIS ON FUNCTIONAL EFFICIENCY

We have only been raising elk as farmed livestock for twenty some years - are we ready to start judging them? As if we really know what makes a good elk? I think we are ready, and more than that, I think it is essential that we begin. Only by actually making some strides will we move ahead on this path.

Where to start? As with so many other aspects of our business, we can learn a lot about our elk by applying what we know from other livestock, particularly cattle. Cattle are, of course, judged every year at thousands of shows and sales around the world. But anyone who has been in that business and has stepped back occasionally and really looked critically at cattle judging and showing, and selling for that matter, will admit that the cattle business is as trendy as the high fashion clothing world. Remember the short legged, stubby Grand Champions of the 1950's? Some of those show toppers were incapable of natural breeding! Then we went full circle, and in the 1980's we were rewarding cattle that looked more like thoroughbred racehorses than beef producers. Somewhere in the middle of that thirty-year period, we must have been raising good beef cattle.

What did they look like?

One of the most influential and ultimately successful cattle breeders was the South African, Professor Jan Bonsma. In the process of developing the Bonsmara breed of cattle, Bonsma studied many different individual cattle, antelopes and deer, measuring such characteristics as: adaptability, body temperatures, respiration and pulse rates, tick counts, hide thickness, hair counts per square centimeter, average daily gain, weight for age, fertility, milk production, longevity, heritable defects or inferiority in function of any organ that resulted in lower resistance to stress or disease.

It is also interesting to mention some of his measuring techniques:

- Measurement of growth was by monthly mass determination.
- Milk production was determined by measuring the calf's growth and weaning weight.
- and also by measuring the actual milk intake of the calf, by weighing it before and after suckling.
- Fertility was measured by keeping a record sheet for every female kept in the herd and any cow that skipped two calves in 8 years was slaughtered.
- Body conformation was based on a subjective evaluation by careful observation, and fourteen body measurements were taken on each animal from birth to maturity or until it was eliminated from the herd.
- Temperament was measured by doing tractability tests on free-gazing animals, by approaching animals and recording the distance between you and the grazing animal before it walked away. The behaviour of the animals in the measuring pen where they were intimately handled gave a very good indication of an animal's temperament. Temperament increases in importance as the amount of handling increases and pen size decreases. Animals that live with constant stress will **never** thrive.
- Longevity was measured by how long the cows were kept in the herd. As long as they could produce a good calf annually and did not lose too much condition - no more than 20 percent of their weight at the time of calving. Their young cows often lost less than 10 percent of their weight during the lactation period.

Obviously, the weigh scale and record book are essential in assessing these characteristics.

Returning to visual assessment, once Bonsma had identified the superior producers in his herd, he began observing and recording the consistent characteristics they expressed. I have adapted his comments to more accurately describe elk livestock.

First, males and females must be different. A masculine female and vice versa, a male that appears feminine will never be functionally superior.

For cows, look for these traits:

- Average size for the breed, neither overly tall, short, large nor small.
- A lean, very feminine head and neck with limited mane development and a well defined lower jaw.
- Smooth, slick hair, particularly on the neck, shoulders, spine and sides. In spring, the winter coat should be shed readily and completely. The hair coloring should not be darker at the same places mentioned below for a highly functional bull.
- Small brisket (front of chest) with little filling of the dewlap. A full brisket indicates excessive fat deposition.
- From every angle of observation, the body should increase in size and volume from the front to the rear, with a noticeable "barrel" shape around the rear of the chest and sides of the abdomen.

Fig. 7 Excellent conformation in a fast growing 4-month old heifer calf

- From the back and side, the rump area should be full and rounded, with the tail positioned below the crest of the rump and pointed straight down.
- Well - developed and at least average sized udder for the breed, with small, sleek and very shiny teats.
- The vulva should be normal sized for the breed, and should not protrude or point upwards.

Functionally efficient bulls must look and behave really masculine. In elk, most of these characteristics will be particularly visible during the rut. Look for:

- Very noticeably darker and coarser hair at the opening of the penis sheath and on the neck, crest, upper front legs, along the brisket, the lower rib area and the flank.
- Muscling should be clear and well defined on the shoulders, upper shank, thighs and rump.
- Neck should be full and thick with a well-developed mane.
- Testicles should be large and of a uniform size during the rut.
- To produce fertile, feminine females, look for very fertile and masculine bulls with a high libido combined with a level and respectful temperament.

Both males and females of all ages must have sound feet and legs. The hooves should be straight sided and hard, with no excessive length, incurving or crossing of the tips. The front feet may have a slight outward "set" to their placement, but not an inward set. The front legs should be straight up and down from every angle, but the hind legs will have a shock-absorbing bend at the "knee".
Elk may be slightly "cow-hocked" from behind, but this should not be so extreme as to allow the knees to "knock" together. Only a slight outward set is acceptable on hind foot placement, but not an inward set.

Fig. 8 Very good masculine bull, in full velvet

Fig. 9 Good feet and legs are essential. This young bull lacks good length of body.

I hope this presentation of ideas will begin an ongoing discussion of what conformation traits are desirable in our elk. So many of you have accumulated so much experience and knowledge - we need to collect and analyze more data, then improve on this presentation.

INTRODUCTION

Elk farming is fortunate to be developing in parallel with new concepts in the approach to disease management and enhanced welfare of domesticated animals.

The term "herd health" focuses on health status rather than disease and is accurately described as preventative medicine. This procedure identifies the known and potential risks of disease and establishes procedures that reduce the risk and the effect of disease, or treatment that contributes to disease prevention.

Good animal management, feeding, handling and care, are basic to animal welfare, but some aspects, seen as central to the welfare issue, are not always consistent parts of animal production requirements - the objective of farming.

The single most important factor in a herd health program is the prevention and reduction of STRESS.

WELFARE ISSUES

The Canadian Venison Council's (CVC) Code of Practice now forms the basis of industry philosophy and directions, and significantly, is among the first in the world to have achieved such established guidelines.

The embodiment of these principles outlined in the Code for every aspect of deer and elk farming gives new emphasis to the industry direction and credibility, as well as an acknowledgment of its responsibility in a modern farming and marketing environment. As a consequence, herd health programs are emphasized rather than treatment of disease. This enlightened stance aids to the enjoyment and reward of farming, and at its heart is the common reality that farmed deer themselves are highly adaptable healthy animals in the first place.

PREVENTATIVE PROGRAMS

HEALTH RISK FACTORS

To assess risk, accurate records must be kept of all management practices and production, e.g. mating programs, feeding regimes, supplementary feeding, weight gains, etc. Management factors that pinpoint health risk concerns may be:

- Growth rates and weight for age data
- Reproductive rate
- Clinical health records
- Mortality records and slaughter data
- Antler production
- Specific diseases
- Chemical, Biochemical, Immunological and Physical Disorders

The most important general animal health principle is that elk are fed to their requirements of nutrition as it is influenced by seasonality.

A stress free environment under good feeding is further enhanced by knowledge and prevention of:

1) Parasitism
2) Yersiniosis and other bacterial and viral diseases
3) Mineral deficiencies
4) Predators, mechanical damage, accidents
5) Disease testing and surveillance of existing herds and particularly additions to the herd

The art of preventative management is to practice routine programs according to a seasonal program. Variations on the routine are dependent on the stock class and age and any particular local influences i.e. climate, predation risk, mineral deficiencies and health threats. All farms have potential problems that can suddenly arise under conditions of stress or poor management, or because of low resistance brought on by other factors.

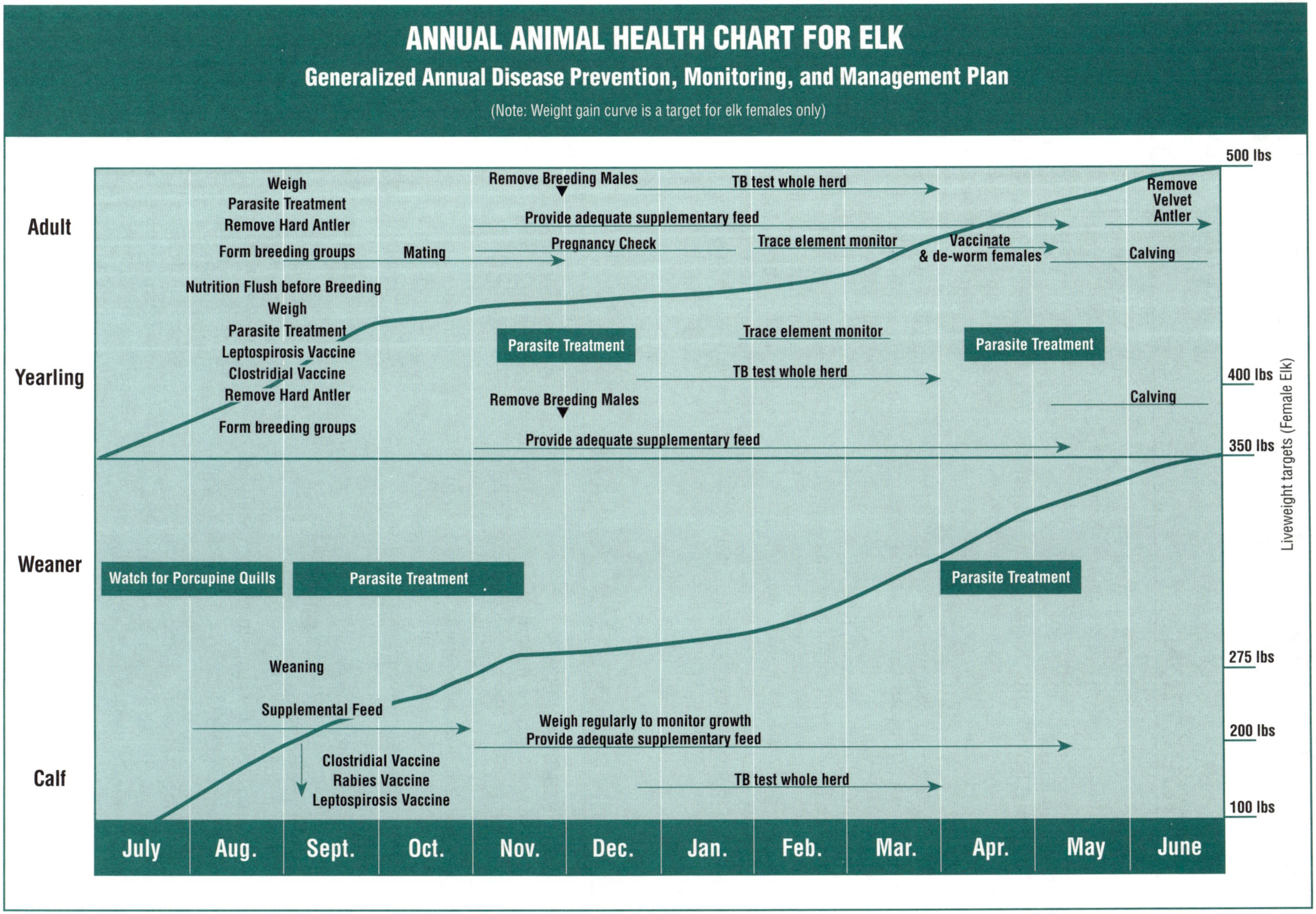

Fig. 1 presents a generalized procedure chart - for routine programs

THE ANNUAL HEALTH PROGRAM

As in many programs the emphasis is on young animals as elk have a great capacity to develop resistance and immunity to many of the parasites and bacterial threats that affect young animals. Young animals, properly treated, develop long term viability and top health status which requires only annual treatments against the most common health threats.

The most significant of these are the normal gut and lung parasites common as management problems to all farmed species. Control of these through routine parasite treatment programs is the primary emphasis in young stock.

SPECIFIC HERD HEALTH ISSUES

PARASITISM

Lungworm (*Dictyolcaulus viviparus*)

General Lungworm is the most important parasitic worm in farmed deer and can cause ill thrift, poor growth and death due to secondary problems in condition loss or parasitic bronchitis in chronic infections, particularly in young weaner deer. The first fall and spring are the most critical periods for control.

Elk and hybrids appear to be at greater risk than red deer, although the risks are major in all three.

Adult worms live in the windpipe and lung bronchi and infections can be so severe as to completely block air passages.

Eggs are laid and hatched here or in the gastro intestinal tract after they have been coughed up and swallowed. In the gut they develop into the first stage larvae and are passed in the feces. Signs of coughing and discomfort after yarding young stock are often indicative of heavy infestations.

Lungworm larvae thrive with moist warm conditions on pasture and in spring, summer and fall conditions will rapidly pass through two moults to a third stage larvae (L3), the infective stage. In warm conditions (18 - 21° C) this can take only 3 - 7 days. L3 larvae may migrate up grass stems where they are eaten by deer in normal grazing. Once in the gut they migrate through the stomach wall to intestinal lymph nodes, undergo a fourth moult and move to the respiratory system. Here they rapidly develop into adults and may reach sexual maturity (patency) and begin the shedding cycle in 20 - 21 days.

Adults have been reported to continue to shed for up to 10 months and even with good parasite treatment programs an infestation can be maintained over the wintering period to arise as an infectious source on spring pasture, both for these hosts and newborn stock. Deer with lungworm lose appetite, are listless with a dry rough coat, and show weight loss. This can be quite severe in young weaned deer.

Adult deer do build up an acquired immunity to lungworm, but nevertheless should be treated for control of potential infectious sources to young stock.

Calves are the major sources of shedding from as little as two months of age. If unchecked by parasite treatment, infestation buildup peaks in autumn with a secondary peak in spring, and as both these times are critical in terms of deer growth and development, they are the focal points of control. However, winter is no time for complacency even in a well managed program, as anything that contributes to stress - lack of shelter, competition for feed, or poor feeding, overcrowding or simply the stress reaction to indifferent handling can induce a parasitic response and ill thrift.

Routine fecal testing to estimate larval counts are easily done through local animal health laboratories. Fresh samples of 15 grams per animal can be collected either from the ground following yarding, or manually via rectum, and bagged and submitted for analysis. Samples should be stored chilled and identified for stock age, and gender, with any treatment history or symptoms recorded.

Treatment Lungworm is not difficult to control and reduce to levels that encourage the immunity response but have no impact on productivity or health. However, due to the short cycle of the parasite, in the right climate conditions a control program must be established rather than treatment becoming the response.

Ivermectin has become the drug of choice because it eliminates all life stages of the worm and remains effective in the blood stream for a 5 - 8 day period after parasite treatment thus increasing the available time between applications from 20 days to 26 - 28. However, that is not to say that other anthelmintics are not effective, although they may vary in their efficacy and required dose rate and time between treatments.

Due to the small size of the industry some common animal remedies may not be licensed for use in deer, and any decision on parasite treatment policy and anthelmintic choice should be done in consultation with veterinarian advice and the licensing status in mind.

Ivermectin at 200 mg/kg dose (normal cattle) is effective in red deer, but the NZ experience is that hybrids and elk need 2 - 3 times the dose rate, particularly if stomach and abomasal parasites are involved as well.

New products such as moxidectin, eprimectin, and doramectin are being tested with excellent results and some cost advantage in the larger species. Eprinex® has been proven effective and is the only one of these treatments registered for use on elk in North America. Table 1 highlights considerations in parasite treatment that should be routinely considered in a treatment program.

The spring and autumn conditions in Eastern North America are very favourable for maintaining the lungworm life cycle and this parasite treatment practice needs to be a central focus for deer farmers in this region. Frequency of treatment should increase with stock densities.

There have been some reports of significant irritation, muscle effect and discomfort for pour-on application in summer - possibly due to inflammation caused by biting insects and irritation. Pour-on treatments may also be less effective when applied on heavy winter coats.

The importance of good practice and routine is clearly demonstrated by the liveweight gain chart from recent work at Invermay. (Fig. 2)

GASTROINTESTINAL PARASITES (GI PARASITES)

As elk farming has become more and more intensified, the usual risks of farmed stock to abomasal and small intestine parasites like ostertagia and trichostrongylus increase. Elk are susceptible to both although there have been very few reports of clinical disease. These worms can cause clinical diarrhea, loss of appetite and ill thrift.

Fortunately all GI worms are susceptible to the normal broad spectrum anthelmintics registered for use in sheep and cattle. Ivermectin or moxidectin control give excellent results.

There is some evidence that heavy GI burdens may appreciably change the stomach conditions for the best uptake of oral parasite treatment. In such cases injectable or pour-on treatments are indicated.

Practical Approaches to Deer Parasite Treatment & Vaccination

- Different parasite treatment (and vaccines etc.) should never be mixed.
- Do not store parasite treatment in clear and unmarked containers.
- Read the label, or on veterinary advice, treat at the correct dosage.
- Check the accuracy of the treatment gun. Under-dosing is a common fault and may encourage parasite treatment resistance.
- All parasite treatment/veterinary drugs have a withholding period - check and observe strictly, especially for animals destined for slaughter.
- Parasite treatment is a management technique not a test of speed. Accurate placement of the gun in the side of the mouth, over the back of the tongue allows time to swallow and avoids damage to teeth and jaws, or treatment being forced into the lungs.
- Record the event - date, group, and dosage and develop a system when treating groups to prevent over-dosing or missing animals.
- Store parasite treatment as the manufacturer suggests. Some have a shelf life and are light sensitive.
- Injectable Ivomec - can have a painful or stinging reaction about 30 seconds after dosing - stags and adult hinds can react with aggression to each other, or the handler as a response.
- Correct placement of the needle between the skin & muscle is important. Tent a fold in the neck skin area (lower neck) for delivery. Injecting is not appropriate in the rump, loin or shoulder areas.
- Pour-ons - It seems important that the animals remain dry 5 - 6 hours following treatment. If raining or threatening rain, deer should be kept under cover but in well ventilated conditions.
- If possible after treatment, shift treated young deer onto new pasture after 24 hours to minimize risk of infestation and if possible rest the last pasture for as long as possible before reintroduction of stock. Use adults as follow up groups.
- Application should be done with care - the whole midline back region used to spread the liquid applied from tail to shoulder against the lie of the hair on the skin.
- If practical - begin the parasite treatment programme three weeks prior to weaning - lungworm burdens will already be evident and it is good training for young stock.

Table 1

Such problems are often related to management conditions such as high livestock densities, stress factors or the introduction of an animal from extensive range conditions to more intensive farm conditions. Transportation can often lead to that stress and susceptibility. As a general rule on arrival at a new property, incoming stock should be treated to prevent new parasites coming onto the farm and to override the stress effect.

MENINGEAL AND TISSUE WORMS

- Brain worm *(Paraelaphostrongylus tenuis)*
- Tissue worm *(Elaphostrongylus cervi)*

Both these worm groups are important parasites to different species of deer in their normal and natural settings and have had a huge impact on the developing deer industry - by default in effect. Tissue worm is naturally associated with red deer in Europe, and brain worm is endemic in Eastern North America in white-tail deer.

Both worms have significant impact in quarantine and interprovincial trade of deer.

Both are the normal parasites of their own deer species, and are non-pathogenic in these deer.

In white-tail deer in Eastern North America, adult worms (generally 4 - 6) are found on the meninges of the brain in a high percentage of wild animals with virtually no signs of any effects. These deer are not considered diseased or sick, nor is any restriction placed on their use as a hunting resource.

The problem lies in the potential for transference of brain worm to other deer and domesticated stock and the clinical effects that may develop in an abnormal host.

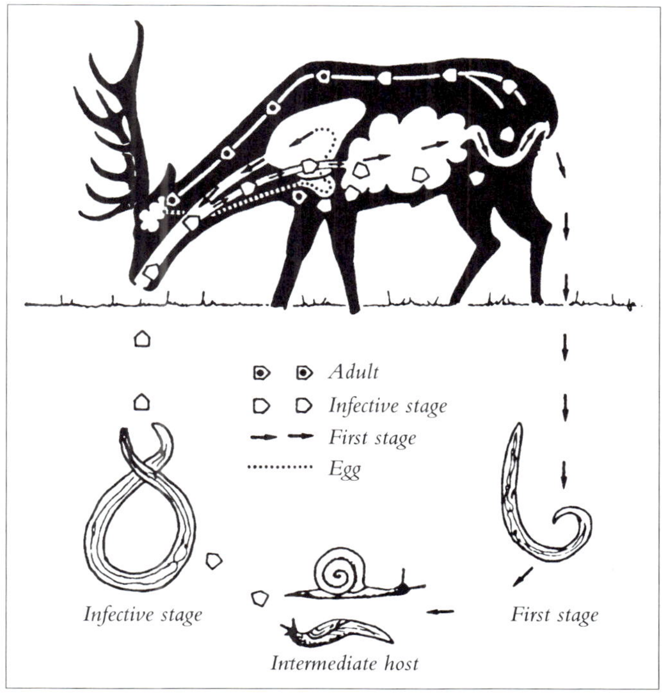

Fig. 3 Lifecycle - Ingested with the host slug, the worm enters the deer gut, and migrates to the tissues

It is important to distinguish between these two worms, and to acknowledge that based on the Ag Canada import quarantine procedures for red deer from NZ and the UK, there is no record nor evidence that E. cervi has been introduced in recent years and it currently does not exist on mainland Canada.

The life cycles of the two worms are similar, and they differ from that of lungworm in that it involves an intermediate host, a non-specific slug or snail commonly found in most parts of mid and eastern North America.

The adult releases eggs which are moved by the blood system to the lungs where they are coughed up, swallowed and passed in the feces. These first stage larvae will

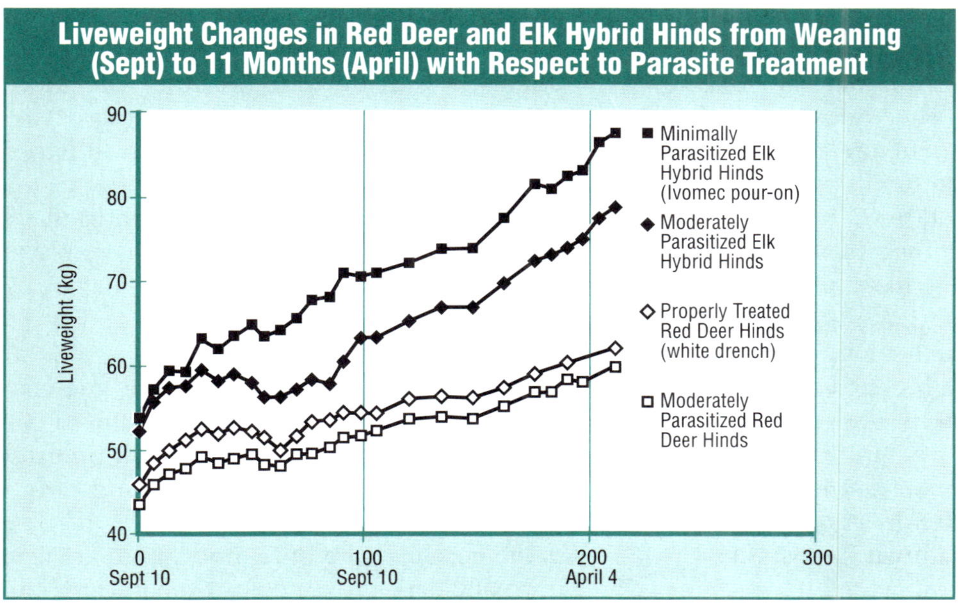

Fig. 2 (After Waldrup 1994)

opportunely burrow into the fleshy foot of molluscs, mature to a third larval stage and can remain at rest in this host for up to 2 years. When eaten accidentally by deer, the brain worm larvae leave the snail in the abomasum and slowly migrate through the spinal cord and move towards the brain. During this travel they continue to mature, and in abnormal hosts at high infection rates can cause damage - hind limb unco-ordination, paralysis, circling behaviour. Once located in the brain they live in the meninges and develop to reproductive maturity (patency) in 90 - 110 days from the initial ingestion. In heavy experimental infection, severe central nervous system disease can result in death, listlessness, stiffness and lameness, hindquarter collapse, circling, blindness and abnormal head and neck posture, paralysis, or death.

The possibility of farmed red deer and elk being infected by P. tenuis from white-tail cannot be ignored although little evidence has been found in deer farming to date. Of the estimated 15,000 red deer in North America, almost all farmed in the recognized brain worm zone, anecdotal reports in 1993 suggest that there may have been four isolated cases of a brain worm effect in farmed deer, with symptoms of blindness, arched neck and a circling gait. Recovery was shown by two animals after rest and there were other contributory factors that may have complicated the infection. To date in 1997, no documented evidence exists of the presence of meningeal worm in well-managed farmed elk anywhere in North America.

Once the adult brain worm is mature, its presence can be detected by fecal pellet examination similar to the lungworm analysis technique, and distinguished by the presence of a spiny tail. There is suggestion that at low numbers and in the early stages of shedding there is variability in shedding frequency and detection techniques which has had significant impact on quarantine regulations and tests.

A number of anthelmintic treatments have been tested although none have been considered as completely effective against adults. The normal parasite treatment routine for lungworm control may be contributing to a lessened susceptibility for farmed deer, combined with the pasture environment and heavy stocking rates that make this an unsuitable area for snails and slugs to survive. Shedding rates are greatly reduced in experimentally infected animals by parasite treatment regimes.

WINTER TICK

Of the four genera of hard-bodied tick reported in elk in North America, only the winter tick *Dermacentor albipictus* has been reported to be a problem. This tick is generally recognized more as a serious parasite of moose than of elk, and it has also been found in large numbers on the hides of dead reindeer from a zoo. Up to 178,000 ticks have been recovered from a single moose hide, but one of the reindeer had over 400,000 ticks on it! This tick can also cause problems in mule deer.

The winter tick is a one host tick that completes three stages of its lifecycle on a single animal during the winter months. Long warm autumns and early spring snowmelts enhance survival and transmission of this parasite. The warmer the autumn, the longer the larvae remain active and able to infect passing hosts. At environmental temperatures below freezing the larvae seem to be inactive. The larval ticks climb up vegetation in the autumn, and clumps of them can be found on the tips of plants. When a host walks past the larvae leave the plants and attach to the animal. They will also attach to other objects, and researchers routinely use a flagging technique, which involves dragging a flannel sheet attached to a pole through the underbrush. After attaching to the host the larva feeds for a short while, and then molts into a nymphal stage. It goes through a rest period before feeding again, and again molting, this time into the adult stage. In spring, the adults take another meal, and the engorged females, after mating, drop off onto the ground and lay their eggs. A single female can produce many thousands of eggs, which hatch into the larval stage and then wait out the summer before again attaching to a passing host that brushes past the vegetation upon which the tick is waiting. The length of the wait varies markedly with location. In southern regions such as California, Texas and Oklahoma, a waiting period or "diapause" of over 4 months may occur. The period in British Columbia may only be two to three months, and in Alberta it may be as little as two weeks.

Clinical signs Heavily infested moose may lose much of the hair on their bodies, due to constant rubbing stimulated by intense irritation caused by the ticks, especially in late winter and early spring.

The hair of moose is white near the base, and as the coat is rubbed the moose changes from its normally dark brown colour to appear almost white over much of its body. It has been suggested that this tick has led to the legends of the "ghost moose". Certainly, the parasite can and has created such severe hair loss and debility that population die-offs have occurred.

The lesions in elk are generally similar to, but less severe than, those in moose. Severe debility has been reported in wild elk, but most often infested animals are seen in the spring months with a loss of hair around the base of their necks.

Diagnosis The distribution of the lesions in elk will give a strong clue as to the nature of this problem, but a definitive diagnosis would require the examination of the ticks themselves and their positive identification. Ticks have been seen on parts of the body other than the neck, especially in debilitated animals, and a careful examination may be needed to confirm a diagnosis.

Fig 4 First sign of winter tick - "bite marks" on the hair coat in autumn

Differential Diagnosis Any condition in which loss of hair occurs, especially in winter, must be considered as a possible alternative diagnosis. Lice can cause itchiness and coat loss through rubbing, but the lesions may not be as localized. "Grooming" due to dominance behaviour can also cause an animal to lose hair, and trauma, either inflicted by other animals, or due to repeated handling, will affect the look of the coat. It must be remembered that in spring the heavy winter coat of deer species is shed over a fairly extended period, and they can look scruffy for a short period of time.

Treatment Persistence is the key to satisfactory treatment of animals infested with winter tick. While it may be argued that they are not a serious disease problem in elk, they have the potential to exacerbate other conditions and certainly increase costs of feeding. If allowed to build up on a farm they could develop to severe levels. The loss of hair is unsightly, and heavy infestations will cause reduced insulation. The main challenge is that for much of the time that the tick is present on the animal, it is in a dormant phase, so that effective drugs do not reach the parasite. The only effective regime might be to treat regularly over the period in autumn or spring that the ticks are likely to be feeding. The best recommendation at this time is at least two or more treatments with topical acaricides such as "Coopers DeLice" in late November and March.

Control Paddock management has been tried as a method of tick control on elk farms. The rationale is that if the larvae are not picked up in the autumn they will not survive the prolonged period of inactivity and cold over the winter, and so the cycle could be broken in a single year. In practice, where winter tick has been seen in elk, this method of control has not been a success. Whether this was due to an insufficient distance between the animals and the infested pastures, or an inadequate number of "clean" pastures has not been adequately investigated. Check with your veterinarian for the most effective control methods for your area.

GIANT LIVER FLUKES

Flukes can cause liver damage in free-living and captive populations of elk, white-tailed deer and other ungulate species. They rarely cause direct mortality but may contribute to slower growth and reduced productivity.

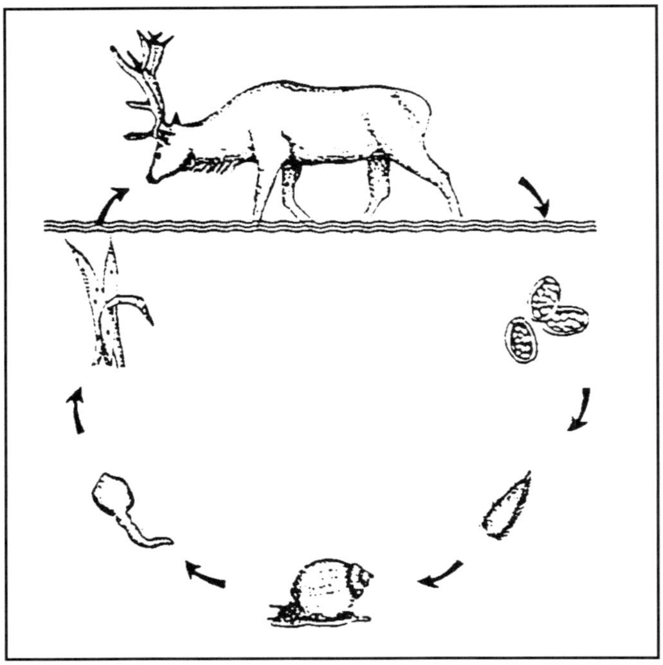

Fig 5 Lifecycle of Facioloides magna

Cause Giant liver fluke, *Fascioloides magna*, is the primary cause of worm-like infections in the liver of deer farm species. It is a large (30 x 80 mm) flat worm that migrates through liver tissue in search of another fluke. When two flukes find each other, they stop migrating. This allows the infected host (i.e., elk or deer) to surround the pair of worms in a fibrous tissue capsule.

Spread and Transmission Flukes release eggs into bile in the liver, which drains into the intestine. From there, the eggs pass out of the host along with the fecal pellets. The eggs hatch in water and the fluke larvae enter an aquatic snail. After four to six weeks, larvae leave the snail and encyst on aquatic plants. The life cycle is completed when an ungulate eats plants contaminated with fluke larvae. The spread of flukes is focused around wetland areas because aquatic snails are required by flukes to complete their life cycle. Infected animals can easily spread giant liver flukes to previously uninfested wetlands. Liver flukes are found in association with non-alkali aquatic habitats throughout southern Canada and much of the U.S.

Signs It is unclear whether animals infected with giant liver fluke can be identified by specific clinical signs. Capsules, flukes, and tissues damaged during the migration phase interfere with normal liver functions, but this often is difficult to detect by looking at the animal. Heavy fluke infections cause the liver to increase in size and fluids to build-up in the abdominal cavity. A distended belly and sagging bellyline are signs of this level of infection. Such animals often are slow-moving and somewhat depressed, however, the majority of liver fluke infections are not apparent by visual inspection. Infected livers will be condemned at slaughter.

Diagnosis An infection of adult liver flukes may be detected by looking for eggs in fecal pellets. Fecal samples should be collected fresh and kept cool if they will be evaluated within three or four days, otherwise the samples should be frozen until they can be tested.

Veterinarians use a combined sedimentation and flotation test to detect liver flukes. The eggs of giant liver flukes are heavy and a 2M Sheather's sugar solution is required to make them float.

Post mortem diagnosis relies on recovery of worms or evidence of damage to the liver. Adult flukes are found within white fibrous capsules in the liver. Immature flukes (5 - 50 mm long) often are found within bloody tracks as they migrate through the liver tissue.

A principle feature of giant liver fluke infection is a deposition of black pigment, often in streaks, through the liver tissue. The flukes feed on blood and the pigment is likely an iron compound, resulting from the destruction of red blood cells. The presence of black pigment is considered a sure sign of giant liver fluke infection.

Treatment Triclabendazole has proven to be effective against giant liver fluke in elk and white-tailed deer. Currently, it is not licensed in Canada but is available to veterinarians as a medical emergency drug. Triclabendazole is marketed by Ciba-Geigy as Fasinex®.

The recommended treatment for elk involves administering an oral parasite treatment of Fasinex® at 50 to 60 mg/kg body weight. It is effective against immature and adult flukes and thus treatment in spring and fall (after the primary periods of transmission) is likely the most effective.

Albendazole, often used to treat fluke infections in cattle is of limited use in treating elk and white-tailed deer. In Texas, albendazole-treated corn is used to limit (but not eliminate) fluke infections in white-tailed deer. Albendazole has only limited effect on immature giant liver flukes.

Control Infections can be prevented if farm animals do not have access to waterbodies where snails and plants abound. It is impractical and undesirable to drain wetlands in spite of their importance to the transmission of giant liver flukes. Controlled burning of grasses and rushes in spring may help to reduce the number of larvae available to grazing elk and deer when the use of wetland areas cannot be avoided.

BACTERIAL DISEASES & ASSOCIATED VACCINATION PROGRAMS

Clostridial Disease With on-farm intensification, a small but significant risk of ill health due to a variety of bacterial species develops. Cases are rare, yet it is a common practice to vaccinate against clostridial bacterial disease on a timed program as a cheap and effective insurance policy.

The associated diseases are the same as found in other domesticated ruminants, and may be referred to as blood poisoning, or other local names. Some of the most common are:

- **Enterotoxemia**- Pulpy kidney (Clostridium perfringens - Type D) Associated with sudden access to high carbohydrate feed (grains) or a sudden change in diet.
- **Black leg** (Clostridium chauvoei novyi), "gas gangrene"
- **Tetanus** (Clostridium tetani) - associated with trauma or wounds
- **Malignant Oedema** (Clostridium specticum)

Clostridial diseases are most commonly associated with young stock and cows during late pregnancy and calving. Often the animals are simply found dead. Under stress or injury the bacteria which are normally present in the environment and in the animal's intestinal tract gain entry to the blood system or trauma site, multiply rapidly and cause toxic excretions and changes that overwhelm the animals defences.

Producers may face local variation in the type of clostridial bacteria that challenges their elk. Redwater disease, also called bacillary hemoglobinuria, is caused by a proliferation of *clostridium hemolyticum*. Appetite, rumination and milk flow all suddenly cease. The elk will exhibit a high temperature and rapid breathing. Both urine and feces are usually blood tinged (hence "redwater"). Death occurs in two days or less. Redwater has caused significant problems in some poorly drained pastures. It is possibly associated with liver fluke damage.

Once contracted, there is no practical cure for any of these clostridial infections. Prevention is accomplished by vaccination. Many multivalent vaccines, i.e. vaccines that cover a total spectrum of these diseases are available as standard livestock treatments. In some cases, the multivalent vaccines are of insufficient concentration to give elk an adequate protection level. In such cases, specially designed vaccines (e.g. "Redwol" for redwater disease) must be used, and vaccination must occur at more frequent intervals.

The vaccination procedure is to give long term protection to the animal and to be properly effective, a sensitizing and booster program is used.

Elk calves acquire some passive protection at birth via colostrum from the mothers and this lasts 2 - 3 months. After this time they must make their own antibodies for protection. Considering the stress imposed at weaning and the potential for feed changes, minor cuts and injuries, ear tagging etc., this is an ideal time to begin a vaccination program with a five way or eight way vaccine depending on local conditions and risks associated with other livestock in the area. Veterinarian advice is important for local conditions.

A second dose, to fully develop the protection, should be given 4 - 5 weeks later.

Following this the animals need an annual booster dose. Adult cows should receive this 2 - 3 weeks prior to the calving period - as they are then at their greatest risk, and this will significantly boost the protection offered the newborn calf (end of April).

If a farmer is unsure of the cow's status then a sensitizing dose should be given first in late March.

Bulls should receive an annual booster pre-rut, although many prefer to handle this at velvetting time when the animals are quieter.

To avoid infection risks, and to preserve meat quality, vaccines are given subcutaneously and should be done in the loose skin above the heavy neck musculature.

As a general rule farmers should dedicate one side of the neck to such treatments as injectable parasite treatment, vaccinations, velvetting tranquilization and trauma cures and leave the other side for TB testing to minimize confusion of casual lumps and injection site reactions, allowing accurate TB testing.

Needles should be short, sharp and frequently changed and kept sterilized, and care must be taken to place the vaccine correctly. The skin can be pulled upwards in a small tent with a free hand and the needle carefully placed in the fold away from muscle with the dose delivered quickly and smoothly.

Calf Scours Elk calves are susceptible to many of the same organisms which cause scours in beef cattle. As a preventative measure, particularly if your herd or property has a history of problems with scours, vaccinate your cows with one of the following products, or another equivalent product as advised by your veterinarian:

- **Ecolan RC®** - for protection against E. coli bacteria, Rota virus, and Corona virus or;
- **Scourgard®** - for protection against E. coli, Rota virus, Corona virus and Clostridium perfringens type C.

Some farms with recurrent scour problems may need to develop a unique vaccine - talk to your vet.

Leptospirosis In New Zealand, cases of leptospirosis in deer are being increasingly reported particularly when animals may be associated with dairy cattle and pig effluent.

It can cause septicaemia and death, and has been associated with abortion in fallow deer and occasionally red deer.

Skunks have been reported in some areas in Canada as being a reservoir for the type (serotype) "pomona", although it seems that because of the highly synchronous breeding of elk they are not pregnant in late summer when risks are at their greatest. A simple vaccination program is very effective if indicated.

Yersiniosis This bacterial disease is the most important management related disease that affects deer and elk, particularly young stock after weaning and throughout winter.

It is caused by the organism *Yersinia pseudotuberculosis*, but its naming has nothing to do with the disease tuberculosis.

Invasion of the bacteria is promoted by a variety of on-farm stressors such as underfeeding, cold wet conditions, transportation, weaning in bad conditions, mixing unfamiliar groups of animals, etc. in the first year of life. The bacteria are present in the environment and it seems that a stressor is needed to promote the disease. Massive reactions in the intestines cause a smelly green loose watery scour often noticed as a stain or loose fecal material under the tail or on the hocks. As the infection takes hold massive hemorrhaging from the gut occurs and death can follow quickly about 24 hours from the first clinical signs.

Calves in particular are more susceptible. In bad weather fawns can lose a lot of body heat very quickly because they have very low fat reserves and their coats are not well insulated. Their natural response is to stop eating, seek shelter, and lie low. If bad weather continues, their lack of eating is compounded by heat loss and their intestinal movements tend to slow down. This slowing of intestinal activity allows Yersinia organisms to multiply excessively and produce toxins which damage the lining of the intestines. This leads to rapid loss of body fluids, bleeding into the intestines, dehydration and shock, and frequently results in the death of the affected animal.

The disease is quite treatable but very dependent on the speed of detection and response. Outbreaks affecting up to 40% of a group have been recorded. An early sign is often that the affected animals stand away from a group being fed. Infected animals are depressed and listless and often in discomfort and commonly will show signs of the scouring. These symptoms are accentuated by bad weather, but will always be associated with some stress activity, although this may be quite subtle, that has occurred 48 - 72 hours previously. As the animal stops eating the gut conditions become more and more favourable for bacterial invasion and reaction, so there is a strong time constraint associated with treatment.

Treatment is via antibiotic or antibacterial therapy using a terramyosin long acting injection, or tetracyclines. It is often prudent to extend the treatment to a number of animals - smaller, weaker animals, or anything showing lethargy or unusual behaviour. Occasionally entire groups have been treated with positive results. While the yarding and handling itself may be a factor, removing the predisposing factors is critical. Most commonly, along with treatment, the elk must be provided warmth and shelter, and the group size reduced if over crowding is suspected.

Severely affected animals will also need electrolyte fluid therapy to combat dehydration and removal from the group to warm quiet conditions is advised. Quality hay should be provided and high energy/high risk grains avoided.

In any cases of scouring in young animals, Yersiniosis should be suspected and immediate veterinary attention sought. The bacteria can be confirmed by laboratory analysis although this takes time so a full discussion and treatment program should be considered with expert advice.

A vaccine has been developed in New Zealand, requiring a two shot campaign - doses given 4 - 6 weeks apart for full protection. This is best done at weaning and four weeks later so this fits a program of early weaning and worm control well. The vaccine has been tested successfully and used commercially since 1992 and while it will not totally prevent the disease, as often the first outbreaks occur associated with weaning stress and the onset of bad weather, it has significantly reduced losses. Good management, however, is the key to prevention.

"Foot Rot" *(Necrobacillosis, Fusobacterium necophorum)* Outbreaks of "foot rot" have been reported closely associated with conditions of poor management and stress. The disease has a variety of effects, including lesions of the mouth and jaw causing extensive damage and swollen face. The condition can lead to generalized infection and massive lesions in the lungs, liver and body cavity and lesions of the limbs and feet with potential for loss of toe digits and lameness created by heavy swelling of the lower foot and open discharging abscess.

In elk, characteristically, outbreaks have been associated with management factors i.e. overcrowding, changed feed or poor quality feed, reduced resistance, and commonly, some injury which may only be slight grazing or cuts to the lower legs. The bacteria are normal environmental bugs, and live anaerobically, often associated with poorly drained feed lot situations where buildup of mud and fecal material is common. In these conditions, any physical defects in yards or pens that can cause injury (protruding nails, sharp stones, projecting concrete ledges or roughly built feeders that can cause mouth injury, etc.) can predispose the animal to casual infection.

Any signs of lameness, swollen mouths or depression are indicative of the need for immediate further attention.

Treatment may require veterinary assistance. Early cases respond well to antibiotics, but as the disease advances the blood supply, particularly to the lower limbs, is disrupted and makes it difficult for circulation of antibiotic drugs. The affected foot needs to be carefully cleaned and dressed and hydrogen peroxide can be used for cleaning wounds and abscesses. A locally applied long acting antimastitis treatment commonly used in treating dairy cows has been tried, with good success, by applying it directly into the wound site along with routine antibiotic treatment. Affected animals should be rehoused in safe, warm, dry conditions, and it is most important where there are several cases to revisit the feed pen or handling yards and try to identify the source of injury and contamination and fix this at the same time.

Fallow deer have been increasingly reported as showing the disease in other forms - the killer fawn disease - which has been associated with significant losses in newborn fawns or recently weaned fawns. Entry to the body may be gained via the navel cord and can cause massive internal lesions and death in apparently healthy animals. There is a move towards research and vaccine development in North America.

OTHER BACTERIAL RISKS

The normal array of disease risks faced by all livestock producers should be considered by elk farmers and veterinarians in developing a herd health management program. Although rarely found in farmed elk or deer, producers should be particularly aware of the risk of Salmonella, Brucellosis, Listeriosis, and the *Mycobacteria*, especially Johne's disease and Tuberculosis.

The *Mycobacteria* include some of the most difficult diseases both human physicians and livestock veterinarians have faced. At least two of these may infect humans, posing a disease transfer risk to livestock managers. The two of greatest concern to elk farmers are *M. bovis* and *M. paratuberculosis*. *bovis* is the well - known cattle TB. The testing and maintenance of a Negative Status or Accredited TB - free herd is critical to maintain the integrity and value of all cervid herds. Most important for elk farmers is to realize their essential role in maintaining the hard-won TB - free status of our entire livestock industry. This status has been achieved at great expense and with substantial impact on the financial success of other livestock producers. Maintaining that status requires uniformly high testing and record-keeping standards for all producers and managers, as well as for the Canadian Food Inspection Agency, the United States Department of Agriculture, and all other national, State, Provincial and Territorial Animal Health Management groups.

TUBERCULOSIS *(Mycobacterium Bovis)* TB

Tuberculosis has always been associated with man and the interaction with farmed livestock worldwide, and has been reported virtually everywhere where deer are farmed. TB in livestock has the potential to limit markets for all meat products, depending on the extent and prevalence of the disease. In the deer industry TB concerns have been responsible for undermining confidence, reducing live sale prices, causing extensive and expensive eradication programs and attracting enormous bureaucratic and political pressures to rest public safety concerns. Accredited free status is hard won, and is valued and important.

In New Zealand, where TB eradication and testing is the dominant focus of herd health issues it is also worth considering this perspective: In 600,000 annual tests on breeding stock and 350,000 annual animals slaughtered for quality venison production, the reactor rate (confirmation of TB) is only 353 cases and declining annually - i.e. 0.03% of the national herd. These reactors are 85% from "endemic areas", i.e. areas where a known non-farm host, the introduced Australian brush tailed possum, is a disease reservoir and continues to reinfect deer and cattle in the area. The balance of reactors are almost always due to lack of care in relocation of deer from these areas, or some breakdown in test recording and stock control.

In Canada, recent outbreaks are fully under control, i.e. the eradication is complete. A new understanding of test procedures and the development of a whole herd approach with a combination of tests has been developed. With a TB free cattle status, and no known feral reservoir of disease source, the Canadian elk industry is poised to have and maintain like the cattle industry, a TB free status, and add this important indicator of quality farming to its merits.

In the U.S., an effective system of uniform methods and rules for the control and eradication of TB in farmed elk is being implemented.

TB Testing

There are a number of variations on a common theme amongst the regulations for various jurisdictions. The common theme involves the testing of specific animals selected for sale and relocation, and a background whole herd history, generally maintained on an annual or two year basis.

Before any testing program can be undertaken, a reliable and readable individual animal identification and recording scheme must be in place, as testing requires each individual to be tested and the test read.

The primary screening test is the midcervical intradermal test, known as the skin test ST, or the Mid-Cervical Test (MCT), or the Single Cervical Test (SCT). This involves the injection of a measured amount (0.1 ml of a fixed strength) of a tuberculin extract into the skin (Purified protein derivative of Mycobacterium bovis is used for elk and red deer) - hence bovine PPD.

If the animal has the disease, or has been exposed, the body's immune system will recognize the antigen and form a localized reaction at the injection site. This takes 72 hours - 96 hours to develop to a reliable "readable" stage, and a swelling will result at the site. This is often highly visible as a lump, and a measurable difference between normal skin and the injection site is seen. For clear testing and reading there are a number of prerequisites.

Site Preparation

Traditionally, cattle have this test routinely given in the area of clear skin at the base of the tail (the caudal fold), used because it is easy to administer, needs little site preparation and was accurate. In elk and deer the caudal fold is not prominent, and early evaluation of it and other sites showed it to be a poor test site and the midcervical area on the side of the neck region to be the site of choice and greatest accuracy.

For testing, all the hair at the test site must be removed by close shaving to leave a square patch 100 x 100 mm (4" square). As elk and other deer, especially at wet times of the year, often have dirty muddy coats the common practice is to use a horse clipper and coarse blade to remove outer hair, and a fine grooming blade (Oster no. 30 or 40) for final preparation. Deer have a very fine down undercoat and this can be difficult to work through. In winter blade life can be quite short. The combination of clippers works well. Site preparation with scissors is not good enough. Because there is usually some noise with clipping, the head and neck should be restrained during clipping. The ideal site is halfway along the neck and about 75 - 100 mm down from the mane area. Once clipped and cleaned, if necessary the site must be examined for any abnormalities - previous injection scars, healed cuts, etc. and these noted.

The veterinarian or approved testing officer will inject a small amount of tuberculin with a fine needle into the layers of the skin to leave a small raised lump or bleb.

Good lighting is essential so that all operations can be completed accurately, particularly site inspection and the injection process, and finally for reading. The small cut ends of the hairs in good light accentuate any reaction, and the area must be clearly visible for reading. Reactions are not always a prominent lump, and may be a diffuse broad area of skin thickening occasionally painful or hot to touch.

Recording

All elk must be tagged and a master list of stock presented for testing and checked as the animals are injected and later read.

Good Management and Quiet Handling

As the test involves a measurement of the immune system, any stress that affects its application and reading should be minimized.

It is not advised to test bulls immediately prior to the rut, or cows in late pregnancy, both for the difficulty of handling and the chance that results are compromised because of the altered physiology and immunology of the elk at those times.

Good Communication and Understanding

Because no single test of this nature can be 100% accurate, and because of the nature of the group of bacteria involved in tuberculosis there is a strong possibility that some elk will show some reaction. For farmers and veterinarians this is always a difficult time and feelings can run high on the discovery of a lump.

The issues of finding a lump are complex and prior to testing a clear understanding of the procedures involved in retesting, comparative testing, sensitivity, specific and false negatives, false positives etc., should be had.

No reaction should be ignored no matter how slight as this animal is a valuable source of information of herd status and the worth of testing. The reaction animal adds to herd test history and focuses future testing procedures. Many factors can affect the test reaction.

RESULTS & TERMINOLOGY

Bovine TB, Avian TB: *Mycobacterium bovis* belongs to a large group of related bacteria present in the environment, soil and in other species. Avian TB (*M. avium*) is an important group, associated with birds, although infection in other species is common. It causes no disease in elk, but a heightened immune response and cross reacts with *M. bovis* tuberculin. Some farms usually report 7 - 15% avian reactions at annual testing and there is a trend that the number usually increases over winter if grain feeding is practised and birds - sparrows, pigeons, etc., are common in wintering pens and barns. Summer and autumn testing also may yield high number of reactions. December to February is the preferred time to test. The same elk do not always react each year. The cross reactivity response can be resolved by:

The Comparative Cervical Test - CCT In accredited herds, or where there is good evidence of herd history to support its use, injection of both tuberculin antigens can be done at testing. Two site preparations are used on the neck, advisedly, each separated by the usual hair. Avian PPD is injected into the uppermost site, and bovine PPD in the basal site. The body's immune response will be to form a swelling or reaction in 72 hours at the site of injection that is specific to the invading bacteria, or commonly there may be no reaction at either, or a slight reaction at both. Differences in the site reaction are measured, against normal skin thickness, in that area and contrasted to each other. Some authorities consider that if the swelling at the avian site is greater by 2 mm than the bovine site then clear avian reactions are confirmed, while some may consider any swelling at the bovine site warrants further investigation irrespective of bovine reactivity. If changes at the bovine site are equal to or greater than the avian site a bovine reaction is confirmed.

Sensitivity and Specificity In a disease test that is complicated by cross reactions and variations in animal responses, the perfect test would show 100% of all animals infected, i.e. 100% sensitive, and would not show reactions to any animals not infected, i.e. be 100% specific (for *M. bovis*). The comparative test is a lot more specific for the reaction type, than the single test, but all evidence suggests that its sensitivity declines quite sharply.

CCT is not advised as an initial screening test if there is any risk or previous history of TB in a herd, and where a ST has been first used, a minimum of 60 days delay is needed before a CCT.

False positive Non-specific reactors, or avian TB reactions to a skin test ST.

False negative An animal that has exposure to the disease and does not react. In suspected TB cases these animals represent a danger element to the herd. Either an animal is in the very early stages of exposure and has a poor immune response to the test, or is an animal that may have severe disease either contained, or in an advanced stage although there may be few outward signs. These animals are considered as anergic, and a small tuberculin test has little chance of showing a response against the huge background of immune response the animal shows as it tries to fight advanced disease.

False positives can also be obtained by poor site preparation, infections during injections, or site damage through previous events that are not noticed at the time of testing.

False negatives are also due to poor site preparation, incorrect injection of tuberculin (under the skin rather than into it), inaccurate measurements, or using sight alone rather than feeling for a lump reaction and human error in determining what is a lump and what is not.

BLOOD TESTS FOR TUBERCULOSIS

A number of blood tests have been developed as ancillary tests to skin testing as a screening process, and for use in repeat testing, and a variety of further test procedures are under development. The blood tuberculin test, or BTB may be used after an initial screening of either ST or CCT. It currently uses three elements of testing an animal's immune response to a disease challenge at the blood cell level.

The ELISA detects specific antibodies and has its major application to determine severely diseased animals. These are often the false negatives in a herd that show other test signs of disease, but that skin testing alone may fail to find these critical animals.

The LT - lymphocyte transformation The white cells circulating in the blood that focus on infection are collected from blood samples, cultured and treated with the bovine antigen in the laboratory. The response of those white cells to *M. Bovis* is measured after culturing and labelling with radioactive material and counted. Response to *M. avium* and other antigens can also be measured. The test involves complex laboratory procedures and reagents and is relatively expensive to run, but extremely important in adding to the compilation of knowledge on disease profile, and defining the herd status.

Antibody measurement & inflammatory protein response To these procedures is added a variety of activities that give information on the background status of the animal - any unusual stress events, disease response, etc. that may be present and influencing or be influenced by the TB disease status.

Results of the TB in Elk Project completed in 1993 - as summarized by Dr. Elizabeth Rohonczy, Canadian Food Inspection Agency, Ottawa:

- At least 4% of elk in "TB in elk" project were Johne's positive (Mycobacterium paratuberculosis). 10% were positive to Mycobacterium species other than bovis, avium or paratuberculosis.
- The MCT, or mid - cervical test we currently use for initial screening, was estimated to be 81% sensitive (i.e. 81% accurate in detecting the presence of antibodies to Mycobacteria) and 80 % specific (i.e. 80 % accurate in differentiating Mycobacterium bovis, which causes TB in elk, from other species of Mycobacteria).
- A combined interpretation of a lymphocyte stimulation test (LST) and ELISA test (using the antigen MPB70) was estimated to be 88% sensitive and 76% specific.
- The Comparative Cervical Test (the CCT we now use for followup testing) was estimated to be 39% sensitive and 97% specific.
- The BTB test (a set of blood based tests

patented and performed by a company in New Zealand) was estimated to be 52% sensitive and 90% specific if "equivocals" (results that were not clear) were classed as negative, and 59% sensitive and 84% specific if equivocals were classed as positive.

- The "gold standard" is considered to be the culture of cells collected from apparently infected tissue. It is close to, but not 100% sensitive.

LST and ELISA still both measure antibody response (i.e.) detect animals that have been exposed to Mycobacteria, not those that currently have a Mycobacteria spp. infection. Confirmation by Culture indicates a current M. infection. ST is also an "immune response" test.

Ideally a combination of tests would be used and add to the initial skin test screening. In New Zealand the BTB operates as being 98.5% sensitive, and 90% specific, compared with the ST at 87% sensitive and 70% specific, and the CCT at 40 - 60% sensitive and 85 - 95% specific. The use of all three under strict control forms the basis of the whole herd accreditation scheme and test and slaughter program. The Canadian Food Inspection Agency, the Canadian Venison Council and other Livestock groups are currently co-operating on research and development for more effective blood-based tests for Bovine TB and other mycobacteria.

In any blood testing it is important that samples are drawn from animals under normal handling conditions without stress or unusual activity and that arrangements for blood storage and transport (chilled - not frozen) to reach the laboratory within 72 hours are made well in advance.

POST MORTEM EXAMINATION/ CULTURE

The final diagnosis of tuberculosis in deer relies on a thorough post mortem examination of the whole animal and culture of any suspect lesions found. Often in the early stages of disease no visible lesions are seen in the lymph nodes of the gut, neck, and head region and these are pooled and cultured to grow the bacteria for positive identification. This is a difficult process that can take 3 - 4 months for final diagnosis. New technology involving DNA typing may speed this process.

TB TESTING - HERD ACCREDITATION PROGRAMS

The development and maintenance of whole herd accreditation status underpins most regulations that affect sale and relocation of animals across international and provincial borders, and in time will be mandatory for any movement of stock between elk farms. Gaining accreditation status is not rapid, taking 2 - 3 years but once attained is easily maintained by sound stock purchase policy and annual whole herd testing.

In general the procedure is as follows:

1) All animals six months and older are presented for testing.

2) The whole herd should be tested within 60 days from the first group tested to the last. Mid winter is an ideal time as weaners are old enough, and bulls and cows at their most easily handled. Testing should be prior to late pregnancy, and antler growth, or pre-rut for cows with calves at foot, or bulls that are difficult to handle.

3) All animals must be uniquely identified.

4) Tests should be at least 6 months apart, and not more than l year apart.

If nonspecific reactions, and *M. avium* are a problem some authorities may grant a CCT as the initial screening test after verification of bovine-free status via BTB, and culture tests.

Herd status is maintained by routine ST testing with BTB or CCT testing of reaction animals.

In the United States, the US Department of Agriculture has established a set of recommended Uniform Methods and Rules (UMR). These UMRs establish the procedure for establishing an Accredited herd. This is defined as a herd of captive cervids that has tested negative to at least three consecutive official tuberculosis tests of all eligible cervids that has tested negative to at least prescribed standards. The tests must be conducted at 10 - 14 month intervals. Each individual State Animal Health Board may choose to adopt and apply or modify these UMRs. Currently, Canada's Negative Status and teh USDA Accredited Status are recognized as being equivalent for purposes of trade.

To ensure their own herd status is not compromised, farmers should only buy stock from herds with an equivalent or better TB status than their own herd. Routine sale testing for individual animals within an accredited or whole herd tested farm is also often requested within 60 days of transport.

Currently in Canada the finding of a positive TB reactor will mean the whole herd eradication procedure will occur as the ultimate exercise in risk prevention. Compensation at capped levels for the various species of deer is paid. There is a major effort in consultation to implement aspects of a test and slaughter positive animals program using the new testing procedures combined with traditional skin tests and this is under extensive evaluation at present.

The concerns expressed echo the importance the threat of this disease poses both to farming and wildlife concerns and the responsibility in elk farming should be no less.

MYCOBACTERIUM PARATUBERCULOSIS – JOHNE'S DISEASE

Johne's is very similar in nature to M. avium, or avian TB. Most non-specific but non-bovine reactors to the ST are avian reactors, but some of these may actually be Johne's.

In 1993, 2600 farmed elk were slaughtered in western Canada to eradicate bovine TB that had been imported into the farmed herds. At least 4% of these elk were determined to be Johne's positive. 10% were positive to Mycobacterium species other than bovis, avium or paratuberculosis. These latter are probably one or another of a large group of M. organisms that live as saprophytes in soil. These are not a concern, and Avian TB is virtually never an actual disease problem for elk farmers, but Johne's definitely is.

Johne's is referred to as a "production - limiting" disease. It is called this because it typically takes many years after infection before an animal actually begins to show symptoms of the disease. Infection is through exposure to manure or other excretions from an infected herdmate.

Johne's is not uncommon in cattle, especially dairy cattle. It is estimated that close to 30% of dairy herds in North America are somewhat affected by the disease. In cattle, less than five per cent of infected animals develop clinical signs of illness. These infected but non- clinical animals act as a source of infection to all other animals on the farm. Clinical signs seldom develop in cattle under two years of age, but elk and other deer appear to be more susceptible and will show symptoms at an earlier age.

Clinical signs may be difficult to pin down. Generally, you should consider Johne's in any elk that is not doing as well as you think it should, and you cannot figure out why. It may have a good appetite and show no other outward symptoms, but it does not thrive. It continues to lose weight, and its hair coat becomes rough and unhealthy looking. It may develop intermittent bouts of diarrhea that gradually become more frequent. Their rumps become soiled with greenish coloured fecal material around their tail, hindquarters and hocks. The disease can run from days to months, but generally the younger the animal, the quicker the progress to emaciation and death. In red deer, the most common age of onset is eight to twenty months. Sporadic cases may appear in older animals, but they appear to be less likely to develop clinical symptoms than younger individuals. Most often, clinical signs develop in stressful situations.

Once Johne's develops, there is no treatment. If you already have the disease on your farm, the most important factors in managing it are cleanliness and a low stress environment. Since it is younger animals that typically develop the disease, it is absolutely essential to maintain clean calving and weaning areas, with low population densities. The best option is to avoid establishing the disease.

Start avoiding Johne's and many other possible problems by building your elk farm on new ground, and avoiding the use of buildings that have been used for other livestock. The more intense and long-established that use, the more potential for problems.

Jurisdictions such as States and Provinces may follow the International Animal Health Code and require a Health Certificate attesting that the animals proposed for import:

- showed no clinical signs of paratuberculosis on the day of shipment;
- were kept in a herd in which no clinical signs of paratuberculosis was officially reported during the five years prior to shipment;
- showed negative results to diagnostic tests for paratuberculosis during the thirty days prior to shipment.

You can also avoid many problems by the way you approach and examine a farm where you are considering buying livestock. These precautions apply not only to Johne's, but also to many other disease concerns, so they are included in the summary at the end of this section.

VIRAL DISEASES

Rabies Rabies is a risk in elk held in areas where this disease is common in wildlife that may gain entry to farms. Farmed deer may be particularly at risk because of their curious nature and any small animal, especially showing unusual signs, that enters a paddock will be investigated. It has been recorded in red deer in Southern Ontario by the typical unusual behaviour; high stepping gait, wood chewing, and salivation. In risk areas vaccination as for cattle can be considered.

PRION DISEASES

Chronic Wasting Disease (CWD) is one of the most difficult challenges our industry currently faces. Although only a very few elk or deer have ever been affected by this disease, its association with the "Mad Cow" situation in Europe makes CWD a potential disaster for our industry. CWD is one of the group of diseases referred to as "Spongiform Encephalopathies" or "SE's". SE's are found in cattle, sheep, elk and deer, cats, mink and humans, but we must be very clear that not all SE's are the same. No evidence exists to suggest that CWD may be transferred under natural circumstances to any species other than cervids. SE's appear to be caused by abnormal proteins called prions. Something triggers normal proteins in the brain of affected animals to become abnormal, to the point brain functions are disrupted and the animal dies.

CWD was first observed in 1967 in mule deer at a Colorado wildlife research facility. It has since been reported in elk, mule deer, and white-tailed deer in free ranging animals in northern Colorado and southern Wyoming. Extensive survey work has failed to identify CWD in any other wildlife population, although the disease has persisted in the endemic area described for at least twenty years.

In the early 1990's, CWD was identified on elk farms in the United States, and since then, close to ten farms have been affected. In some cases, only one animal has been found positive, but in others, a larger number have succumbed. No one has been able to explain why.

Canada considers itself free of CWD, however, four cases have been found in the country. The first case occurred in 1974 in mule deer imported from the United States to the Toronto Zoo. These animals had originated in one of the research herds in Colorado now known to have the disease. The zoo killed all the remaining deer from the same source, and no other cases have been identified there.

Three other cases have occurred in farmed elk in Saskatchewan in 1996, 1998 and early 2000. Actions to control the disease were taken on the basis of the information known about each of the cases and the disease at the time. To date, all animals that have been killed as a precautionary measure were examined and found negative for CWD at that time.

All of these cases appear to trace back to one U.S. herd, and that herd may trace back to the Colorado endemic area. CWD remains largely a mystery, as most of the other SE's are. However, we know we do not want it on our farms, and measures are and must be taken to avoid and eradicate the disease.

How does one recognize the presence of CWD? Here are some guidelines:

1. **CWD Suspect Profile:**
   - at least 18 months old
   - emaciated
   - abnormal behaviour – social dysfunction
   - neurologic symptoms
   - drinking/urination excessive or abnormal
2. **First symptoms:**
   - rolling, vacant eyes
   - abnormal behaviour, dull, unresponsive
   - pacing, stumbling, tremors
   - weight loss
   - salivation
   - drinking and/or urination excessive or abnormal
   - pneumonia, respiratory problems

Incubation period after exposure until symptoms develop appears to be from twenty to thirty months.

Any producer that observes these symptoms should immediately contact a veterinarian and isolate the animal from the herd. If an elk or deer dies with any of these symptoms, the entire head and upper neck should be removed and submitted as quickly as possible to a diagnostic laboratory for examination for CWD.

SUMMARY

Prevention is always the best approach to managing disease. When you buy livestock for your farm, don't be shy about asking questions before you hand over the money. Look at the farm environment. Is it clean and well- ordered? How crowded are the animals? Is there plenty of feed, water and shelter available? Look at the whole herd, not just the elk being offered for sale. Is the level of management and condition the same throughout the herd? If not, which group of elk looks the best? And the worst? Is the difference explainable - for example, bulls just after the rut? Are there individual animals that look particularly poor, or particularly good? Is there any evidence of diarrhea, rough hair coats or emaciation? If you see any of these signs, you should either start asking questions or leave. When were the elk last dewormed? How was it done - all the elk at the same time? What product? What dosage? Has there been any deaths in the herd in the last year or two? From what cause? Was there a proper postmortem done? Can you have a copy of all of them? Did the postmortem include examination for CWD and Johne's disease? Was the death sudden or a slow wasting? Who is their regular veterinarian?

No answers? Then politely thank them for their time and leave. Good managers never have a problem bragging about how well they manage their elk. They are proud to share such information. Remember, it's your money, and the problems will be yours if you bring them home. Prevention is always more painless than treatment.

After all this discussion and precautionary warnings, it must be emphasized that elk are not often sick, and death losses on a well managed farm average less than 2% per year. The health management problems have proven to be very small, and on well - managed operations, negligible. Nature has provided us with livestock that are resistant to most disease or parasites, and our care and planning will ensure that situation continues.

A comprehensive review of health concerns and issues is presented in *Farming Wapiti and Red Deer*, by Drs. Jerry Haigh and Robert Hudson. This text is a most authoritative and relevant reference.

MANAGEMENT AND HERD HEALTH

In a recent survey of deer on farms in Ontario, researchers identified that of over 2,000 animals during the previous three years there had been only 16 reported deaths, the majority (14/16) due to either accident (3), misadventure (2), calving difficulty (l), Yersiniosis (2), or grain overload (6) i.e. management related problems.

Grain overload is the most significant concern and frequently occurs when animals are offered a change in diet which includes high energy supplements like corn or barley without adequate introduction and gradual increase of the ration.

Farmed elk, once accustomed to grain feeding, respond very quickly to reintroduction and if given the chance can eat large amounts very quickly. Even if a small amount is fed, there is a chance that some animals can eat to excess by bullying and dominating the feeding station. Introduction to full rations ideally takes 7 - 10 days with a gradual increase daily. When feeding, grain should be spread out either in small piles, especially suitable for males, or in long troughs. Feeding space of about 0.6 m per head is ideal, though may not be all that practical. Nevertheless care must be taken.

Wheat and barley are considered particularly "dangerous" grains with corn and oats less so. Cracked or crushed grains are greater threats as the grain starches are more available to the rumen bacteria. These undergo population changes under a sudden influx of carbohydrate and produce large quantities of lactic acid which damages the intestinal lining and causes massive scouring, discomfort, depression and often death through many side effects, including tendonitis and other foot and leg discomfort, particularly in adult bulls. Treatment is often difficult because of the severity of damage and the speed of onset of the response to overload, but chalk based preparations can be given by mouth, the grain removed from access and treating with electrolytes to restore fluid loss and revitalize the pH balance.

However, the cure is careful management of dietary change and ensuring there is no risk of accidental access to grain stores. The problem can be very significant in young stock on their first exposure to supplementary feeds. There is some merit in feeding limited amounts to cows and calves in late lactation, to boost the condition of the cows and to familiarize young stock with feeding and the presence of the farmer, and to have a teaching experience while the risk is low. Farmers in feedlot winter feeding should recognize that there are groups of shy feeders that may be forced from feeding areas during the daily feeding round i.e. some animals gain more than their share while others may have little opportunity. If these shy feeders are removed for separate management the initial precautions should be observed.

Sudden changes in diet can also be upsetting when the animals are first released from feedlot conditions onto lush grass in spring conditions. Any transition should be adjusted by continuing to feed, but at reduced levels, the hays or supplements being fed previously over the course of 7 - 10 days.

ACCIDENTAL DEATH

Yarding or handling misadventure

This type of accident, when an elk hits a fence or a gate and breaks the neck, is hard to avoid once an animal has taken that turn of mind. Occasionally and often out of character an animal will become very agitated and may leave the group, become separated or panic. The quickest way to calm such an incident is to allow the herd to regroup around the individual and remove the stressing factor. High visibility of fences in pressure areas aids prevention, but in time it is ultimately stockmanship that prevails. Accidents can also be associated with loose wire, untidy pastures and yard hazards that are all preventable. Elk are very inquisitive and will investigate vehicles, machinery or working areas on a farm, so litter from fencing is potentially dangerous as it can either be swallowed, picked up in feet or wrapped around antler stubs. Basically, if there is a hazard on an elk farm, the elk will find it first unless it becomes common practice to check and remove these obstacles.

PORCUPINES

Elk are particularly inquisitive and have a natural aggression against small animals that may invade the herd territory. Encounters with porcupines are often recorded, usually with a bad outlook for the elk. Their habit of sniffing or nuzzling curious objects often causes multiquill attacks in the nose and muzzle region and in response, elk often stamp and kick with the front and hind feet.

Fig. 7 Porcupine quills in hind leg

Thus a porcupine encounter can result in many quills that should be removed before infection and migration occur. Veterinarians recommend antibiotic treatment and a tetanus shot as part of treatment.

PREDATORS

Elk are particularly defensive of their young and will drive, or surround and kill most predators up to coyote or dog size, although predation still occurs. Other than reinforced fencing solutions, i.e. mesh fencing and the use of electric outrigger wires, vigilance in high risk areas is the only remedy. In risk areas a few adult cows as sentinel animals in a group of newly weaned calves may help. Similarly, donkeys have been used successfully in a protection role, often in adjacent perimeter pastures rather than in with the elk.

TRACE ELEMENTS

We know from experience that herd health and productivity can be greatly compromised by some mineral deficiencies, particularly copper and selenium, and occasionally cobalt and iodine.

As trace elements are involved in a complex interaction with physiology and other mineral levels, recognizing and correcting imbalance is a job for expert analysis and remedial recommendations.

Copper Deficiency has proven to be the most common trace element deficiency in farmed elk and deer. Although all farmed species can be affected, elk and deer require much higher dietary levels than cattle or bison, and the copper levels in feed supplements commonly fed to elk would kill sheep. Copper is necessary for the proper function of many parts of the body, and is critical in enzyme function. Deficiency leads to improper growth of bones, hair and nervous tissue and many other abnormalities.

Deficiency can occur in two ways: Primary copper deficiency occurs when there are low levels of that mineral in the soil and therefore in feed. Secondary copper deficiency occurs when there are adequate levels of copper in soils and feed, but there are excessive levels of other minerals such as molybdenum or sulfur that bind to the copper molecules and make them unavailable for absorption and use. Many areas in Western North America have excessive sulfur in water sources or high levels of molybdenum in certain soil types. In particular, peaty and swampy soils tend to have levels of molybdenum sufficient to cause copper deficiency.

Typical deficiency affects young, rapidly growing stock rather than adults, and the effects are most often observed during rapid growth phases in autumn and spring. Copper is transferred from feed and stored in the liver, then released as required to the blood. As stores are depleted and blood levels drop, the signs are variable in different herds, but often include:

- slow growth rate
- hair coat is rough, kinked, orange or lighter in colour, lacks the luster of healthy elk
- staggering gait
- uncoordinated action of the hind legs
- swelling of the hock joints
- increased rate of bone fractures
- diarrhea

As this is a nutritional deficiency, most animals in the herd will commonly be affected to some degree. Diagnosis of a copper deficiency problem requires several approaches and analyses. The first indication is background information regarding the levels of copper, molybdenum and sulfur commonly found in you farm area. This information is available from your neighbours, agricultural specialists and laboratories. Next step is to actually sample your soil, feed and water and have the analysis done and considered by those same assistants. Then, blood samples can be collected from a number of elk in your herd and the levels of copper determined. Blood sampling is rather unreliable except on a whole herd basis, but the **copper levels in blood should be 10 to 15 micromols per kg. Dry Matter Basis, or 1 mg/kg DMB.** Liver sampling is much more accurate in determining copper levels, but this requires a biopsy done by surgery on a live animal, or more likely, sampling of animals slaughtered or dead from other causes. **Copper levels in liver samples should be in the range of 120 to 150 micromols per kg DMB or 10 mg/kg DMB**, with less than 100 micromols or 7mg/kg suggestive of deficiency. However, levels may range as high as 100 to 200 mg/kg (ppm) without risk of toxicity, and samples taken from neonates or very young calves can be amazingly high. The elk cow mobilizes copper to the fetus in late gestation, elevating the reserves in storage in the calf's liver to very high levels. Reports of levels up to 18,500 micromols/kg or 1200 mg/kg DMB are considered normal and not toxic in red deer, elk, and whitetail fawns. Males in the antler growth phase also require high levels of copper - hard antlers may be 36 mg/kg (ppm) copper.

Once a copper deficiency situation has been identified, treatment is relatively easy by oral supplementation. In New Zealand, copper oxide boluses are administered, but most elk farmers in North America go with increased copper in a supplemental grain and/or alfalfa- based pellet. The target is 15 ppm DMB total ration, 365 days a year. Analysis of pasture, stored forage and supplemental rations will allow the copper level in the supplement to be increase to target levels. Higher levels may be fed at some times of the year with the expectation that reserves will be built up in the liver and used during peak pasture season when little or no supplements are being fed. Browse contains higher levels of copper and other minerals than grasses and legumes, and elk will happily use this feed source.

Selenium Similarly, selenium is just as important a trace mineral in terms of management and animal physiology and is closely linked to Vitamin E. Poor health, growth and reproductive rates are linked to selenium deficiency and occasionally white muscle disease has been seen in deer. Feeds may be low in selenium, so there is a dietary risk with bought in feeds unless they are balanced during processing. White muscle disease and selenium problems are often linked to young animals that have been in-housed over winter and let out to fresh pasture in spring. The unaccustomed exercise, diet change and stress associated with the change and handling can predispose the animal to this condition.

Routine treatment of animals in deficient areas is practised, either by mineral supplementation in a lick, or in injectable form. Commonly, liquid forms of the trace mineral are added under veterinary supervision to the common parasite treatments. It is an easy and effective procedure although care must be taken to avoid overdosing.

Other mineral imbalances are applicable more to specific local conditions and usually redressed by an advised soil fertilizer correction program. However, **cobalt** is important for the production of **Vitamin B12**, critical in ruminants for aspects of metabolism.

Lack of B12 causes starvation, weight loss and emaciation. Treatment is via injectable Vitamin B12 or cobalt application in fertilizer to pasture.

The whole area of mineral deficiency and correction is complex and should not be approached without expert help. Equally there are a great number of mineral additives and supplements available for farm use that are prepared with regard to toxic limits and used on elk farms that may prevent marginal conditions becoming worse. An investigation into local conditions and practices is a necessary step in operating an elk farm.

DISEASE RISKS TO ELK FARMERS

Some of the common disease risks to farmed elk also have a risk to deer handlers, particularly Yersiniosis, tuberculosis, tetanus, leptospirosis and brucellosis infections. Yersiniosis is the most likely to be encountered particularly with the handling of young elk during the operation of determining sex (by braille!) and in parasite treatment. The bacteria can easily transfer from hands to mouth and cause appendicitis like symptoms, bowel upsets and a flu like condition. Care must be taken when dealing with these infectious diseases and good hygiene practices followed. Good washing facilities should be mandatory in yards and care taken to avoid contamination with fecal matter.

On the same theme, visitors to elk farms, or farmers visiting other farms for field days, etc. should also observe hygiene and wear clean boots, etc. uncontaminated by soil or manure from other properties.

Trauma - for the animal and the farmer

In elk farming there is no such time as "being too early" to involve veterinary advice. Elk are incredibly hardy and tough animals and can survive almost any crippling accident, calving difficulty or disease threat provided it is treatable and treatment is made early enough.

Because of the strong herd instinct, a sick or distressed animal represents danger to the herd and especially in cow/calf groups, dominant animals will bully and drive an animal in distress, or suffering from a potential illness away, often quite brutally. Animals that are shy feeders, or the least dominant animals in a group, especially in overcrowded situations may fail to feed and often are seen with severe coat plucking or beating marks down to raw skin. These animals then come under greater stress and are predisposed to disease risk (parasitism, Yersiniosis, pneumonia) and should be removed into a smaller group - a so called hospital group, and preferably fed in small groups in warm sheltered conditions.

Most importantly, if such signs are observed, it is time to reassess the management and feeding levels. Often such cows will be non-pregnant and may have resorbed their fetus under stress, or are being driven from the group because they are not pregnant. It is prudent at this time to treat for parasites, check for physical damage and even run tests in critical health areas if there are no other apparent reasons.

It is the same for injury - broken limbs, antler injuries, etc. These should be treated at first sighting with veterinarian care.

Fig. 8 Be sure to remove all twine and wire - or you could end up with serious damage

Fig. 9 Shelter is essential - this calf lost the tips of his ears to sunburn

Elk respond well to casts/splints if treated early, although often the compromise must be made between recovery of an individual and the risks of leaving an animal by itself. Both objectives can be reached if the injured animal can be housed or confined with two or three quiet animals of a subordinate group (e.g. calves).

It is useful to remember that generally with elk, only stressed animals become sick, and only sick animals get diseased. By knowing the limitations imposed by management requirements the likely stress areas become obvious. Planned animal health programs reduce the potential for reaction to that stress, and consequently disease risks. Farmed elk remain one of the most healthy livestock farming options available, and their products inherit that healthful status that underlies the principles of quality production.

WELFARE AND NORMAL BEHAVIOUR

To fully assert themselves as farmed animals there are some behavioural features of elk that may seem undesirable yet are critical to wellbeing.

Access to shade is important particularly for newborn stock and bulls at the height of their pre-rut condition. Often a single tree, well protected from bark chewing and antler rubbing, is a boon either centrally located in a pasture or fenced off and incorporated as part of a laneway - gateway system. Some farms fence off small areas of woods, but in close confines elk are extremely hard on undergrowth and softwood trees. Early protection of the trunks with well secured netting, or slab wood strips is advised to preserve this valuable resource.

If it is not possible to use natural shade, effective shelters can be provided cheaply by building a 20% porosity wind fence between a series of posts planted in the pasture - about 5m x 2m x 2m high is effective for groups of 20 - 30 animals or to construct roofed shade shelters around feeding stations, and hay feeders.

SUMMARY

Farmed elk have proven to be particularly healthy with relatively few production limiting diseases threatening animal welfare. In general terms, the minimizing of health problems relies simply on avoidance of management related stress and implementing a pre-planned animal health program. On-farm stress in elk can be extremely subtle and often remains undetected until serious and threatening problems emerge.

In order of severity, or frequency of occurrence the major stresses are:

- undernutrition - particularly in winter and fall, although poor pasture management in summer can also create quality pasture shortfalls in spite of appearances.
- bad weather - combinations of both cold and wet conditions - ice storms, cold spring rains, thunderstorms.
- heat exhaustion - lack of shade and shelter, particularly in newborn calves.
- overcrowding - mixing unfamiliar animals or groups.
- grain overload - acidosis, caused by overfeeding of "hot" rations (grains or formulated pellets) without an appropriate gradual introduction period, sudden changes in diet, or feed dominance by individuals and gorging.
- management related - poor handling, excessive disturbance, unfamiliar routines, anxiety and hassle in yards or on pasture.
- physical - poor fencing and facilities construction posing injury risk, obstacles or hazards on farm (foot injury, bruising).
- hard antler damage - fighting, aggression, poor temperament.
- disease risks - tuberculosis, leptospirosis, Yersiniosis, parasitism, rabies, clostridial diseases, Johne's, necrobacillosis.

Most diseases are preventable, either by following strict testing procedures prior to stock purchase, or as part of a routine vaccination and testing program.

Under any of the management related stresses, elk can become very susceptible to other invasive diseases. Often in new operations the problems can compound and present a major crisis. Typically under stress, parasites or invasive bacteria can gain a threatening hold with great speed. Failure to act on intuition or first signs of a problem is the greatest cause of loss in farmed elk, and frequent and open consultation with veterinary advice is essential. In most instances elk are no different from other farmed ruminants and the same sound basic principles of preventative animal health programs and emergency treatment apply. Any elk under threat of stress or disease or injury must be able to be yarded quietly and kept in darkened conditions with food and water until veterinary assistance can be obtained. No farm should be in operation without handling facilities. These are as essential as the boundary fence.

For the most effective treatment and control it is vital that good records of management practices and production are also available.

Other known limits to production critical to elk involve a complex of essential and trace minerals, particularly copper and selenium, and to a lesser extent cobalt, phosphorous, calcium and sodium. Some areas may be deficient, but a combination of soil testing in a local area, relevant veterinary advice from other livestock producers and a preventative approach eliminate most problems.

Finally, a planned animal health treatment program is simply established and administered at critical times of the year. Animal welfare on-farm then only requires minimizing any imposed stresses due to farming practice, and feeding to seasonal demands.

SEASONALITY

Temperate cervid species exhibit highly seasonal patterns of reproduction that are profoundly influenced by the changing conditions of photoperiod, or regular seasonal changes in the ratio of night and day length. These patterns have evolved to maximize survival of offspring within highly seasonal environments, and on the whole, fit the requirements of elk farming extremely well.

The seasonal patterns, however, impose advantages and disadvantages on the routine farming calendar as management tasks and animal behaviour changes with the season.

All these changes are ultimately initiated by the release of the hormone melatonin in the body. Its release rate and circulating levels change in response to increasing and decreasing night time length. Such photoperiod changes are detected by the pineal gland in the brain. Red deer, fallow and elk are short day breeders; i.e. the influences that induce oestrus cycles in females, and the physical changes in males are prompted by an increase in melatonin secretion as night lengths increase towards autumn following the summer solstice in June. The effects of seasonality differ outwardly in cows and bulls and for both are a critical part of the annual cycle.

MALES

Adult males show great changes in annual liveweights reflecting normally controlled seasonal changes in feed intake and in rutting activity.

Melatonin, via the endocrine pathway, operates primarily on the testicles and the production of the male hormone, testosterone. Liveweight changes, particularly the huge loss (up to 30%) over the rut are linked to high testosterone levels. Testicles are at their maximum size and development, initiated and controlled by the pituitary gland secretion of LH (luteinising hormone) which increases in frequency of pulse and extent under the short day autumn stimulus. As testes increase in size, sperm production and development of the accessory sex glands involved in sperm maturation and storage develop. By the onset of the breeding season large numbers of sperm are present in relatively low volume : high sperm count ejaculate and the bull is physically and behaviourally at peak activity for a six week rut or breeding period.

The testes remain mildly active over winter secreting low levels of testosterone and large numbers of spermatozoa. Behaviourally, the bull slowly loses much of the aggression and dominance activity, and is at its lowest weight and most vulnerable physical condition with no fat reserves, and in physiological survival mode.

In spring as day length increases, again under the same pathways and endocrine influence, LH secretion diminishes, the testes dramatically reduce up to 60% in size and very low levels of testosterone are produced. Under this "biochemical castration" condition, sperm production halts, the animals become infertile and docile, aggressive behaviour ceases and hard antler is cast to begin the annual antler growth cycle. The males remain infertile for two to three months during velvet growth and the cycle repeats once the summer solstice is passed.

The annual changes in testosterone secretion also cause the antler to begin calcification in July and August. This is followed by the mineralization process with antler becoming hard and stripping the velvet covering in August. Hardening and stripping are often quite rapid processes as autumn melatonin level and the consequent testosterone level increases. To support antler weight and prepare for mating activity there is a marked change in carcass water balance and noticeable change in body development. The neck muscles can hypertrophy, massively increasing neck muscle mass and body size. A thick mane develops and the voice box enlarges to allow the typical bugling of competing males during the rut.

Behaviour too is markedly altered. Bulls of equal rank - as shown by antler size, age and familiarity spend much time in aggressive behaviour, sparring and fighting to establish dominance. Bulls also create and seek out wet areas to mark territory by wallowing and scenting. As the intensity of the rut increases, noise level, sparring and sharp savage fighting increases. Bulls

Fig. 1 Rutting bulls sparring

that are normally reasonable and pleasant to handle at velvetting become quite unpredictable and can easily cause injuries to themselves, or in an intensive farm situation to fences, buildings, trees and vehicles. In general they need space, and rough or broken terrain (rocks, woodlots, hilly pasture) to establish their own territory and remain more settled over the rut, if they are not required for mating.

FEMALES

Females too have their reproductive status controlled by the same day length change and melatonin secretion endocrine pathway with the ovaries being the ultimate target organ of hormonal action. The onset of the breeding season occurs late in summer as photoperiod decreases and is preceded by a period of deep anoestrus and no ovulation in early summer. Ovulation cycles begin after approximately 85% of lactation and milk production is finished. The first detectable oestrus or heat that can be detected is preceded by one or more short cycles and silent ovulations. These produce a single 10 - 12 day duration corpus luteum (CL) which may serve to naturally synchronize the first normal cycle in the herd. A short concise calving pattern is a real advantage in the wild, and equally allows greater flexibility in pasture management for lactation on-farm.

If cows are not bred, then the reproductive cycle of ovulation, CL formation and breakdown can continue for four to six months all under typical hormonal control until the springtime daylight changes cease the cycling activity. The oestrus cycle takes 19 - 21 days for farmed elk.

There are few outward physical signs of oestrus in female elk. On-farm, oestrus cows, when handled, may stand and, if pushed on the rump, squat slightly and can show a vigorous tail wagging, but this behaviour is too imprecise to establish timing for an artificial breeding program. Crayon marking harnesses to detect oestrus have been tried with reasonable success in fallow, but much less so in elk because of greater difficulty in handling during the rut and their low mount to service ratio.

The female reproductive statistics for elk are presented below.

Reproduction Characteristics for Female Elk

Characteristic	Elk
Cycle length (days)	19 - 21
Onset of 1st oestrus	early Sept.
Time in oestrus	8 - 12 hrs
Gestation length days	247 + 10
Breeding ratio, mature males	1 : 30
AI technique	laparascopic or cervical
Success rates	40 - 80%
AI timing post progesterone CIDR withdrawl	56 - 64 hrs

Table 1 Features of reproduction for female elk

Fig. 2 Males become very aggressive during the rut

MALES AND MATING

On most elk farms, farmers will have the option of choosing one or a small number of sires for breeding, and to pass on their superior background, the very reason they will have been selected and purchased in the first place. Bulls can only pass 50% of that genetic merit to progeny, the other 50% coming from the cow, but the high ratio of bull : cows at mating means an individual male's influence is large.

Bulls must be prepared for mating, if nothing else for their own protection, and to allow them to get on with the task of mating instead of continually establishing and defending territory and controlling the female harem. The typical mating behaviour of wild populations is much modified on-farm because of fencing limitations and the close proximity of similarly ranked bulls in adjacent pastures, and the often much greater numbers of animals in a small and somewhat artificial setting.

PREPARATION

Hard antler or any re-growth following velvet removal should be cut off as early as possible for protection and removal of injury risk. The time of velvet stripping or rub is ideal, although if the animal is to develop a full hardened antler set with sharp and polished points and some colour, cutting should be delayed but not at the expense of safety.

Make no mistake, in general terms there is no place for hard antler on intensive elk farms where handling of males in the rut in yards is required.

Bulls that are relatively settled on pasture have established their hierarchy and space. In yarding the rules change even in small groups. The later in fall this occurs the greater the reaction as males are compromised in their typical behaviour, and confused as the balance is artificially altered between harem control, challenge from other animals and the interaction of the farmer. Even a simple changing of the herd from one pasture to another can upset a previously established balance. The most extreme pressure and unusual behaviour however, occurs if the mating group is yarded.

Fights and charges can happen without warning and penned bulls have been known to also turn on cows, calves and people. That combined with a full head of hard antler is a lethal combination.

Removal of hard antler is painless although sawing can create discomfort with heat transference and the simple act of trying to confine the animal. A drug tranquillizer may be indicated, which also allows checking of the mating equipment. Check for damage or injury to the penis and prepuce, palpate the testes which should be large, firm and well descended in the scrotum. Check for foot soundness and teeth condition. This is also the time when copper and selenium supplementation is useful and a parasitic treatment given to allow healthy post rut recovery.

If breeding bulls are being shifted onto the farm, or being sent elsewhere this should be done as soon as possible after velvet growth completion and harvest, and prior to the rut. Bulls in late summer are prone to heat exhaustion and stress during transport and ideally must be transported in individual pens with frequent watering if travelling over large distances.

Bulls should be well settled on a new farm six to eight weeks before breeding begins so that as many cows as possible will be mated in the first cycle. This allows easier handling and bulls to settle into a new hierarchy and dominance pattern prior to breeding. Experience suggests it may take up to six months for an animal to fully settle into a new farm and management system.

FEEDING

During late summer the males lay down huge reserves of body fat that will be mobilized during the intensive six to eight weeks of rut activity, when intake and appetite are virtually non-existent. Some luxury feeding in early fall will ensure that peak weights are achieved, but more importantly feeding levels after the rut are critical. With up to 30% weight loss, all fat reserves are gone, but for a short time bulls will recover some lost weight and condition. A liveweight gain at this point may have a positive effect on future velvet antler production (perhaps as much as eight to ten percent) although this is variable between years and age groups.

Single sire mating reduces the effects of weight

loss through constant competition. Managers notice that, if velvet bulls and non-breeding males are removed well away from breeding groups and given as large and as rough an area as possible they settle quite quickly after establishing their own territory and comfort zone, and typical weight loss is reduced with a positive carry over effect for velvet production.

Elk Bull : Cow Ratios	
Single sire mating	
Bull Age	Ratio Bull : Cows
15 - 18 months	1:5, max. 1:10
27 months	1 : 15
3 years	1 : 20
4 years	1 : 30
5 years & over	1 : 40 - 60

Table 2 Elk bull : cow ratios related to mating age

MATING RATIOS

Rapid genetic progress is best achieved by mating selected top young sires, as young as possible, provided they have some sort of performance recorded backing and not just an informal guess or dramatic photo of their parents as a pedigree. In multisire matings an older bull can dominate and hold 80 - 90% of cows in any one area and may not necessarily be able to mate all in the first cycle. Subordinate bulls will operate on the fringes of large groups and take opportunity as it presents. There are trends in the NZ system as the industry focuses on low cost venison production systems to use large numbers of young stags together for saturation mating and a narrower mating/calving range. There is, however, no requirement to identify progeny or retain them for breeding or livesale. The objective is simply numbers of calves on the ground in as short a space of time possible.

To succeed as breeders, elk bulls must feel dominant over the cows in their harem, and over nearby bulls. Isolation in a pasture, removed from other bulls, may be enough to accomplish this, but spiker bulls are always suspect breeders. Young bulls should be isolated as far as possible away from competition, and run with younger cows if possible.

TIMING OF MATING

Years of selection and heritage in the wild mean that superior adult males, i.e. the dominant male, tend to be most advanced in antler growth, both size and stage, and the seasonal reproductive season. These will breed first, coinciding with the peak seasonal first overt oestrus of cows. Younger males lag this peak by three to four weeks and will mate younger lighter females, or recycling adults. No matter how early prior to the onset of the breeding season bulls are joined with cows, the arrival of the first calf is generally within two or three days every year, indicating a very fixed start to successful mating. On-farm, the start of calving can be moved forward a week or two by pre-rut weaning and better nutrition.

MATING BEHAVIOUR

Mating is not often seen on-farm, either because of the speed of the final event or that 25 - 30% occurs at night, with the balance during the day only when the animals are undisturbed. The cow comes into oestrus and initiates most of the pre-mating activity and will only stand for the bull at the peak of oestrus.

Just before and during rut, males become exceedingly aggressive. They will charge fences and vehicles, although seldom pressing home their attacks. Bulls on opposite sides of a fence will spar repeatedly and do considerable damage to it. From a management point of view, it is advisable to either separate males by at least two fences or use a strand of electrified wire on both sides of the fence. This should be about a meter off the ground spaced out 25 - 30 cm from the posts. If antlers are left uncut for any reason then bulls will make short work of these electric lines.

If males are handled during the rut then it is essential that the dominant position of the farmer be maintained. In particular, the handler should not infringe on the personal distance of the bull, which will often be greater than at other times of the year.

Fig. 3 Making Flehmen - glands in the upper lip and nose detect hormone levels that indicate estrous status

Rutting male elk will spend a good deal of time rounding up their females. They may even chase them away from water or feed troughs. They will frequently stretch their necks and grunt or bugle. At this time, they often perform the so-called "thrash-urination" display. Throughout rut, bulls can be seen periodically checking cows, both by testing their hind ends and sniffing their urine and performing "Flehmen" (lip curl)(see Fig 3). If a female is coming into oestrus the male will pay more attention to her and a series of interactions between the two will take place.

First, the bull chases the cow for short distances and then appears to lose interest. Length of the chases increases and finally the female will stand for the male for a few seconds. At this time, he will show great excitement, his tongue will often flick in and out and he will arch his neck over the back and neck of the cow. He will also perform so-called low mounts. During these mounts, he will mount the cow and lay on her back. His penis will come out of the sheath, but no intromission occurs. The low mounts may be seen as much as 12 hours before mating, but usually this time is less than six hours. As many as six low mounts have been recorded by a single red deer stag before actual mating. Interspersed between low mounts, the cow may actively solicit attention from the male and even mount him or other females. She will also rub her head and neck over the rump or under the neck of the bull.

The actual moment of breeding lasts only a few seconds. The female stands after a brief chase and the bull mounts and then thrusts upwards in almost a single motion. He places almost no weight on the cow's back, but the force of the thrust usually drives the cow forward a few paces. Just after mating, the female may adopt a squatting stance and urinate. Urination in this position is not seen at other times.

On small farms where there is only one adult male often no bugling and little mating activity is observed. Total infertility is rarely seen, so as long as some behaviour is seen there is no reason to doubt that breeding has occurred and that the bull is fully fertile.

ARTIFICIAL BREEDING TECHNIQUES

Techniques now exist to extend the capability of an elite bull by synchronizing cows either side of the 20 day first cycle period and offering five or six cows in oestrus on alternative days. The technique is not difficult but is still under some refinement at present. Weaned hinds are fitted with progesterone impregnated CIDR's (Controlled Internal Drug Releasing Devices) worn in the vagina for a seven to ten day period prior to normal mating or for a longer 11 - 12 day period in the height of the season. The intravaginal device (9 - 12% progesterone weight/volume) is a formed plastic matrix that slowly diffuses the contained progesterone into the body and simulates a constant mid oestrus condition inhibiting follicle development. On removal, plasma progesterone levels drop sharply within two hours and the normal hormonal events of ovulation follow predictably. In red deer and elk an intramuscular injection of PMSG (Pregnant Mare Serum Gonadotrophin) at removal ensures ovulation occurs. Response is dose dependant. High levels will induce twinning or multiple ovulations which have been proven a distinct disadvantage under normal farm management conditions.

This technique also forms a reliable basis for fixed time artificial insemination (AI) procedures. For natural mating, removing CIDRs from five or six cows treated every second day gives the bull a small number of cycling cows to mate and thus spreads the potential to mate a large number of females in a relatively defined period.

Artificial Insemination of Elk There are several reasons why the use of artificial insemination may become more common in elk. Artificial insemination provides a method of exploiting the genetic superiority of the male. Superior males are just now being identified but more records on growth and antler production will be required for concise, less biased evaluations of individual genetic merit. Through artificial insemination and semen preservation techniques, a sire can continue to produce offspring after his death or in his advanced years after his health has declined. Artificial insemination also allows more females to be serviced and increases the number of offspring produced. These techniques also provide an easier method of movement or transport of genetic material.

A problem associated with artificial insemination is that it is labour and handling intensive. During handling, some animals may be injured. The collection of semen and the artificial insemination of females requires skilled professionals.

Methods of Artificial Insemination Detection of oestrus (heat) followed by artificial insemination is impractical. The stresses associated with removing oestrus females from the herd for artificial insemination would decrease conception rates and potentially cause injuries to other females. This method would be highly labour intensive and only have application in a limited number of herds.

Oestrus Synchronization is a term used to describe methods of treatment to induce oestrus and ovulation at a precise time.

The method that has been used the most for elk is a CIDR (controlled internal drug release) implant that is inserted into the vagina. The CIDR is removed after 14 days and PMSG (pregnant mares serum gonadotropin) is administered to promote ovulation and improve synchrony. Females are inseminated 60 hours after CIDR removal. This procedure can be conducted at the beginning of the natural breeding season.

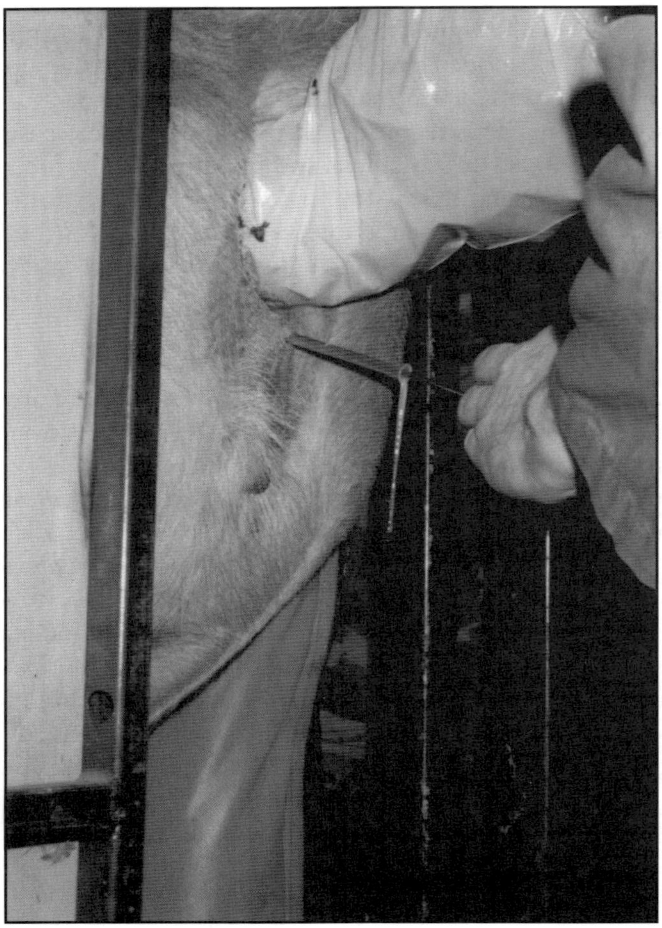

Fig. 2 Elk Cow Artificial Insemination

Insemination Techniques The techniques used to artificially inseminate females will vary with cow size. Females should be handled in a manner that minimizes stress at this time. Hormones released in response to stress inhibit the hormones that are released to cause ovulation. An increase in animal stress will result in a decrease in conception rate. The large body size of elk will permit rectal palpation of the reproductive tract. This permits insemination techniques that are similar to those used in cattle. One of the technician's arms is inserted into the rectum and the cervix is located. With the other hand an insemination gun is passed through the vagina and is manipulated through the cervix. The semen is deposited at the anterior end of the cervix, which is also the posterior end of the uterus. A person experienced in cattle artificial insemination will find that passing the insemination gun through the cervix of an elk cow is possible but more difficult than inseminating cattle. A person experienced in cattle embryo transfer procedures should find that manipulating the insemination device through the cervix of elk is similar to transferring embryos into cattle heifers (at the time of cattle embryo transfer, the cervix is constricted down and extremely narrow).

The smaller body size of red deer or elk cross breds will not permit rectal palpation of the reproductive tract. The techniques used are procedures that have been modified from use in sheep. Some of the first artificial inseminations in red deer were conducted using a vaginal speculum and attempting to visually locate the cervix opening to manipulate an insemination device into the cervix. As in sheep, the success of this technique depended highly on the competence of the inseminator and, consequently, the results were highly variable.

More recently laparoscope procedures have been developed for depositing semen into the uterus of deer. First the animal is anesthetized and placed on a cradle, on her back with the hind quarters elevated. An area ahead of the udder is washed and disinfected and two or three small incisions are made. A laparoscope with light source is inserted into the peritoneal cavity through one of the holes. A probe to manipulate the peritoneal contents is inserted into another of the holes. Once the uterus has been visually located, an insemination device that is tipped with a small, short, fine bore needle is inserted. The needle pierces the wall of the uterine horn and semen is deposited into it. This system is widely used and results in conception rates ranging from 50 to 75%.

Semen Collection The techniques and equipment used for semen collection are similar to those used in cattle and sheep. Collecting semen from elk requires properly designed handling facilities. Animals can be collected while under general anesthesia, but the quality and concentration of the semen will be decreased.

Semen is usually collected by electro ejaculation. In this procedure, a skilled technician first removes the fecal material from the rectum of a restrained standing or anesthetized male. Next a probe is inserted into the rectum. Elk bulls will require a probe designed for elk or small cattle bulls. The probe is positioned with electrodes directly over the internal male reproductive glands. A low current is passed through the probe for two to three seconds. The current is stopped for five to ten seconds and then a slightly higher current is applied. The currents are thus cycled until first an erection and then an ejaculation occurs. It is usually best that the semen be collected in several fractions so that if urine or, in the case of elk, a dark seminal fluid is collected that it can be discarded, as both kill sperm cells. The semen should be kept warm from the time of collection until processing. This procedure can be conducted twice (10 to 15 minutes apart) on a collection day and once weekly during the breeding season. The semen can then be used fresh or frozen for latter use.

The procedures used to freeze elk semen are similar to those used to freeze cattle semen and usually use an egg yolk/tris buffered semen diluent and glycerol to allow freezing. The quality and concentration of semen is highly variable. Time of the year, age, breeding status and health can all influence semen quality. When freezing semen 10 to 15 million live sperm at thawing are usually targeted for. Semen can be frozen in .25 cc or usually .5 cc semen straws. They are frozen in the vapour phase of liquid nitrogen until use. If the animals require general anesthesia, then weekly collections may not be possible as anesthesia procedures may be harmful to the animal. In general, 20% of all semen collection attempts will not produce freezable semen, and 15 to 150 units of semen can be frozen from successful attempts.

END OF MATING

Even in times of an expanding industry and reasonable returns for live sales, gaining a few extra late born calves by extending the mating season is a management headache at weaning and for the average growth of the young stock.

It is recommended that primary sires be given a four to five week period from late August to early October covering two full cycles in most cases, and then sires exchanged for younger bulls or chaser bulls for a further 20 - 30 days, and then all breeding bulls removed to allow a definite end to the breeding season. After primary sire mating, individual groups can be pooled and multisire mating with young stock used. There may be a reduction in conception in the first year of adapting from extended mating, but from then on calving patterns will continue to concentrate in the first cycle and thus with better lactation and feeding, weaning weights and cow performance improves.

This management best suits an early weaning program, but also has application on smaller properties where calves may be left unweaned by choice.

FEMALES - MATING, PREGNANCY & CALVING

THE OESTRUS CYCLE AND MATING

Once minimum threshold weights have been achieved in the first 15 months of growth, elk females should always be either pregnant or lactating. Reproductive success, measured by calves weaned to total number of cows mated is generally very high in North America; (adults 92 - 96%, yearlings 84 - 92%), but problems have occurred if yearling heifers are not well grown.

Once born, most calves survive unless victims of opportune predation by coyotes, bear, etc. There can be a 2 - 4% loss of young calves through difficult birth problems, mismothering or misadventure. This figure has increased up to 7% in intensive situations (e.g. the enforced requirements of research or overcrowding).

Conception rates, and the spread of calving, can be negatively affected by mismanagement both at calving, and at mating.

Table 3 represents a calving/mating planning

ELK CALVING PLANNER
Mating and Calving Dates

Date of Mating	Mating Event	Expected Calving Date
		Elk 247 ± 10 days
Aug. 20	Begin Weaning	April 23
Sept. 1	Synchronize Females	May 5
Sept. 4	**Primary Sires In**	May 8
Sept. 10	Silent Oestrus	May 14
Sept. 20	Peak of 1st cycle	May 24
Sept. 30	AI/ET	June 2
Oct. 10	Peak 2nd cycle AI/ET	June 12
Primary Sires Out, Backups In		
Oct. 30	3rd cycle	July 3
Nov. 10		July 14
All Breeding Bulls Out - Finish Breeding		

Table 3

calendar and anticipated calving dates for elk cows and Table 4 the milestone dates in the life cycle of breeding elk females. Mating success can be improved through the following management techniques:

(a) Early weaning 10 - 14 days minimum prior to joining with a bull. Late weaning, i.e. removing young stock in late October when the bull goes out, will not significantly reduce the number of cows that are ultimately bred and become pregnant, but there is a tendency under farmed conditions for levels of 60 - 65% only to conceive to the first cycle instead of the 80% plus possible. Weaning stress at the time of first oestrus seems to delay or upset this first cycle and as a result mean calving date can fall by 8 - 10 days.

Early pre-rut weaning gives immediate relief from the high feed demands of lactation and after a 48 hour settling period cows are quickly able to regain lost body condition and be on a slowly rising plane of nutrition over mating until early winter. Any sudden change or restriction in diet during the three cycles may increase embryonic mortality and further reduce calving percentage. There is anecdotal evidence that dry cows, or cows that have lost calves may not conceive if they are too fat at mating. Far better to lose weight over the summer and regain during the fall.

(b) Keep cow groups as stable as possible in terms of mixing animals that haven't been together before, and if purchasing new animals, have them introduced well before the mating season begins (particularly for yearling heifers).

STAGES IN THE LIFE OF AN ELK COW

AGE (in months)		
39	Aug	
38	Jul	
37	Jun	
36	May	
35	Apr	
34	Mar	Mixed Age Cow
33	Feb	
32	Jan	
31	Dec	
30	Nov	
29	Oct	
28	Sept	
27	Aug	
26	Jul	Calving
25	Jun	
24	May	
23	Apr	
22	Mar	
21	Feb	Yearling Heifer
20	Jan	
19	Dec	
18	Nov	
17	Oct	
16	Sept	
15	Aug	
14	Jul	Rising Yearling
13	Jun	
12	May	
11	Apr	
10	Mar	
9	Feb	Weaner or Heifer
8	Jan	
7	Dec	
6	Nov	
5	Oct	
4	Sept	
3	Aug	Heifer Calf
2	Jul	
1	Jun	
	May	
	Apr	
	Mar	
	Feb	
	Jan	
	Dec	
	Nov	
	Oct	
	Sept	

Table 4

(c) Keep mating groups away from newly weaned calves and as free as possible from disturbance.

(d) Yearling heifers: It is often more practical to run yearling heifers together. Once male yearling bulls are separated from the young animal group (by June at latest) yearling females may be slow to come into oestrus in sympathy with the adults. The introduction of one or two adult cows per 10 or 15 yearlings seems to bounce these yearlings into earlier oestrus, provided they are well grown. If yearlings cycle late and a defined end of mating date used, the chances of missing mating altogether increase. This induced synchrony and advanced ovulation appears to work well.

The common techniques of oestrus synchronization, using vasectomized teaser bulls, CIDRs or running entire bulls in adjacent pastures immediately prior to mating all seem to work very well in adult and yearling females. The common objective is to reduce the calving interval, start to finish and have as many cows as possible calving in the first 20 days of the season.

PREGNANCY

Cows are pregnant for 65% of the year, but emphasis on critical feeding and animal health treatments occupy only the last six weeks, provided the females are maintained to requirement during late fall and winter. Under conditions of stress and underfeeding, embryonic mortality may occur. In tough conditions cows are known to resorb fetuses even at quite advanced stages. Under such stress, gastro intestinal parasites, lungworm and the threat of disease present some management challenge. As a general rule, cows should maintain or increase their weight by 15 kg over the first 100 days and then slowly increase a further 15 kg for the final third. In effect cows enter and finish the winter in the same condition they started and the approximately 15% weight increase is simply the weight of fetus and placenta. Signs that cows are being underfed include fighting and beating of weaker or less dominant individuals. Excessive hair loss is an indication that something is wrong. The bullied animals should be removed and preferentially fed, and ration of the main group increased.

If cows are overfed they quickly lay down surplus energy in the form of fat through the pelvic channel and around the kidneys and rump, all of which can contribute to difficult birthing, especially of large male calves. In late pregnancy the fetus gains most of its weight in the last six weeks, so excessive restriction of diet at this time can negatively impact on fetus size.

Over fat cows can have their weight controlled during early pregnancy by restricting the diet. However, the offered ration should be increased in energy content in the face of bad weather or other potential stressors with a supplement that the cows are adjusted to previously.

Exercise is as much an important factor in preparing for calving as is controlling fat deposition. This can be hard to achieve during late winter and the spring thaw, but larger pastures will allow some exercise for cows confined to feedlots until April. Where possible cows could be given the run of more open and hilly land at least for two or three hours per day to maintain fitness.

PREGNANCY DETERMINATION

The simplest and most effective method of pregnancy determination in elk is rectal palpation at 6 weeks or more post breeding. The uterus of a non-pregnant elk lies entirely within the pelvis. The pregnant uterus feels like a smaller bovine equivalent, with less than ten cotyledons. The technique of rectal probe ultrasound scanning to determine pregnancy at an early age, or to fetal age to distinguish cycles of pregnancy, e.g. pregnant to an AI treatment, or pregnant to the backup bull is becoming quite common, and forms a back up to many advanced breeding programs.

The primary purpose is to distinguish pregnant and non-pregnant cows. Scanning is done by a professional operator using a linear rectal 3 or 5mHz probe and a portable scanner.

These machines, or particularly the probe head itself, are very expensive. For safe and effective use the elk must be fully restrained inside a mechanical or hydraulic squeeze so there is no chance of damage to it or the probe.

Ideally this would be done early December, at least 10 days after the finish of mating and the backup bull has been withdrawn.

In natural mating there will be considerable cross over as the cycles of individuals overlap, but some fetal aging might be possible. The technique that gives a simple pregnant/non pregnant response is extremely quick and reliable and will in time become a common marketing tool, or a prerequisite for female live sales in the early part of winter.

The more subtle use is for pregnancy confirmation in an AI program to distinguish between pregnancy from AI, and pregnancy from later backup bulls. Two scannings can be used, the first at 29 - 35 days after the initial fixed time insemination, which will only confirm treatment pregnancies and the second at 54 days where fetus can be seen and the backup bull pregnancies and dry cows identified. Alternatively a scan at 40 - 45 days will allow distinction between treatment and chaser backup pregnancies.

An alternative, and very accurate pregnancy detection technique is a blood scan for PSPB, Pregnancy Specific Protein B, available at specialized laboratories.

LATE PREGNANCY: VISUAL CHECKING AND UDDER SCORE

From late winter in the third trimester of pregnancy a practised eye can pick pregnancy from the thickening and swelling flank region, just in front of the stifle joint of the hind leg. The insight "she looks pregnant to me" has a high confidence level, so that buyers of pregnant cows should be able to select pregnant animals with some certainty from February onwards.

For management reasons it may also be prudent to separate early and late pregnant cows so that they can be split into two groups for calving. This is particularly necessary to ensure that late calving cows do not overfeed in late pregnancy under spring flush conditions and run risk of fatness and large calves potentially causing dystocia or calving problems. Udder checking and a simple scoring system are relatively reliable indicators, although a delicate balance exists between accurate scoring, the closeness of first calving and risk of physical damage to cows in late pregnancy through the yarding process.

Preparation for calving

To ensure maximum calf survival and the best health for milking mothers it is recommended that cows are given a pre-calving health treatment. This includes an anthelmentic treatment for parasites, selenium and copper treatment if indicated and clostridial vaccination for the normal risk bacteria of the area.

These treatments allow a level of protective clostridial antibody to be passed to the calf via colostrum feeding in the first 48 hours after birth, and maintain high copper and selenium blood serum levels in the new-born calf. A parasite treatment reduces the fecal egg output from any lungworm for the early part of lactation and reduces the risk of any chronic parasite infestation until later in the autumn when hinds and calves can be safely yarded and the annual program initiated.

It takes 10 - 14 days prior to calving for an effective immune response to build up from the annual booster vaccination, so cows should be yarded and treated two to three weeks prior to calving, and certainly no less than seven days before the anticipated first birth. From the gestation chart this is likely to be no earlier than the 1st of May, unless a pre-season AI or synchronization program has been used.

Cows are intensely irritable at this stage of pregnancy and may react quite differently to injection, parasite treatment and handling. Ideally, yarding early morning or in the cool of evening assists handling ease. Group size should be reduced and care taken not to rush animals through narrow doorways or risk any sort of hard contact on the flanks or belly. Gate catches that project out into handling pens are a classic physical risk. Handling should be kept to a minimum and the operation conducted as quickly as possible. If cows have been single sire mated now is the time to separate them into calving groups based on the sire so that some identification of calf-to-dam-to-sire is possible.

If intensive recording that involves collecting birthdate is required, cows must have clean visible management eartags.

Calving Pastures

Because elk are such strongly herding creatures the stability of the herd structure, and the successful

Fig. 3 Pregnancy Checking - notice the "pointed bulge"

rearing of a calf by its mother within a group, relies on establishing a very strong cow-calf bond. In the wild, cows will leave the herd only shortly before birth (perhaps 24 hrs) and select a safe and solitary calving spot, staying with the calf after birth for 10 - 12 hours constantly feeding, grooming and establishing a smell and calling sound pattern unique to the pair. Following this period she will not be far away for the next few days, but calves are natural hiders and for the initial post birth period seek cover and protection and stay put until called for feeding or protection. Disturbance of the cow/calf within 24 hours of birth can disrupt this bonding and can lead to mismothering or, at worst, desertion and calf loss. This is particularly true in first calving cows.

Calving pastures must be selected with this behaviour in mind. They should have the following features:

- quiet - minimal disturbance from outside influence.
- adequate feed for six week calving period.
- good clean water.
- shade, shelter, and space. Hiding cover is essential for calves.
- easily observed from a distance, either around perimeter or high spot in an adjacent pasture.
- connected to a lane system, or adjacent pasture to allow a simple one paddock shift for new feed (drift system).
- predator proof.
- have some simple access to yards in case of difficulties.
- secure and calf proof perimeter fence.

Because of the huge increase in feed demands that occur once calves are born and lactation begins (a more than double increase) a decision has to be made on stocking rate of a pasture over the calving period.

Management is greatly assisted if cows selected for a group calve as closely as possible together, ideally in a 10 - 14 day period, hence the value of pregnancy scanning and aging, or more practically udder checking. If a tight calving pattern exists, cows can be safely moved into new pasture to meet their rapidly increasing feed demands seven days after the last calf is born and establish the routine of rotational grazing for good milk production and calf growth. This may involve simply harvesting the spring growth of a pasture next door for silage or hay just prior to the birth period and letting it recover to a clover/legume dominant green leafy re-growth and allowing the group to drift next door at 21 days. After 24 hours cows can be locked in the new pasture and a final check for any calves that may have been left behind. If all are of roughly the same age they will be running with the herd and a deliberate round-up and shift is effective. If older calves are left behind, the relevant cows will quickly leave the herd group and begin calling for their babies. These cows are best returned to the original pasture to collect their calves and regroup rather than trying to move the young stock. If there is a very young calf it can be left with gate open, for later collection or can be moved and left at the gateway entrance. By calling the cows, the mother will come and collect the youngster quite happily. This may require a little experience just to be confident that it is okay. The drift and collect system works well provided conditions that allow undisturbed calf-dam bonding at birth are observed and enhanced.

Some managers continue to feed a token amount of supplement (corn, pellets), just a mouthful per cow, to maintain some sort of quietening contact with the group. This can be an advantage if part of a routine. It allows the settled animals to feed and any animal that needs calving assistance or special attention to be worked as an individual without disturbing the rest of the herd and new-born calves. Such feeding can also be used to entice cows to shift pastures particularly up laneways etc. Gates to and from the new and old pasture must be locked in position and be hard against a fence line. Young stock have an uncanny knack of finding trouble spots inevitably with bad results. The cows will return for the calves and after a day can be contained in the new area.

Any calving pasture should have direct laneway access leading to the facility so that anything in trouble can be quickly and easily taken to the yards for treatment without disturbance to the herd.

This is one of the few times elk can be handled as individuals. Cows having calving troubles generally respond positively to assistance. They will be isolated from the group, running the fence line in obvious distress and generally will follow the fence and out of the pasture into a laneway without fuss.

However, in order to make the right decision whether to interfere or not requires a basic understanding of calving behaviour and recognizing when and if a cow is in trouble.

Calving Behaviour

In general calving is a rarely observed event and commonly occurs in early morning or late afternoon, and fits with the typical behavioural patterns of grazing and resting. Mostly the actual birth process, from the appearance of the water bag or amniotic sac until the calf is born and on its feet takes 25 - 60 minutes. The longest "normal" birth should be completed in 4.5 hours, so in broad terms anything still trying to calve after 5 hours should be assisted unless there is a much more obvious problem and help should be given earlier.

As the majority of births are early morning, a cow observed still trying to calve during the 9:00 - 11:00 am quiet resting period has probably been trying to calve 5 hours and should probably be assisted. An observation routine that occurs between 8:00 - 9:30 am and 6:00 - 7:00 pm is likely to cause minimum disturbance yet cover any problems.

Observation and establishing a routine are as important a part of calving as any assistance given to animals in trouble. These will be less than 5% of the herd, and while they can be an obvious concern, management to provide the other 95% with the best possible conditions is much more important.

CALVING SEQUENCE

Cows will attempt to find a calving site some time in advance of the birth sequence. Commonly, cows choose a small area in the paddock away from the normal resting area of the herd. This can be on an open hillside, edges of gullies or in patches of natural cover in areas where a cow can avoid immediate disturbance by farm staff or the general movement of the herd.

Obviously if pastures are small, flat and without some sort of relief in terms of shelter, cover or terrain this behaviour is compromised. As elk settle into an on-farm routine they show an amount of tolerance and will calve in plain view of normal farm activity. If some attention can be paid to minimizing disturbance and allowing space to spread out and find privacy it will help calving patterns and reinforce the important calf-cow bonding process.

Fig. 4 Fence Pacing

Fig. 5 Appearance of Amniotic Sac

Fig. 6 Amniotic Sac broken and one forefoot showing

Fence pacing

The continual walking up and down a fence line is frequently observed in elk (approximately 48 hours before birth). In some cows this activity can also be seen up to seven days prior to birth and often also

Fig. 7 Lying down during contractions

Fig. 8 Final birth effort

Fig. 9 Mother cleaning calf *(Photo Credits: Dr. John*

occurs when cows are shifted to the final calving pasture after handling and pre-calving treatments. (Fig. 4) Observers believe this behaviour may reflect a need for isolation, or perhaps be a response to overcrowding and a lack of ideal calving sites. However, on-farm this can be considered usual behaviour and shouldn't be considered as a major concern even if cows appear quite agitated for a short period. Undue pacing may indicate overcrowding or some other imposed disruption and some remedial action should be taken.

Once settled the balance of the herd will generally be showing normal behaviour - sitting or grazing in groups and the about to calve cow seeks some distance from the group before moving to the isolated calving site.

The best indicator of imminent calving is an increase in udder size, which also becomes quite firm with teats very pinkish and prominent. 24 hours prior to calving the pregnant hind also loses the sagging appearance of the abdomen as the calf moves up into the pelvic girdle.

The Birth sequence

The following events are quite characteristic as described from the first appearance of the amniotic sac signalling labour onset.

(a) Appearance of amniotic sac: Cows often continue to graze until the sac bursts (30 - 45 minutes average) and then will lick up the expressed fluids and eat the grass in the area. They may resume pacing or lie down at the calving site. The time range, however, is highly variable (1 - 240 minutes). Cows are very vulnerable to disturbance at this time. Other females rarely approach the site. If cows are disturbed, particularly young females, they may rejoin the herd and return to the site, or find a new site but are very agitated at this time. (Figs 5 & 6)

(b) Contractions: Cows can often be restless and will alternatively lie down and stand. Contractions occur during either position at an initial rate of one per minute for five seconds over the final 30 minutes prior to birth and the emergence of a nose, head and the front feet. The cow will frequently lick her teats and vulva and the emerging hooves and forelimbs of the calf. If lying she will lie on her flank with the limbs extended and hold the head up looking back over the flank. The tail is raised and some vocal sounds may be heard. (Fig 7)

(c) Final efforts, birth: Once the head is out very intensive closely spaced contractions normally expel the shoulders with forelegs extended out past the head. Often then the cow will stand and the final pushes quickly expel the calf past the rib cage and the calf normally slides to the ground with a final push. (Fig. 8)

(d) Calf cleaning: Elk cows are active mothers and generally commence cleaning immediately. The calf is usually quite vigorous and responds with head shaking, and first attempts to stand can follow as close as 10 minutes after birth. If birth has been difficult cows may rest before commencing cleaning. The cleaning process is interspersed with eating the membranes, licking amniotic fluids from the ground, grazing while standing or sitting, and resting. (Fig. 9)

Variations:

(i) Backwards presentation: Observations suggest that presentation of the hind limbs first is a minor, but quite normal form of presentation on-farm (up to 30% in some herds). This may be in response to an over disturbance of cows at the point of calving, creating an unusual turning of the calf as it is lifted to engage in the pelvic region. Often calves will be seen with both hind legs extended straight out with hooves pointing skywards. If these animals are left undisturbed there is little reason to interfere and aid this birth unless this is again observed four to five hours later.

(ii) Disturbance: If a cow is disturbed when the calf head and shoulders have presented she may stand and move off with the calf dropping to the ground some distance from the calving site. The cow generally returns, but young or nervous and flighty animals may not, and return to the original calving site, or join the herd.

Occasionally, especially in crowded calving pastures, other cows may approach and show interest in a new-born. These may be tolerated but more often are vigorously driven off by the new mother.

More rarely, older cows that have lost a calf may be quite invasive of new-borns and can cause disruption and calf loss. They can either attack new-borns or attempt to steal the calf and should be removed from the herd if positive confirmation of this activity is seen.

(e) Suckling: After being cleaned the calf will generally suck and stand within 30 - 45 minutes of birth. It can be assisted to stand by the cow with nose and head nudging. Once the calf stands, sucking and feeding follow almost immediately. There is something seriously amiss if the calf has not fed by 2 - 3 hours from birth. The calf very quickly gains control of legs and co-ordination once standing and the cow will often try and lead the calf from the birth site soon after first feeding. At this time (1 - 2 hours after birth) the placenta detaches and is expelled quickly and almost invariably eaten immediately by the cow.

(f) Following and hiding activity: Typically the cow will encourage, by nose to nose contact and repeated calling, the calf to follow her from the site within 2 hours. Eventually the calf will leave her, seek cover, and hide on its own accord. The mother will graze and rest close by for periods up to 4 - 5 hours and she will initiate short suckling feeding bouts. As it feeds she licks the ano-genital region, ingesting any feces or urine shed. The calf concludes the activity by seeking cover and hiding. During this period the cow is quite vocal in contrast to other times during the year - both with a quiet calf calling and the 'squeak' response of the calf and occasionally a calving bugle - a long soft squeal.

Note: This time and behaviour is very important for the calf-cow bonding process and any disturbance is negative and can risk mis-mothering or desertion. Observation is a greater tool than interference and wishing to help. The apparent deserted calf, perhaps lying on bare ground or behind a single thistle is almost certainly not deserted and interference should only be considered if the calf itself is showing obvious distress; i.e. frequently standing and moving, calling with no response or lying unnaturally. The hiding calf may not have cover in intensive farming situations but will naturally lie quietly - ears down and motionless.

CALF/COW BEHAVIOUR

Bonding Elk calves are hider types and individual recognition is important to both as they are separated for long periods. Bonding begins in the first few hours following birth. There is some belief that the cow labels her own calf with her saliva when cleaning after birth. The onus for recognition appears to be predominantly smell and is centred on the mother, although the calf call and response also seems to play an important role.

If a hiding calf is disturbed especially at two or three days of age and calls and squeals, any number of cows that have recently calved will leave the grazing herd and race towards the calling calf, calling to it themselves. They will sniff the calf to confirm identity and only claim their own. Under normal conditions, as the mother returns to the calf she approaches and calls with a soft mewing. The calf will rise and may move to the cow, or remain on the bedding site. Nose to nose contact is made, identity confirmed and the cow begins to groom the calf which suckles standing up. The calf will stand nose to tail alongside the mother and the mother will lick the ano-genital region ingesting the milk feces and urine for some weeks after birth. Calves appear much slower to adopt their own bond to the birth mother and initially are quite indiscriminate in their approach to other cows and will suckle from any cow that does not repel them. These calves will often feed from behind or from the side and there is little cow-calf interaction. This cross suckling behaviour is very rare in the wild and seems to increase on-farm as stocking rate increases.

Feeding Behaviour For the first two days after birth calves suckle every 2 - 3 hours while their mothers are present, with the cow returning from grazing or resting periods and calling the calf to feed.

Cows which have calved and those due to calve are frequently grouped together while late pregnancy and barren cows form a separate sub-group.

Most calves join the group at 7 - 10 days of age and run with the mother close at hand. Feeding frequency is about 8 times per 24 hours. Calves initiate feeding from their bedding sites by squeaking/calling and if this fails to attract, will stand up attracting a positive calling response from the dam, and any other cows in the area until identity is clarified. Mothers then initiate suckling by grooming an active calf and calves will also nudge the cow's flanks with their muzzles for stimulation. Feeding itself is often accompanied by vigorous bunting of the udder and can take 30 seconds to 2 minutes per bout. By 3 months, feeding frequency declines to 4 or 5 times in 24 hours with a 60 second duration. Calves will often attempt to feed at times of disturbance, and many show a distinctive mounting to the cow if they are prevented from suckling or they become agitated or excited. The cow may try and stop feeding attempts during distraction or disturbances, but suckling is usually terminated by the calf. After 8 - 10 weeks the cow is more reluctant to continue feeding, lifting her back legs or moving away.

Hiding behaviour Elk are hider type species and for the first 4 days calves are initially unresponsive to disturbance. The calf will select its own bedding site with attempts made to conceal its position. In pastures with short grasses and little natural cover calves will often be seen, agitated, often panting and occasionally exhausted walking fence lines and attempting to push through fences into longer undergrowth outside the farm in search of a hide. This then presents a problem if the calf needs to rejoin the cow for feeding and protection, as they can find time to crawl out of a calving pasture, but cannot find their way back quickly. They may panic and run blindly off, or jump and become hung up in fence and wire with death or injury resulting.

Calf losses through misadventure and mis-mothering are usually a consequence of insufficient natural or artificial cover for calf hides. A distressed fence walking calf (because it is unusual and perceived as a threat to group stability) can also be attacked, kicked and bitten by other cows or pushed through fences into other groups where it is at even greater risk.

Cover can easily be provided in calving pastures, and has the additional advantage in hot humid conditions of providing shade. Heat exhaustion through exposure is proving to be a significant source of calf loss - a management related problem that is easily prevented.

Shelter and cover can be provided simply in a number of ways:

• Natural woodlots contained with pastured calving areas.

• Leave unmown strips of grass in the corners of pasture, or unmown strips in the centre. Protect these with temporary fencing (electric) until calving when the grass will be at full height and these form a natural attraction for hiding.

• Tree branches can be cut and anchored to stakes or hay feeders to form a reasonable sized jumble of cover that acts as a focus for calf hiding.

• Placing large round hay bales in groups will allow the calves to hide in gaps between, or in the shade of hay bales.

• Artificial housing - some success has been had by placing small triangular shaped wooden slatted shelters in calving pastures, or larger shelters placed on short poles and covered with slatted snow drift battens, although these may be used more for shade at older ages than for cover for neonatal calves.

Post calving behaviour Cows with calves tend to be a lot less comfortable with human presence than prior to calving. Usual flight distance of 40 m increases to up to 240 m with calves of two weeks of age, but in time this reaction decreases, especially if food is used as an enticement and a regular inspection routine established. Cows are a lot more wary of people on foot, and may react quite aggressively to dogs, small children, and interference at tagging near birth time. Inspection is more comfortable for the animals if a vehicle or ATV motorbike is used. Care should be taken to follow a consistent route and existing wheel marks - even to the extent of a pre-mown grass circuit so that there is no danger of running over hiding calves. Once the calves are three weeks old they become quite independent and will form kindergarten groups, watched over by one or two sentinel cows, especially during the morning and afternoon rest periods. Early mornings and late afternoons are important periods of play activities where chasing and fleeing and play over tree stumps or high points in the pasture often occur. Once this stage is reached cow/calf groups can be shifted to new pastures, although the distances should not be great or involve complex laneway or gateway transfer.

For easier management, pastures adjacent to the calving pasture can be prepared with vigorous palatable new pasture growth, either re-growth from previous pasture conservation, or by grazing and a three week rest period. An open gate and drift policy teaches calves about gateways or lanes without pressure. Shifting cows and calves through gates or laneways at any time requires some caution and preparedness for things going seriously wrong. If the group is too large or moved too quickly, there is a huge risk of calves being knocked down and trampled, or in laneways these young animals can be forced into laneway fences by the pressure of the group and broken limbs can easily result. If the herd is spread out, calves tend to hold back and can become separated. With no calming influence they can race off blindly into fences or attempt to charge the fence rather than seek the unfamiliar gateway.

In this situation cows will probably stop and return, but in any case a group of cows should be returned to the previous pasture to collect the calves and moving attempted again. If this occurs in a laneway, groups of agitated calves may leave the herd and charge back along the lane. It is much safer to move away and allow them to return to the pasture, move the cows back again and try again, than to control a nervous group of calves only. If this is not an option, a roll of synthetic barrier cloth that reaches across the lane can be used as a mobile fencing barrier to turn and direct the calves back to the mob. Discretion is advised in either case, and experience suggests it may be easier to shift groups early morning or late evening when the calf group is relatively active anyway.

MANAGEMENT IMPLICATIONS

Assisted Calving Generally, elk have little difficulty in calving although it might be expected that a few may need some assistance in the farm situation. Observation is the greatest tool, although there is a delicate balance between unnecessarily assisting, and leaving help too long. It is rarely possible to approach a cow in the pasture to help, so a trip to the handling facility and some form of assistance should be anticipated. A calving cow in trouble will often be well separated from the herd and fence pacing, and it is often possible to quietly work her out of the gateway into the laneway and handling system without disturbing other cows and calves. If the group is used to regular supplementary feeding, the group can be fed away from the gate and the target cow will separate and move.

A common problem is when one front leg is prevented from emerging as it is caught behind the head, or in large calves, does not extend properly and is trapped in the pelvic arch. The frequency of this condition seems to increase if cows are moved from sparse feeding conditions into luxuriant pasture growth at the point of calving. A sudden increase in stomach content volume may restrict the ability of the calf to turn and re-engage for birth.

These problems are relatively easily fixed by manipulating and the cow generally only needs restraint and light sedation. It is far better to manipulate the calf position and let the cow quietly finish calving herself rather than assisting for the whole procedure. Some other observations regarding dystocias:

- The majority of cows with dystocia are overfat or unfit, or both.
- In paddock, calf loss seems to be a consequence of disturbance, lack of cover or overcrowding in calving pasture.
- Posterior presentations - both hind legs back are relatively common but take longer and are a contributor to assisted calvings.
- Most problems are with large male calves.

In many new situations the farmer will be faced

with the difficult decision of when to act and assist, balanced against the disruption and potential risk to other newly calved cows and their hiding offspring. The importance of a routine of observation and some sort of familiarity with the calving group is a distinct advantage, and it may be easier to collect the cow in trouble with a small group of other cows for yarding rather than concentrating on the individual. If you instinctively feel you are doing more harm than good in trying to salvage a calving problem, you probably are and it may be time to leave and observe and try a short time later. People in the past have tried to rectify calving problems in the paddock with dart guns etc., but in reality the risks and disturbance may outweigh the gains. At best - an orphaned live calf results, at worst - two burial sites!

The Assisted Birth Once the cow is in the handling facility, veterinarian, equipment and facilities should be on hand so there is little delay. Ideally cows should be given minimal chemical restraint, or preferably, held securely in a crush. On the whole they seem to respond to assistance, but depending on the severity of the dystocia, assistance can take from minutes up to 45 - 60 minutes and major discomfort.

Good hygiene conditions and an ample supply of water and a natural lubricant are essential, and also the realization that manipulation of the calf can be quite physically demanding because of the nature of the cow's anatomy and simple size limitations.

If the cow is heavily sedated, the calf will also be. There is merit in reversing the calf first and more slowly reviving the mother so she recovers and focuses on the very alert calf rather than the surroundings.

The recovery area needs to be a well ventilated but smaller pen with clean fresh bedding, and quite subdued in lighting in a quiet area of the facility. The cow should be left with the birth membranes and free to clean and ingest her placenta. It helps to aid bonding and focus attention on the calf rather than the facility by rubbing the birth membranes on the cow's mouth and nose. While the cow is sedated colostrum can be collected via a human breast pump, or oxytocin used to induce let down and the calf supplemented with tube feeding. Frozen dairy colostrum can also be collected and used as an effective substitute.

Early success in mothering will be shown within an hour if the cow licks and cleans the calf, although she may show no apparent interest if disturbed and in such cases release can be that evening or following overnight confinement. Ultimately it becomes a common sense judgement call, and a decision made whether or not to treat the calf as an orphan.

Routine antibiotic therapy and oxytocin are used to combat infection and stimulate milking. Oxytocin injection also stimulates the passing of the placenta. Once tranquillizer is reversed the cow and calf are left alone for 12 - 24 hours before release into an adjacent small pasture with some cover, before rejoining the herd. The success rate is variable depending on stockmanship skills, level of birth trauma and the relative calmness of the cow. Often she will rush off and ignore the calf when first let out from the shed. If the calf is taken first and placed near the access gate to pasture in full view it will remain prone on the ground and the hind will quickly return and collect it. It certainly pays to I.D. both cow and calf to keep some observational notice of success, and also to highlight future calving problems or production limitations. Veterinarians and more experienced producers should refer to Haigh & Hudson, ***Farming Wapiti and Red Deer***, for more details on assisting calvings.

Feeding and Care of Orphaned Calves Once a decision is made to remove a calf from a cow, then that act is irreversible. Rejection cannot be overcome in time and the calf returned, but in rare situations a weak calf can be supplemented in the field with replacer milk or colostrum. If you do provide this supplement, be careful to give just enough - if the calf does not nurse at all, the mother may dry up, and you'll have the calf to feed. There are two scenarios that logically justify bottle raising an elk calf: 1) the cow has died and cannot raise her calf herself. 2) The cow is guilty of mis-mothering her calf.

Mis-mothering can manifest itself in several ways: not allowing the calf to nurse; physically mistreating the calf; not producing enough milk for the calf to thrive; paying too much attention to the calf and causing physical harm by excessive grooming (creating sores where she has licked the calf excessively) and, as often happens when you have to pull a calf, rejecting the calf.

Those are the logical reasons for bottle feeding a calf, but we often fall prey to illogical behaviour when confronted with a newborn elk calf nestled in the grass and yield to an irresistible temptation to pick it up, bottle feed it, and make a pet of it. On the other hand, there are perfectly good management reasons to have a bottle-fed elk. The presence of a bottle-fed elk in a herd usually calms other elk; a bottle-fed elk can be used effectively as a "lead animal" to aid in moving other elk into working facilities, trailers, or new pastures; and they are a great public relations animal to have on your ranch.

Whether you embark on the chore of bottle feeding a calf because of choice or of necessity, the first thing you must realize is that you are making a major commitment in terms of time and money. Once you pick up that calf, there is no turning back. You are now its mother, and its survival depends on you. This is not a chore to be taken lightly. You will be committed to that calf for a minimum of 12 weeks.

The first major rule of bottle feeding is **calves must receive colostrum**. It is essential for the calf's survival. Antibodies formed by the mother's body are passed to her calf in the colostrum which is present in her milk during the first few days after giving birth. Colostrum is thicker and more yellow in appearance than normal milk. The antibodies allow the calf to fight off infection and disease; without colostrum its immune system is defenceless. Antibodies cannot be passed from the mother to the calf through the placenta; they must be absorbed through the calf's intestinal walls. It is generally believed that ingestion of colostrum must occur within the first 24 hours of life. After that time the openings in the walls of the calf's intestines shrink, and larger particles which contain colostrum cannot be absorbed into the calf's system.

In the case where the mother dies, if you are on hand early enough, try to milk out her udder and feed the colostrum to the calf. If the calf has not nursed and you cannot milk out the cow, you must feed artificially a prepared colostrum solution. Time is critical; colostrum should be given during the first 12 hours if at all possible. If you are pulling a calf from a healthy mother to bottle feed, let the calf stay with the mother and nurse for at least 48 hours to insure that the calf receives the needed colostrum.

When removing a calf from its mother, the best results occur when the calf is taken in the evening, an hour or so before dark, and put in a box stall by herself. Any similar enclosure would work provided it has solid sides and a covered top with adequate ventilation. It should not be larger than 10' x 10' and should be located in a quiet place, with a bedding of deep straw. Leave the calf alone to get used to her surroundings and check on her once during the night, but do not offer the bottle until the next morning. The calf is calmer by that time and is usually hungry and more apt to take the bottle.

Rule two of bottle feeding is **be prepared**. Assemble all the necessary equipment, have the formula on hand, and make sure the pen is ready before you actually remove the calf from its mother.

NOTES ABOUT EQUIPMENT

Nipples Elk calves seem to prefer a softer nipple, so you may need to boil a hard nipple for 15 minutes to soften it. Some calves prefer baby nipples while others will not suck anything but a calf or lamb nipple. Have all three on hand, and let the calf choose which one it prefers. You may need to enlarge the hole in the nipple for the first week or so. You want the opening large enough that the calf does not have to work too hard to get the milk and get discouraged, but not so large that the flow is too much for the calf to swallow comfortably.

Bottles Baby bottles are easy to handle at first, but you will soon need several bottles each feeding to keep up with the calf's consumption. If you can get the calf started with the larger calf or lamb bottles, you will have less to carry later on.

Bottle Brush This is a necessity; clean bottles are a must! A regular size will work for baby bottles, but you will need a larger one to adequately clean calf or lamb bottles.

Formula Container To store mixed formula in the refrigerator, a half gallon with a large mouth is a minimum.

Record Book A spiral notebook or note pad is adequate. Keep it by the stall or pen and use it religiously.

Rule three in bottle feeding is **establish a feeding schedule and stick to it**. Erratic feeding times result in undue stress for the calf. Feed the calf five times a day for the first seven to ten days: 6am, 10am, 2pm, 6pm, and 10pm. Start out offering four ounces at each feeding. When the calf is consistently eating four ounces at every feeding for three days, increase all feedings by one ounce. Then when it is eating five ounces at every feeding for three days, increase all feedings by another ounce, and so on.

Get into a routine and follow it. Mix your formula, fill the bottles, and refrigerate any unused formula. Warm the milk. (Do not use a microwave to heat bottles; use a bottle warmer or pan of water on the stove.) Take the bottles, paper towels, and record book to the pen. Go in with the calf, talking to her all of the time, and get her to stand up. (Never feed a calf while it is lying down.)

Kneel down with the bottle in your left hand, put your right arm around her neck, and hold her lower jaw in your right hand. (She will be under the crook of your right arm.) It is easier to get her to hold still if you can back her up against the wall or back her rear into the corner until she gets the hang of what you are trying to do. Put the nipple in her mouth; you might need to open her mouth using the fingers of your right hand. Move the nipple slowly in and out. Do not squirt milk down the calf's throat. You might need to drip a few drops on her tongue so she can taste the milk, but forcing it down her throat will only result in choking. If the milk is not swallowed properly, it can run down the throat, accumulate in the trachea, or be aspirated into the lungs and result in pneumonia.

If the calf refuses to drink after five or ten minutes, give up and try again at the next scheduled feeding. Patience and persistence are the keys in the early stages of introducing the calf to a bottle. The calf will eventually get hungry, and with enough encouragement will take the bottle. Some calves begin sucking the first bottle; some take 36 hours. Stay calm, be patient, and work with the calf. You need to experiment with different nipples and bottles until you find the right combination for that particular calf.

Each time you feed you will need to stimulate the calf's rectum with a moist paper towel. When the mother nurses her calf, she licks the calf's rectum so that the calf will defecate. You are now the calf's mother, and you need to see that she is stimulated to urinate and defecate. You will need to do this at every feeding until the calf is voiding on her own, usually by the end of the second week.

Once she has eaten and voided, leave the pen, get out your record book and write down the time of feeding, the amount of milk consumed, the colour and consistency of the stool (firm, runny, pasty, brown, green, yellowish), the general attitude of the calf (sucked readily, listless, etc.) and the calf's temperature and respiration rate if you have taken it. (Normal average temperature for an elk calf is 101°F, normal pulse rate is 80, and respiration rate is 26.)

If you do not record the information in your record book, you have no way of knowing when to increase the feedings, if the calf is getting enough to eat, if she is constipated, etc. In the event that the calf should get in trouble and you have to call a vet, this information will be invaluable in making decisions for proper treatment of the calf.

Immediately throw out any formula in the bottles that the calf did not eat. Never put it back into the refrigerator to be stored for later use. Clean the bottles and nipple thoroughly with the bottle brush, using warm water and a mild detergent, and allow the bottles to air dry. Sterilize the bottle at least twice a week. It is very important that you wash the bottles carefully as soon as possible after each feeding and keep all your equipment clean so that harmful bacteria does not enter the calf's system through the milk. If you are feeding several calves, use one nipple for each calf, and label it. Do not "share" nipples between the group.

Commercial milk formula may not exactly match the composition of elk milk. However, specifically designed formulas are improving constantly. Fresh goat's milk is very close and is the choice of many breeders who bottle feed regularly. You may wish to purchase goats and milk them yourself or purchase milk from a local goat dairy. **Note:** Goats must be tested and shown to be free of EHD, MCF, Johne's and any other disease before they, or their milk is used to feed elk. As an alternative, you can used canned goat milk. Some people train the goat to let the calf suck directly from her. This can be quite a time saver, but the calf is usually not as tame as one raised on a bottle. As the calf gets older, it will usually take two goats to produce enough milk for one elk calf. This is a particularly good choice for male calves. You will also need to wean the calf from the goat before she convinces the elk that it is a goat and you find your calf standing on top of hay piles, vehicles, etc.

There are several commercial milk replacers that have been used successfully to bottle feed elk calves: Maple Lake Second Nature Exotic Game Milk Replacer, PMT, Brown's, and Carnation LAMA. Use a lamb or goat formula rather than one designed for cattle. You may also wish to use one of the proven formulas shown in Table 5, or refer to section following re: Milk Replacers for Elk Calves.

NOTES ABOUT FORMULAS:

The commercial milk replacers will be the most economical to feed in most cases. Do not use egg whites in any formula because they inhibit the utilization of B vitamins. Use human electrolyte formulas. Probiocin powder helps to prevent scours by keeping the bacteria in the gut balanced. You can use Karo syrup to help regulate consistency of the calf's stool. Too much will produce runny stools; too little will produce hard stools. You might wish to add 1/2 to 1 teaspoon oat or wheat bran to the formula to help prevent scouring.

Once the calf is eating readily and approaching, wanting the bottle, move her to an outside pen with a shelter, where she can get some sunshine and have grass to nibble on. When the calf is ten days to two weeks old, reduce the number of feedings to four a day: 6:30 am, 12:30 pm, 5:30 pm, and 10:30 pm. The calf will be averaging 12 to 16 ounces per feeding. Stay on four feedings per day until the end of the second month. Then, reduce to three feedings per day: 6:30 am, 2 pm, and 10:30 pm. The calf usually averages 24 to 30 ounces per feeding. At three months, decrease to two feedings per day (32 to 40 ounces per feeding) and then gradually drop to one per day depending on how well the calf is eating hay and grain, etc. It is important not to over feed the calf. The calf should be satisfied but not completely full at the end of each feeding. The recommended fluid requirement is 10% to 20% of the calf's body weight per day.

If you are bottle feeding more than one calf, you might want to train them to a bucket. There are commercially manufactured buckets with several nipples around the bottom that allow you to fill the bucket, hold it in the air or put it on a stand, and feed several calves at once. You can also make your own by drilling a 1 1/2" hole near the bottom of any bucket and purchasing a nipple set with washers that attach to the inside and outside of the hole. Be sure to thoroughly scrub both the bucket and the nipples after each feeding.

Always keep the pen or stall clean and change the bedding daily. Always keep plenty of clean, fresh water in reach of the calf at all times. When the calf is about three days old, put a shallow pan of peat moss and forest soil in the stall. Calves seem to do better with a little soil to ingest. When the calf is two weeks old, keep a small amount of sweet feed and best quality alfalfa hay available in the calf's stall, free choice. Change the sweet feed daily. When you begin to wean the calf, if she is not eating feed, grass, or hay readily, you might

Elk Calf Milk Replacer Formulas

JANEY HAYES FORMULA

- 500 - 800ml fresh whole cows milk
- beaten egg yolk
- 15 - 20g glucose
- 1 ml cod liver oil

BILLY WHITE FORMULA

- Day 1 and 2 feed straight cow colostrum.
- Day 3 begin mixing colostrum with equal parts of evaporated milk and whole goat milk, gradually decreasing colostrum. (If goat milk is not available, use 2 parts evaporated milk to1 part whole cow's milk.)
- Day 10 begin decreasing evaporated milk to pure goat milk.

JERRY HAIGH FORMULA

- 1 litre whole cow's milk
- 1 can condensed milk
- 1 egg yolk (do not use egg whites)
- 2 drops multivitamin B
- 1 teaspoon cod liver oil

TIMBERLINE RANCH FORMULA

- 1 can condensed goat milk
- 1 can Carnation milk
- 1 tablespoon Karo syrup
- 1 egg yolk (do not use egg whites)
- 2 teaspoons Vedco Calf Quencher
- 2 teaspoons Probiocin Powder
- 12 oz. water

Table 5

want to mix infant cereal in with her formula to get her accustomed to more solid food.

SCOURS

Most bottle-fed calves will scour at some point in their lives. Severe scouring (diarrhea) will dehydrate the calf and cause a life-threatening situation. If you cannot control scours with the following method, or if the calf becomes severely dehydrated, call a vet immediately or begin fluid treatment with an IV Ringers solution. You can tell if a calf is dehydrated by pinching the skin on the neck. If it returns to normal by the count of three the calf is probably okay; if it stays elevated longer than the count of three, the calf is dehydrated. Dehydration will kill a calf very quickly, so you must act promptly and not adopt a "wait and see" attitude.

At the first sight of scouring, take the calf's temperature. If it is normal (101°) proceed, but if her temperature is elevated, consult a veterinarian immediately. Begin feeding 1/2 normal formula diluted with 1/2 sterile water, and add an electrolyte mixture containing bicarbonate and dextrose (LifeGuard, Nested, etc.) per manufacturer's recommended dosage. Give the calf 25cc of Pepto Bismol or Correction Fluid orally. Use a syringe without the needle to squirt the Pepto Bismol directly into the calf's mouth. If scouring continues, you may need to dilute the formula further with water until the stool becomes firmer. When the stool is normal, gradually decrease the amount of water until the calf is back on full formula.

When the calf is actively scouring, you will need to wash the tail area and clean off all feces using a soft cloth and warm water. Spread vaseline on the calf's bottom and tail area to keep the diarrhea from sticking and burning the skin.

CONCLUSION

Bottle feeding elk calves can be a very rewarding experience. An adult cow that has been hand reared as a calf is extremely loving and a joy to have around. Once you get the calf to accept the bottle, everything usually goes smoothly as long as you keep an eye out for potential developing problems and treat the calf promptly should any occur. Good luck with all your calves and enjoy your new friend.

Much of the preceding topic was gleaned from an article published in North American Elk magazine (Winter 1997), entitled "Bottle Feeding Elk Calves", written by Sharyon Doshier, Timberline Ranch, Ltd., Manager.

MILK REPLACERS FOR ELK CALVES

When an elk calf if born, the first stomach, the rumen is not considered functional for the first 40 days of life. These first 40 days are most important to elk calves, particularly those calves that need to be hand fed a milk substitute. After 40 days, feeds other than milk can become a major source of nutrients to all calves.

Intake Levels

The level of intake of milk by elk calves is similar to that of other ruminants in the same stage of development. It is common for the calves to consume 15 to 20% of their body weight in fluid milk. For the first 3 weeks, a newborn elk calf weighing 20 kilograms can drink 3 to 4 kilograms (or litres)of milk per day. Intake of milk peaks at 4 litres per day at 3 weeks of age, and then declines as illustrated in Table 6. At the same time, body weight is increasing, and by the 6th week of life (approximately 40 days), the intake of concentrate and forage compensates for the decreased level of milk intake and provides nutrients to the calf.

Digestion of Milk

When calves drink milk, two folds in the rumen curl inward and form a tunnel which causes the milk to bypass the rumen and send it straight through to the abomasum (true stomach). This mechanism allows nearly 99% digestibility of

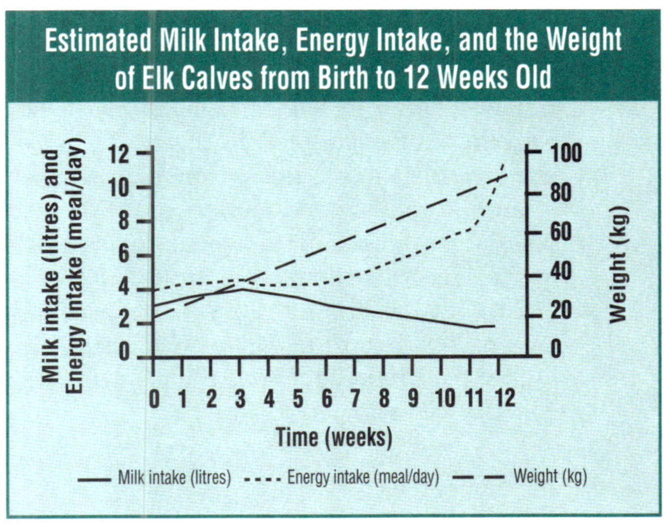

Table 6

the milk that was allowed to enter and bypass the rumen.

Composition of Milk

The composition of the milk received by hand reared calves should have a nutrient composition similar to that of milk from an elk hind. Nutrients in milk that are of importance to elk calves include fat, protein, lactose, calcium, and phosphorus. On average, elk milk contains 19.80% dry matter, 37.87% fat, 31.31% protein, 20.70% lactose, 5.56% ash, and 6.09 kcal/gram gross energy on a dry matter basis. On an as fed basis, elk milk contains approximately 7.50% fat, 6.20% protein, 4.10% lactose, 1.10% ash and 1.20 kcal/gram gross energy. In comparison, whole milk from a Holstein cow (24 hours after calving) contains approximately 12.50% dry matter, 3.70% fat, 3.20% protein, 4.60% lactose and 0.70% kcal/gram gross energy (as fed). Thus the cow's milk is not a good substitute for elk milk.

Energy

Energy intake by an elk will have a direct effect on it's daily gain in weight. Close to 58% of the energy in elk milk comes from the fat.

Protein donates close to 30% of the energy, and lactose provides the remaining 12%. It has been reported that elk calves can gain approximately 790 grams per day when drinking 3 to 4 litres of elk milk. In comparison, a 20 kilogram elk calf drinking 3 to 4 litres of whole cow's milk will only gain approximately 350 to 450 grams per day due to a lower energy content.

Protein

Because the elk calf has little rumen function for the first 30 to 40 days of life, very little to no protein is produced by microorganisms in the rumen. Therefore, the protein supply must come from the milk. The protein that a calf ingests goes into creating muscle tissue, for essential bodily functions, and is used as an energy source. In order to sustain a 0.8 kilogram average daily gain, the protein intake by elk calves should approximate 25% to 26% of the milk (dry matter). Any protein content less than these values will result in a slightly depressed average daily gain due to inadequate tissue deposition.

Lactose

Lactose is the main sugar in mammalian milk. Lactose is composed of glucose and galactose in equal parts. Both glucose and galactose are additional sources of energy and are highly digestible.

Minerals

Because the elk calf can have an average daily gain of up to 1,000 grams per day, adequate levels of minerals must be provided for the formation of bones, enzymes and other components. Two essential minerals are calcium and phosphorus, which are involved with bone growth. If these minerals are deficient in the milk, bone formation is affected. Calves deficient in calcium and phosphorus may have swollen joints, suffer spontaneous fractures, appear unthrifty and suffer from a nutritional disorder called rickets.

In the milk from a Holstein cow, calcium and phosphorus compose approximately 18.7% and 14.4% of the ash content. This translates into 0.13% calcium and 0.10% phosphorus of milk content on an as fed basis (1.04% calcium and 0.80% phosphorus on a dry matter basis). Elk milk by comparison contains approximately 12% more calcium and phosphorus (1.17% calcium and 0.90% phosphorus on a dry matter basis). This suggests that if an elk calf was to be raised on cows milk, it would not be receiving adequate levels of calcium and phosphorus in its diet.

Vitamins

Young calves that are fed milk for the first 30 to 40 days of life require some vitamin supplementation.

Healthy Elk Calf

However, once on a diet of concentrates and forages, the rumen begins to function and the calf will be able to produce its own B and K vitamins due to the fermentation. The colostrum produced by the cow when the calf is born is high in Vitamin A, E and B12, but these levels are reduced thereafter. There is limited information available regarding vitamin levels required by elk calves.

Suggested Nutrient Levels in an Elk Calf Milk Replacer

Table 7 is a compilation of recommended nutrient levels that should be available in a milk replacer for elk calves. This type of replacer should provide enough energy, protein, minerals and vitamins to support an elk calf growing at 0.8 to 1.1 kilograms per day when mixed at a 5:1 dilution rate (e.g. 1 cup of replacer to four cups of water).

Suggested Nutrient Levels in an Elk Calf Milk Replacer

Component	As Fed (%)	Dry Matter (%)
Total Solid (Dry Matter)	20.00	100.00
Fat %	7.60	38.00
Protein %	6.40	32.00
Lactose %	4.20	21.00
Ash %	1.12	5.60
Calcium %	0.24	1.20
Phosphorus %	0.18	0.90
Vitamin A, IU/kg	100,000	50,000
Vitamin D3, IU/kg	1,600	8,000
Vitamin E, IU/kg	15	75

The preceding section was adapted from an article published in The Canadian Elk & Deer Farmer, Winter 1996, entitled "Wapiti Calf Milk Replacers", written by M.S. Feist, D.A. Christensen, and V.J. Racz, Department of Animal & Poultry Science, University of Saskatchewan.

WEANING & WEANER MANAGEMENT

Bring up the subject of weaning to a group of elk farmers, and you'll get a lot of different opinions. That's partly because we are still fairly new at intensively managing these animals in a farmed environment and partly because management philosophies vary between different styles of elk farms.

There are several reasons to consider weaning before the rut, (i.e.) in late August or early September. These include:

- Removal of the need to lactate reduces the nutritional requirements for the cow, and she can then devote more feed resources to improving her body condition. This should also allow earlier conception and easier wintering of the cow.

- Calves are growing rapidly in the fall. Separating them from the cows allows them to be fed exceptionally good feed without competition for that feed from their mothers and away from the stress associated with a bull's rutting activities.

- Artificial Insemination (AI) programs require several sessions of handling the cows. This is much easier and safer to do if their calves have been separated from them.

- Handling at weaning time allows the farm manager to vaccinate, deworm, treat for ticks and weigh the calves. The dams of the calves with the best weaning weights can be identified and selectively bred to the best bulls, either naturally or by AI.

- Weaned calves (weaners) can be handled again in thirty days for their second Chlostridial vaccination, another deworming and deticking and weighing to monitor their growth rates. This is much simpler without their mothers and the breeding bull.

- The suckling of a calf stimulates release of prolactin, a hormone that has a negative effect on the release of hormones governing the estrus cycle. Removal of the calf may stimulate earlier cycling and breeding activity, but observations of wild elk suggest that this effect may not be particularly powerful.

The most important reason *not* to wean before the rut is the stress associated with the process. Detailed research projects have documented that calves, cows and farm managers may be highly stressed by the process of weaning. Both cows and calves pace, call out to each other, and generally remain unsettled for several days following separation. This is particularly so with abrupt procedures that generally involve running cows and calves into the handling facilities, returning the calves to the pasture from whence they came, and moving the cows to the furthest pasture away from the calves. Although this is a relatively simple and efficient process, and research has shown that the calves continue to gain weight at the same rates as with other weaning procedures, there are two other procedures that work very well and that are more humane for all involved.

The first more gentle alternative is "fenceline" weaning. In this procedure, the cows and calves are separated and returned to adjacent pastures, with only a fenceline between. This allows much socialization behaviour to continue, including vocalization, nose to nose contact, and lying together. It does require a well-built and maintained fenceline and gates, to ensure that neither the fences nor the livestock are damaged.

Fig. 10 Fenceline weaning, cows on one side, calves on the other

The second alternative is "soft" weaning, which involves the removal of about ten percent of the cows each day for ten days. This allows a gradual change to the social structure of the group, and allows access to the comforting presence of at least some of the cow herd.

No matter which procedure is followed, there is always some stress associated with weaning. A good farm manager will prepare to minimize this stress by using one of the gentler techniques mentioned, and by following these procedures:

- Begin feeding supplements to the cows and calves in mid-August. Oats and specially formulated pellets or cubes are a good choice. Start feeding at about half a pound per head per day, and increase by half a pound per head each day that the herd cleans up (in a short time, say half an hour) all that is offered. The calves will start nibbling grain at about 60 days of age. Be sure that you spread out the feed enough to allow each animal an opportunity to eat.

- Prepare a pasture for the calves by seeding and fertilizing to produce a lush growth, and rest the pasture from grazing for the month of August. Fall rye or annual ryegrass planted in July makes a perfect autumn pasture for weaners. Try to have this pasture located near your house and yard so you can observe the calves easily. Newly weaned calves are more susceptible to attack by predators or disease, so you need to monitor them more closely.

- Avoid weaning when the weather forecast for the next few days suggests cool, rainy or windy weather. Calves have low fat reserves, and their baby coats do not provide good insulation until the winter coat grows in. The stress of weaning predisposes calves to diseases, and this is magnified by poor weather.

The most common problems encountered in newly weaned calves are injuries due to poor handling, inadequate nutrition and Yersiniosis. All three of these are "management diseases" and the course of prevention is good management:

- Be sure that your handling facilities are in good working order, and that you are prepared to handle smaller animals. If you have a quiet older cow, put her in with the calves for the first week or two to give them a leader to follow.

- If you have followed the procedures mentioned above, the calves will be on high-quality (young, tender, rapidly growing - YTRG) pasture, and they will be eating their fill of oats and supplement pellets, preferably fed to them twice or even more often per day. Feeding several times a day increases voluntary feed intake (VFI) and familiarizes the calves with your presence, making them much easier to handle in later life.

- Yersiniosis is caused by bacteria that are present in the farm environment at all times. Healthy stock can resist the bacteria, but stress, poor nutrition, social instability and/or bad weather can drag down an animal's immune system and allow an invasion of the Yersinia bacteria to occur. Affected calves hang back from the herd, appear depressed, listless, uncomfortable and typically show signs of very watery greenish scouring which will also turn bloody. Treatment must be immediate, and involve fluid replacement therapy and treatment of the affected calf and the entire group with tetracyclines. Consult your veterinarian immediately, as there are other diseases or problems that may affect weakened calves.

Should you wean early or not? In some situations, you may be better to let nature take its course. Elk mothers will slowly cease milk production as autumn turns to winter. Their calves stay with them, and the females will remain together for life as a matriarchal social group, if they are allowed to do so. If you have a small herd, for instance less than 12 calves, they will be more uncomfortable on their own than if the herd is larger and has an older, experienced leader. If 90% of your cows all calve in late May and early June, then you aren't having problems with early and adequate conception. Top-quality nutrition can compensate for not weaning early, but there are other advantages in the points listed above.

These are most of the considerations regarding weaning. Now, you have to decide which procedures suit your management style and on-farm resources the best.

Chapter 2 ***Breeding***

SELECTION & GENETICS

Seasonality is the key feature to remember in managing the breeding of elk. The natural breeding season, the "rut", begins every year in late August and early September, signalled by the readiness of the breeding bulls. The elk females are stimulated by photoperiod, physical condition and the rutting behaviour of the bulls to begin cycling. A good, sensible breeding bull can service and settle thirty to fifty cows in forty days. This allows the selection of a few genetically superior males to gain maximum benefit for a herd improvement program.

An investment in a breeding bull should be carefully considered within the long term breeding objectives of the program. That is the first critical step, defining your objective, "where do I want to be in 5 years?"

Questions involved include:

- "Is the operation at its best producing velvet antler, venison, foundation stock for livesale, or developing breeding bulls?"
- "Are the physical aspects of the farm suitable for this objective: climate, location, feed availability?"
- "What level of investment is required, and what rate of returns are needed? How are these returns realistically going to be achieved, (meat, velvet, breeding stock)?"
- "How much time input is needed in the operation? How big should the animals become? Can I manage and handle the results? What size is most efficient for our production goals?"

Once these questions are clearly answered, the breeding options and choice of bull and management system become more clear. There is a huge variation of size, antler shape and other characteristics that with careful selection allow the breeding of styles of elk with improved production characteristics that suit individual markets, the farming operation or simply the enjoyment of the investment. The final stage in a breeding program requires appropriate recording of animal performance and seeking maximum genetic expression through good management.

In an intensive operation with well defined objectives, single sire mating is the most practical and rewarding technique. Cows can be selected for their superiority as expressed through:

- weaning weights of calves, (the lactational advantage)
- early calving - the ability to conceive early in the season and to feed calves well, taking advantage of spring and summer pasture growth peaks. Actual wean weights, rather than adjusted, reflect this early calving advantage.
- manageable temperament - cows that are

stress free through calving and weaning and can be handled at will. These animals also make excellent candidates for artificial breeding and embryo transfer programs, to fully utilize the base of elite genetics now available through routine reproductive technology.

- Annual mid winter weights.

- 15 month weight and ranking in a herd group. This weight determines conception rate at this age.

Elite bulls should be selected from the top 3% of any age group based on the character, or characters desired. However, it is important to remember that the more characters that are selected for, the less progress is made for each.

In any program, excellence still requires the choice of the best genetics available. A mere increase in size, just because it's bigger is not efficient unless that size increase represents the top merit of that bloodline.

In selecting breeding bulls important records and selection criteria include:

- weaning weight (and herd ranking)
- 15 month weight - spike antler weight
- 27 month weight
- 2 year old velvet weight (including days of growth, grade, antler definition characters, beam diameter, style) subsequent velvet weight and grade per year, annual mid winter lean weight, (peak weight prior to the rut is interesting but often reflects extremes of luxury summer feeding), and hard antler characteristics, longevity, temperament plus "strain or type and style".

Bruce Friedel, from Moose Lake Wapiti in Alberta, has analyzed the relative merits of selecting for antler growth, body size and reproductive performance. Examining the potential profitability of each selection criteria may help to rank the relative importance of each.

SELECTING FOR VELVET GROWTH

In 1993, Alberta producers cut an average of 11 lbs of velvet from 463 three year old bulls. Farmers competing in the Elk Expo velvet competition suggested that a three year old son out of a 30 lb plus sire might fetch from $8,000 - $12,000. Let's examine the financial wisdom of purchasing better velvet production genetics (Table 7).

Let's make the assumption that an average velvet producing 3 year old bull is worth $2,500, and that production costs of rearing average, good or excellent bulls are comparable. Would it pay to buy one of the best three year olds in the industry, cutting 19 lbs of velvet for $12,000? If a producer has a herd of 25 average cows (replacing breeding bulls every 2 years), an 88% weaning rate, and retains both male and female offspring in their herd, they can expect an additional $6,820 worth of velvet production (Table 7).

The bull will continue to produce velvet at better than industry averages, and will likely cut 8 lbs/yr more than average bulls. Harvest an additional 16 lbs of velvet over two years at $50/lb, and the bull returns the producer another $800. The additional $9,500 investment returns you

Velvet Production Genetics

Breeding Bulls	Velvet (lb)	Heritability 1	Selection 2 differential	Generation 3 interval (yrs)	Expected genetic 4 gain (lb/yr)	Sire's genetic value 5 at $50/lb velvet
Average 3 year old	11.00	0.35	0	4.5	0.00	0
Good 3 year old	15.00	0.35	4	4.5	0.16	3520
Excellent 3 year old	17.00	0.35	6	4.5	0.23	5060
Awesome 3 year old	19.00	0.35	8	4.5	0.31	6820

1. Heritability of velvet antler growth has been estimated at .35
2. Selection differential - difference in velvet production between selected and average bulls. Good 3 year olds = (15 - 11) = 4
3. Generation interval - average age of parents when calves are born [3 year old bulls = (4 year old bull + 5 year old cows)/2] = 4.5
4. Expected genetic gain - assumes average cows (no selection differential): Good 3 year olds = .35 x [(4 + 0)/2]/4.5 = 0.16 lb/yr
5. Sire's genetic value - increased velvet yields from mating superior sire to 25 cows/yr x 2 yrs x .88% weaning = 22 males
    a. Retain 22 males in velvet herd (Good 3 year old sire) 22 bulls x 10 years x .16 lb/yr x $50/lb = 1760
    b. Consider the effect of genetic improvement of the female offspring = $1760 x 2 = $3520

Table 7

Economics of Selecting for Increased Body Size						
Breeding Bulls	Body Weight (lbs)	Heritability[1]	Selection[2] differential	Generation[3] interval (yrs)	Expected genetic[4] gain (lb/yr)	Sire's genetic value[5] at $3.50/lb venison
Average 3 year old	650	0.40	0	4.5	0.00	0
Good 3 year old	725	0.40	75	4.5	3.33	235
Excellent 3 year old	800	0.40	150	4.5	6.67	469
Awesome 3 year old	875	0.40	225	4.5	10.00	703

1. Heritability of of selecting for body size is .40
2. Selection differential - difference in body weight between selected and average bulls. Good 3 year olds = (725 - 650) = 75 lbs
3. Generation interval - average age of parents when calves are born [3 year old bulls = (4 year old bull + 5 year old cows)/2] = 4.5
4. Expected genetic gain - assumes average cows (no selection differential): Good 3 year olds = .40 x [(75 + 0)/2]/4.5 = 3.33 lb/yr
5. Sire's genetic value - increased velvet production from mating superior sire to 25 cows/yr x 2 yrs x .88% weaning = 22 males
    a. Slaughter 22 male progeny (Good 3 year old sire) 22 bulls x 3.33 lb/yr x $3.50/lb x .55 dressing % = $141
    b. Slaughter the 22 female progeny (Good 3 year old sire) 22 cows x 2.22 lb/yr x $3.50/lb x .55 dressing % = $94

Table 8

$7,620 in two years after which the bull can be resold for $10,000. That's a 30% return on your investment annually. Use that same bull for three years on 30 cows/year and keep all progeny born on the farm, and that $9,500 investment returns you $13,476.

BODY SIZE ... IS BIGGER BETTER?

In the future, red meat production will play an increasingly important role in the North American elk industry. Would it pay to select for larger bodied elk in hopes of cashing in on the upcoming venison market? Let's examine the potential profitability of selecting for larger body size (Table 8).

The heritability of selecting for increased frame size is 40% in beef cattle. We'll make an assumption that elk are similar. Comparable management practices would result in the same number of calves being born, but producers would likely slaughter their stock at 15 - 17 months for the prime venison market. Progeny from an exceptionally large bodied bull (875 lbs at 3 years) would gross a farmer an additional $703 in those two years.

We can't make the assumption that production costs of average sized and large elk are the same, as bigger animals require more feed. A third to half of the increased venison returns could be expected to be consumed by increased feed costs. If a producer pays $12,000 for an extra large bodied 3 year old, their return over the two years would have been 3% on their investment! The biggest risk the industry runs if they zealously select for bigger elk, are the associated problems of increased dystocia, decreased fertility and longevity, and higher labour costs.

IMPROVING REPRODUCTIVE PERFORMANCE

In 1993, Alberta elk producers weaned an 81.8% calf crop. Results were highly variable between farms as some producers reported problems with open cows, underweight 15 month old heifers, late calving cows and increased dystocia due to overfat females. Examining weaning success is an accurate method in determining the profitability of a herd's reproductive performance (Table 9).

The consensus among most elk farmers is that a 90% weaned calf crop is an easily obtainable goal. Improving weaning percentage involves fine tuning your management program. Increasing cow conception and decreasing calf death loss are the keys to a better weaning rate.

Conception rates of 15 month old heifers are strongly influenced by the onset of puberty which occurs when females reach 70% of their mature body weights. Since body weight is so crucial, producers should attempt everything within their power to ensure these heifers reach 400 - 450 lbs before their first rut. This strategy involves calving early the preceding year (strive for 90% May calves), pre-rut weaning, and excellent nutritional supplementation from pre-rut weaning until after breeding the following year (18 months).

Profitability of Improving Weaning Percentage				
Weaning Success	**Weaning %**	**Calves weaned (from 25 cows)**	**Elk calf crop receipts f.$5000 - m.$1250**	**Returns from improved weaning**
1993 Alberta average	81.8	20.4	63891	0
5% improvement	85.9	21.5	67085	3195
10% improvement	90.0	22.5	70280	6389
15% improvement	94.0	23.5	73474	9584

Table 9

Separating different age classes of stock ensures that young females do not have to compete at the feed trough and are not harassed by older cows further up the pecking order. Breed heifers separately from the main cow herd, and use backup bulls. Finally, remove breeding bulls by Nov 1st to avoid late born calves.

Conception rates of adult cows can be improved by culling aged cows, reducing internal parasite loads, pre-rut weaning; flushing with excellent pasture or grain; feeding mineral & vitamin supplements; keeping bull : cow breeding groups under 1 : 30 and using backup bulls. Dystocia is the biggest killer of elk calves and occurs most frequently in overfat females!

Fine tuning your management system involves better nutrition supplementation in the breeding herd (1 lb/day mineral & vitamin supplement 365 days/yr and 5 lbs/day oats for a 90 day rut) which would cost a producer an extra $1,100/yr/25 breeding cows. Pre-rut weaning of those 22 calves born from those 25 cows would cost an additional $600/yr. Better parasite control involving a mid summer and pre-rut treatment would run an extra $300/yr/25 cows. Increased input costs total $2,000/yr in that 25 cow herd. The economics of achieving a 90% weaned calf crop by fine tuning your management program are 320% return on your investment! A 94% calf crop provides a 48% annual return.

Improving weaning percentage is 10 times more profitable than selecting for increased velvet antler production and 100 times better than selecting elk for improved body size. While the relative profitability of these management factors will change as breeding stock prices, velvet and venison markets fluctuate, they do mirror similar comparisons in the domestic beef industry. Selecting for reproductive performance in a beef herd is five times more important than selecting for growth, and ten times more important than carcass trait selection.

Improvements in Genetic Selection & Recording Systems - Production of Breeding Values

Written by A.J. Pearse, Deer Breeding & Genetics Program, AgResearch, Invermay

INTRODUCTION AND BACKGROUND

Any breeding or purchasing program for genetic improvement relies on clearly defining the long-term objectives at the outset.

The phrase "where do I want to be in five years' time" is ultimately as appropriate for the elk meat finisher as it is now for the velvet antler producer or specialized breeder. All operations rely on making a decision on what animals to use at some critical point in the annual management program. This means deciding on what animals to

a) keep for a purpose, which is selection.

b) reject from your program, which is culling.

c) concentrate to maintain and improve those genetics for the future, which is breeding and selection.

In cervidae farming, the objective can invariably be expressed as improving farm profitability by:

- increasing productivity - per animal or per hectare(acre) with the aim of increasing returns.

- improving quality - increasing price per unit (either elite or surplus animals or the product stream).

- improving the timing of supply through either genetics or management to fit market demands, i.e., increasing unit price.

To translate this to the on-farm situation, a

selection objective must have some market relationship because of the long-term nature of elk/deer breeding. Market relationships require good communication from the industry as to what is needed, when and to what specifications. Clearly, indicators must be stable or be a predictably evolving trend that allows on-farm selection programs to be directed long term along these lines.

The most important aspect of any selection program is being able to measure progress. There needs to be a good understanding of what to measure, and when and how to record this information so that it can be retrieved for analysis and decision making.

Elk farmers today face a huge variety of choices in assessing large numbers of well-promoted animals to fit markets or their own requirements. Often these animals are presented with impressive individual performance figures or source/origin and pedigree descriptions that imply a significant "elite" background. Promotional claims often include some measures of superiority that may not be tightly linked to objective measurement. In making a purchasing decision, critical evaluation of this information is obvious. It is also important to best fit your management objectives and the physical and environmental limitations of the farm as well.

In a small and relatively homogeneous farming climate like that in New Zealand, there are massive environmental and management influences that produce differences in performance from animals of similar genetic backgrounds. These factors will be at greater extremes given the breadth and scope of elk farming in North America, as is demonstrated in the north to south gradients in velvet antler weights. Breeders and purchasers are more aware now of records, repeatability and meaningful production data in making selection judgments. In looking at available information and records, some features are more important than others.

RECORDING MUST:

1. Emphasize pedigree: breed types, strains within breed types, bloodlines within strains, and the influence of spectacular individuals.

2. Take account of the highly seasonal production cycle of cervidae: weight for age and weights at different stages of the growth cycle.

3. Take account of product quality (especially for antler, currently): weight for age, quality grade, weight for a specific number of days' growth, and suitability for the market.

Recording such information need not be complicated if objectives are clearly set and the data recorded are relevant and based on the individual and where it fits relatively in its own age group.

Once the measurements are identified, selection requires consideration of all the following aspects:

- Clear and measurable traits or characters are consistent as to timing, seasonality or market end point.
- Characters must be heritable, i.e., have a genetic basis.
- There must be variation for the character within the population.
- The more characteristics selected for, the less the rate of progress made in any one.

The tools of selection can be summarized in terms of what to measure as:

Bulls (for velvet antler, trophy antler and elk meat production)

- pedigree (sire and dam)
- weaning weight
- 12 or 15 month weight
- 27 month weight
- 2-year-old antler weight (preferably corrected for days' growth and antler grade)
- annual mid-winter weight
- annual velvet weight and grade

Cows and heifers

- pedigree (sire and dam)
- weaning weight
- 12 or 15 month weight
- annual mid-winter weight

For both groups in time, the date of birth, date of weaning and the growth rates of young animals as calves will also be significant recording features as they give a strong indication of cow milking ability and reproductive patterns. It is also very important to have a clear distinction between selection for breeding purposes and culling. Culling poorer performers will improve herd productivity and still requires a decision and assessment process, but may not necessarily be agenetic improvement decision. A number of cull factors are environmentally determined, for example, aspects of handling ability or calf rearing ability in the farmed situation.

The main tools of culling (i.e. features that might exclude an animal - both subjective and heritable features) include:

Bulls

- temperament
- velvet weight relative to mean production
- velvet grade and style
- body weight for age

Cows

- temperament
- reproductive performance
- calving spread
- weaning weight of progeny
- conformation

RATE OF IMPROVEMENT

Having selected and then developed a simple system to measure any genetic based characters, the rate of productive improvement at the animal level, expressed as genetic gain, needs to be assessed. This will ultimately be reflected as profitability.

The first need is that elk management, productivity and health status have all been integrated into the program. For example, it's very difficult to take advantage of an improved breeding program for increased weaning weights if you are farming in a mineral deficient area through a predictable late summer drought at high stocking rates, unless management and supplementary feeding systems are in place to meet the animal demands.

Selection for increased velvet weight will be greatly impaired if bull feeding in late autumn, post breeding or in winter or early spring is restricted through inadequate feeding of quality feed. Sufficient bulk of feed may well be offered; but if it is of high fiber and low energy quality, the bull is unable to eat adequately to meet maintenance needs and velvet growth will suffer.

Genetically, there are two possibilities for genetic improvement in cervidae:

Within a strain or bloodline type

Improvement by selection of the top sires and identification of top cows is real, positive and additive, but relatively slow. It has, however, been quite dramatic in the leading subspecies strains, which themselves have been intensively selected by the early elk farming pioneers and those who were fortunate enough to recognize what are seen as top sires now.

Across strains as divergent as possible

The reality is that for elk in North America, strains such as Manitoban and Rocky Mountain are not widely divergent in the core growth features of velvet antler growth and body growth as an inherent genetic difference. Spectacular individuals do, however, exist and with the advent of DNA profiling and gene discovery, strain differences that make a productive difference will emerge.

The situation is markedly different for red deer farmed in New Zealand and North America. The enormous potential of red deer to freely and successfully hybridize across widely divergent strains has meant huge and rapid progress in body size, hard and velvet antler weight and growth rates in a very short time. The variety of interstrain breeds is increasingly complex, for example: N.Z. red x European reds, N.Z. red x Eastern Europe strains, N.Z. red x elk /wapiti, N.Z. wapiti (a hybrid) upgraded with recent elk imports.

The capacity for improvement is large and much more extreme than for sheep and cattle because of the huge diversity in size and production of the red deer family. The more similar the strains, the less obvious the potential change and effect. Irrespective of the route, there is developing in all deer farming a genetic mix of features that best represents the ability of the animal to respond to a particular farming environment and management system and profit center. This genetic base is distinctive and rapidly improving.

While it would be fair to assess that the wider industry acknowledges this identified farmed strain as a superior source of antler genetics (both velvet and hard antler), it is also obvious that there are strong within-family group origins of several sires and their progeny. Bulls like Reed and RJF117 have made a significant impact in the development of the elk industry in relative terms as much as the leading AI dairy sires have made in milk production. With an initial emphasis on antler weight and body size, and in the future, as family or genetic relative information, promotion of growth rate, venison potential and farmed suitability of the strain are emphasized, the information that is genetically based becomes critical.

For elk breeders to take a next step and move

the "best to the best" mating programs and enhance the line breeding selection programs, some basic objective setting is required, for each individual and for the NAEBA association as a whole. Interestingly, the reluctance to further this has been the major hurdle to advancing industry-based genetic progress in New Zealand.

That is not to say that leading stud breeders have not made huge progress individually. The perception is different. Purchasers of breeding stock cite the inconsistency of performance of young sire stock when relocated to commercial farms as the single biggest frustration encountered in cervidae farming. A positive, objectively recorded genetic evaluation and ranking system is missing. NAEBA's well-supported, DNA-based registration and pedigree registry may well be the flagship in worldwide production recording in the near future.

BREEDING OBJECTIVES

Dorian Garrick (Professor of Animal Breeding and Genetics, Massey University) and Peter Fennessy (formerly of AgResearch Invermay) have completed an extensive review of Deer Genetics in the Proceedings of a Deer Course for Veterinarians, NZVA, Deer Branch, July 1997. In summarizing the review, the following points are pivotal (from Garrick, 1998):

- Genetic improvement enhances farm profitability by mating the "best" bulls to the "best" cows. This involves specifying accurately and objectively what is "best." This is most readily achieved by defining a breeding objective that is relevant to the particular farm. A breeding objective defines a list of traits that influence use (generally an economic or a profit basis), plus the relative emphasis for each trait on the list, i.e. the "5 year plan".

- The Breeding Value Genetic Evaluation is an index of each of the individuals for each trait in the objective. This index expresses relative merit or relative genetic contribution to the profit.

- The mating plan, within the herd or group (or industry), is the breeding structure that best utilizes the highest ranking bulls and cows for cost effective improvement, i.e. superior individuals used via reproductive technology more widely than natural mating allows.

For stud breeders, the industry trends in terms of productivity have a distinct bearing on setting an individual breeding objective. In New Zealand there are two points of focus:

Venison or Elk Meat

If the focus is to produce young sires for commercial farmers to use in venison production systems, the farmer's objective is to select sires that add profit per unit of input (pasture and feed costs, animal health and productivity) to their venison progeny.

Carcass weight gain and efficiency of conversion of pasture to venison are the measurable criteria.

Reproduction rate and mortality rate are measurable and impact severely on profitability if either is compromised through poor management. Therefore, sires that significantly improve liveweight gain, weight for age and reproduction rate become the logical choice for commercial clients.

There is only limited analytical information available on the genetic basis of superior growth. Heritability estimates show that there is modest variation in this area. It is also important to realize that true genetic gain, if selecting on liveweight gain, will be slow, but it is cumulative and permanent.

Recent work with the Pere David x red hybrids as the extreme model to track genes shows that some animals genetically have an ability to maintain high growth rates from late lactation, through the weaning check and autumn, relative to others.

A further QTL (Qualitative Trait Linkage) study also identified another genetic base to superior spring growth, i.e., some cervidae have an advantage in later growth recovery. It's possible that these two different genetic components of growth are related to rate of reaching puberty, which could in turn have some influence on spiker velvet growth. These effects will be strongly evident in red deer, and it is certainly worth tracking and identifying the exceptions in the typical, normally distributed population of growth records.

Velvet antler

The public perception of the elite velvet producers is the growth of outstanding heavy, thick and potentially multi-pointed velvet antler or hard antler as a defining feature. The champion N.Z. red deer head (Warnham Park bloodlines) for 1999 was 11.4 kg (26 lbs.). By contrast, the best N.Z. elk head yet produced weighed 18.14 kg (40.7 lbs.)

The NAEBA International Antler Competition, which now exhibits the top 18 heads, selected on a regional basis by age group and judged under a consistent set of criteria, emphasizes the genetically linked features of weight and circumference as the primary expression. Both countries are using a similar approach to define the product so that we all understand its common economic and significant features.

For velvet antler production, the definition of a breeding objective is probably an easy task, given the dominance of superior antler production portrayed. The product and sire markets pay a premium for size and style on the assumption that the quality of velvet as a medicinal product is not changed. For the sire market, high weight at an early age is also very important if it is a breeder's objective to be selling superior stock to satellite breeders within the group, or to those developing their own breeding nucleus for further sale. Heritability of velvet weight, velvet antler beam circumference and ultimate hard antler weight are all easily and productively recorded. Perhaps there needs to be a further investigation into the link between hard antler weight and velvet antler, and a standardized hard antler/velvet weight conversion.

ESTIMATED BREEDING VALUES

Antler breeding objectives can be simply summarized by concentration on producing high velvet weights at 2 or 3 years of age. Predictions of future production based on a 2-year-old weight corrected to standard days of growth (60-65 days for elk) has a high correlation to future velvet production (weight) and is definitive. (Fennessy 1989) Components of weight (beam circumference, days' growth, lack of calcification, and grade) are not individually as definitive as antler weight itself.

Genetic Trends in Velvet Antler Production
EBVs for 2-year-old antler

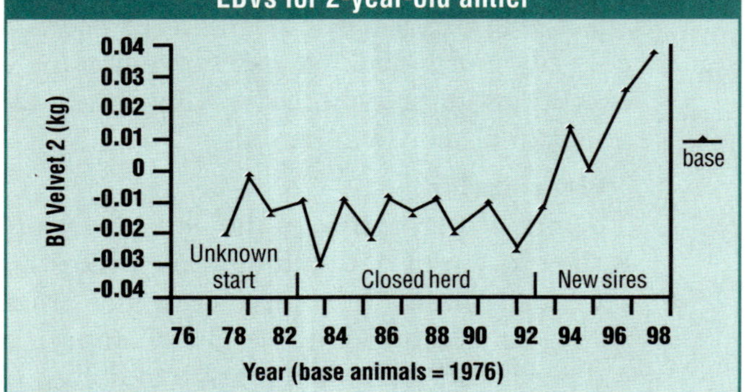

These features are important. Nevertheless, investigation into the objectivity of the CWI index and its relationship to genetic components could be worthwhile if a truly quantitative measure of calcification and style or quality parameters can be found.

It also seems profitable and logical to select for early growth and early maturing antler. Regrowth is an important and additional product when cut properly, with significant medicinal and processing advantages. Similarly, classical typical form or high-scoring non-typical hard antler styles with long tines and multi points are emerging and important traits. Beam circumference, spread, and hard antler weight will also be useful in the trophy business. The longtime established relationship (Huxley 1931) between hard antler and body weight is that antler weight within a specific bloodline increases at a relatively faster rate than body weight increases. This means that selection for antler weight should increase ultimate size and some aspect of the growth profile. Hard antler weight increases at about three times the rate that body weight increases. For red deer, a 10 kg increase in weight (say a 4% increase) increases velvet antler weight by 0.15 to 0.2 kg (5 to 8%). In farmed elk this figure could well be greater.

Velvet antler production at any nominated age has a high variability between individuals. Standard deviations for red deer at 2 and 3 years of age range typically from 15 to 35%, averaging around 22%.

Current Estimates of Values

Weaning Weight	0.20
Weaning - 12 month gain (Co-efficient of variation - 9%)	0.23
12 month weight	0.30
15 month weight	0.30
Antler weight (Coefficient of variation 22%) (2, 3, 4-year-old bulls)	0.043 - 0.45
Antler weight (Coefficient of variation 12%) (2, 3-year-olds)	0.57 - 0.47
*Estimated TB resistance	0.48

*** It is possible that the estimates are a little high because of possible effects of vigor from interstrain breeding within the small groups that have been analyzed.*

*Proportion of variation due to genetic effects - Heritability h2***

Production within any age group is a normal distribution. The challenge is to isolate environmental factors from the genetic component. Variation in elk velvet is likely to be extreme, given the low rate of performance-based culling and the spectacular increase in weights at the top end, reflected in velvet antler competitions over the past five years.

The next major step is the development of estimated breeding values (EBVs) for antler production. These values are the production ranking or expression of relative genetic merit within a herd for economic and productive traits. Ideally, the EBVs will be selected according to their respective economic value. Units are generally the units of the trait and can be centered positively and negatively around zero, or any other productive indicator point.

Features of EBV relating to antler production

- EBVs separate genetic potential from observed performance.

- EBVs correct for systematic environmental effects (e.g. sex, age of dam, cohort pressure and differences between years).

- EBVs are calculated by BLUP analysis (best linear unbiased prediction), using standard computing techniques based on sound and well-established principles of genetics.

- EBVs are expressed as relative genetic merit.

- The progeny breeding value is the mean of the parental breeding values. However, identical pedigrees may have different EBVs because of Mendelian segregation during DNA recombination and the "chance effect." For 2-year-old velvet antler weight, this can range from 0.25 - 1 kg in red deer and at least twice this in elk.

- For accuracy, the calculation of the value should include all pedigree relationships known. Looking forwards and backwards over time to include maternal and paternal half sibling data enhances calculations. Production in different herds in different environments with herd sires linked through AI and ET programs contributes to the accuracy of the genetic component of the animal's relative genetic merit.

For the elk industry, with the widespread use of AI sires, breeders and purchasers are in a strong position to demand and develop accurate and significant EBVs for velvet antler production. Progeny from the star bulls are widespread in a variety of farmed situations and with the advent of the registration system moving to record productive data, accurate estimates of true genetic potential are an immediate opportunity.

For stud breeders the real value in establishing EBVs is to allow comparison of the emerging bulls with the established, so-called leading sires and to allow better selection of the elite females. Velvet traits carried by the elk cow are extremely important, and by ranking females based on the performance of male relatives, rapid, cumulative and predictable gains can be made. The use of EBVs in the whole herd over time allows trends to be monitored, facilitates identification of poorer cows over time, and minimizes risks. (Example: Figure 1)

The figure of real genetic gain based on the Invermay red deer herd shows two factors. Initially, the herd was closed and the best producing males were mated to the "best " females. Beginning in 1975, for 15 years we made substantial weight gain improvement and, increased 2-year-old velvet weight by ~80%. Clearly, this was almost all due to management, feeding and environmental effects. EBV analysis of early data shows little permanent genetic improvement in this closed herd situation. Recent trends are strong, positive and cumulative. The objective was set to improve 2-year-old velvet antler weight through purchase of semen and young sires that were truly genetically superior, but also to objectively review cow EBVs based on progeny performance over time. Recent progress has been almost entirely due to the dam selection process, as few stud breeders have EBV values on the stock for purchase on the objective that was important to our program. We now can also take advantage of a new system, OPTIRANK, that allows the breeding program to put its own bias on the trait values important to our defined objectives and pick the sires on offer that will genetically make the quantum leap in production.

As a first step in this objective approach, gains are positive and encouraging. The future in New Zealand and in the North American elk industry looks exciting and limitless.

A.J. (Tony) Pearse, AgResearch, Deer Research Group, PB 50034 Mosgiel • New Zealand
(64) 03 489 9221 • Fax (64) 03 489 9038
pearset@agresearch.cri.nz

REFERENCES

Fennessy P.F. (1997). *Issues in Genetic improvement of Deer.* Proceedings of a Deer Course for Veterinarians, No. 6, Deer Branch, NZVA. Ed. P.R. Wilson, pp. 118-128

Fennessy P.F. (1989). *Stag selection, progeny testing and recording.* Proceedings of a Deer Course for Veterinarians, No. 14, Deer Branch, NZVA. Ed. P.R. Wilson, pp. 15-22

Garrick D.J. (1997). *Genetic principles for deer improvement.* Proceedings of a Deer Course for Veterinarians, No. 14, Deer Branch, NZVA. Ed. P.R. Wilson, pp. 15-22

Garrick D.J. (1997). *Breeding objectives for farmed deer.* Ibid. pp. 33-37

Huxley J.S. (1931). *The relative sizes of the antlers of red deer.* Proceedings of the Zoological Society, London 72: 819-864, 9-10 Feb. 2000. Produced as notes for the Minnesota Elk Conference for Veterinarians, College of Veterinary Medicine. University of Minnesota. Extension Service. Held in conjunction with the North American Elk Breeders Association 10th Annual Convention. Minneapolis, Minn.

PRACTICAL CONSIDERATIONS FOR SUCCESSFUL BREEDING PROGRAMS

(1) Early weaning: 1 - 3 weeks prior to mating to allow cows to regain condition after lactation, and maintain a slowly rising plane of nutrition for 6 - 8 weeks prior to winter. The primary advantage of early weaning and preferential feeding of calves is to increase the number of cows that conceive to the first oestrus cycle.

(2) Male preparation: Sorting, handling and transportation of breeding bulls should be done 6 - 8 weeks prior to mating to avoid any stress or impairment of performance.

(3) Mating groups should be physically separated to avoid fighting and aggression and left undisturbed if possible.

(4) Primary sires should be used for 40 - 42 days (September and October) and replaced by back up bulls. This insures against any low fertility or breeding problems, which although rare, do occur. Backup bulls should only be offered a limited mating time (November) as late born calves create more management problems than a few additional calves warrant even in times of high valued livestock.

(5) Accurate records of mating groups must be kept where genetic progress is an objective. It is now standard practise to pregnancy check to identify early and late conceived calves and to verify parentage through DNA and blood protein testing. Good records also aid in identifying any underlying management problems, e.g. poor calving performance.

(6) For social and behavioural reasons it is more practical to mate 15 month females separately from adults. However, often these groups first cycle 2 - 3 weeks later than adults, and if restricted mating is used, lower conception results can occur. A small number of adult cows (5 - 10%) run with the group can induce earlier cycling and add some social stability to young animals.

(7) Early weaning also allows the application of synchronized matings and AI programs prior to the main breeding season, so that late calves don't result from follow-up matings. However, success in these requires the development of a stress-free handling routine, good feeding and application of a detailed timed preparation program.

(8) After mating, bulls should be removed and be maintained in a separate group to allow preferential feeding to recover from rut activities and the huge weight loss. This is an important protection of an investment in elite stock and can add an advantage of improving velvet antler growth in the following season.

(9) Post mating, the breeding female herd requires good maintenance feeding as winter approaches. Major diet changes must be introduced slowly to avoid any stress/disease risk and effects on embryonic viability, or acidosis/feed related illness.

Good production, genetic progress and profitability in breeding relies on:

- having clear objectives.
- selecting appropriate elite stock.
- recording objective performance records and animal identification.
- planning a program in advance in keeping with the objectives.

These combined underwrite a long term commitment to herd improvement. To support that commitment, a wider view of the industry and its potential must be taken when considering objectives.

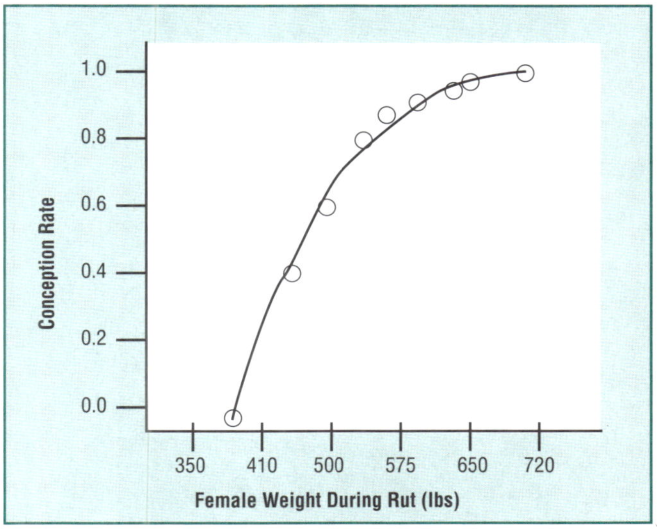

Conception Rates (1.0 = 100%) of Female Elk in Relation to Liveweight during Rut Season

Chapter 1 **Nutrition**

THEORY OF FEEDING & NUTRIENT REQUIREMENTS

Two key points:

1. An elk is **not** a small cow, or a long-legged sheep, or a deer!
2. Feed production and use, and elk feed requirements, vary dramatically with the seasons.

Elk are ruminants, like cattle and sheep, and therefore can make use of fibrous feeds because of the microbes which aid digestion in their rumen. However, ruminants vary in their selection and use of fibrous feeds. Fallow deer and white-tailed deer are very selective eaters, choosing to consume only the most succulent and digestible plants and plant parts. Elk will eat and digest feeds higher in fibre content, but will leave plant parts which bison, sheep or cattle will readily consume.

All deer are highly seasonal in their eating habits. Voluntary intake decreases by as much as 40 - 60% during winter, as compared to the spring and summer peak. This is a considerable advantage in temperate climates, where low-cost pasture growth peaks at the same time as the feed intake of farmed elk.

FEED REQUIREMENTS

The list of feed nutrient requirements for any animal is long and complex. However, these nutrients can be classified in a short list of "major" requirements and a longer list of "minor" ones. "Major" nutrients include: water, crude protein, energy, calcium, phosphorus, salt and fibre. "Minor", but essential, nutrients include: magnesium (Mg), potassium (K), sulphur (S), iron (Fe), copper (Cu), iodine (I), cobalt (Co), manganese (Mn), selenium (Se), molybdenum (Mo), chromium (Cr), fluorine (Fl), nickel (Ni), vanadium (V), tin (Sn), arsenic (As), vitamins A, D and E and essential fatty acids.

It is important to have a basic knowledge of each nutrient's function and to know which feeds provide these nutrients:

Crude protein (CP)

Protein is needed for maintenance, muscle and bone growth and for tissue repair. Elk that are rapidly growing or lactating and bulls recovering from the rut have higher protein needs than when at maintenance or in early gestation. To some degree, greater protein requirements can be met by increased intakes, however, the percentage of protein in diets designed for these animals is usually increased as well. Maintenance rations can be designed at 10 - 12% crude protein, whereas rations designed for lactation or antler growth should provide 14 - 18% crude protein and growing rations from 16 - 20% crude protein. If a single diet is fed to all gender and age groups, a 16 - 17% crude protein level is optimal. Total dietary protein content must be determined (roughage + concentrate) to follow these general guidelines.

Like other ruminants, elk can make use of non-protein nitrogen sources to satisfy part of their protein requirements. Rumen microbes transform non-protein nitrogen into microbial protein. The microbial protein in turn, as it passes out of the rumen with the digesta, becomes a protein source available for digestion in the lower gastro-intestinal tract. Urea may be used to replace a portion of true protein nitrogen in an elk ration with no adverse results. In practice, total amounts of urea or ammonia (non-protein nitrogen) that could be used in a ration depends upon the roughage source.

Feeds naturally high in protein include legumes (alfalfa, clover, trefoil) and vegetable protein supplements such as roasted soybeans, soybean meal, canola meal and linseed meal. The crude protein content of a commercial feed will be displayed on the feed tag as will the proportion of non-protein nitrogen (% ECP/NPS; estimated crude protein from non-protein sources). In general, feeds containing non-protein nitrogen are less expensively priced.

Energy

Energy is derived from the digestion of several compounds, including carbohydrates and fat. Excess dietary protein can also be used as an energy source but is a costly practice.

Next to water, energy is the most essential nutrient, required in the largest quantity in the diet. Energy is required for all bodily functions including maintenance, growth, and activity. Lack of energy will seriously compromise health, production and reproduction.

Energy requirements of elk are commonly expressed as metabolizable energy (ME).

Energy, Protein and Ration Requirements for Larger Elk Cows

	Rut (Sept 1-Nov 15)	Winter (Nov 15-Apr 1)	Spring (Apr 1-May 15)	Summer (May 15-Sept 1)
Cow Weight (lbs)	627	638	665	616
Gain (lb/d)	0.05	0.00	0.10	0.10
Milk L/d				3.00
Energy Requirements				
Maintenance	46	46	45	45
Gain	2	0	4	4
Gestation	0	3	10	0
Laction	0	0	0	41
Total required MJ of Me(= feed units)/day	**48**	**49**	**59**	**90**
Protein Requirements				
Maintenance	268	281	317	401
Gain	12	0	23	23
Laction				
Protein (g/d)	**280**	**281**	**340**	**634**
Efficiency of use	0.45	0.45	0.45	0.45
Crude Protein required (g/d)	**626**	**630**	**764**	**1420**
Recommended Ration				
Components				
Oats lb/day	3.3	2.8	2.2	4.4
Roughage lb/day	7.7	9.0	12.0	15.0
Min/Vit Sup lb/day	1.0	1.0	1.0	1.0
DMI (Dry Matter Intake)	12.0	12.8	15.2	20.4
Energy (MJ/d)	**48**	**50**	**59**	**80**
% Crude Protein in ration	14	12	14	16
Ration CP (g/d)	**743**	**807**	**986**	**1427**

Table 1 Adapted from Haigh & Hudson 1993

Metabolizable energy (ME) quantifies the amount of energy available to the tissues after subtracting energy losses in digestion and metabolic conversions. Total energy requirements are estimated by totalling the appropriate requirements for maintenance, gain, gestation, lactation or velvet production. Metabolizable energy units may be expressed in kilocalories (kcal), kilojoules (kJ) (1 kcal = 4.186 kJ) or megajoules (MJ), (1000 kJ = 1MJ). For purposes of clarity, we refer to 1 MJ of ME as a "feed unit". For example, energy requirements for a 260 kg. elk female during lactation are estimated to be 80 feed units per day.

Elk will voluntarily adjust their intake to meet their energy requirements to the extent that maximal gut fill capacity permits. In a study with white-tailed deer fawns at Pennsylvania State University, minimal feed energy content required was 9.2 feed units; feeds of poorer quality than this could not be fed as a maintenance diet. This corresponds to an overly mature grass hay or good quality oat straw.

Energy requirements of all deer species follow a seasonal pattern. In winter, deer decrease their basal metabolic rate by 40 - 60%, thereby significantly reducing their energy requirements for maintenance. Pastured or free ranging deer increase their maintenance energy requirements by 60% in comparison to penned animals due to their increased activity.

Elk: Average Energy Requirements

Calf	Weight	Rut	Winter	Spring	Summer
Male	210 lbs	35	36	60	75
Female	190 lbs	32	33	42	50
Adult	Sept 1				
Male	1000 lbs	50	62	66	68
Female	550 lbs	46	42	49	82

Table 2 Elk: Average Energy Requirements in "Feed Units" (MJ of ME) "Weight" as of Sept 1, Pre-Rut

As ruminants, elk can meet much of their energy requirements from the digestion of fibrous feeds such as hay, haylage and pasture. Smaller species such as fallow deer have less rumen and fibre digesting capacity than do larger deer such as red deer and elk. Fallow and white-tailed deer require more concentrate (grains and supplements) and less fibre in their rations in comparison to elk and red deer.

Energy fed in excess of maintenance and production requirements will be stored as body fat. Fat deposition in deer is remarkably sensitive to photoperiod. In a Michigan study with white-tailed fawns, well fed fawns essentially stopped growing by mid-November. In contrast, fawns fed a restricted diet lost weight during the autumn. Regardless of dietary treatment, all fawns deposited fat, starting in early autumn and peaking by mid-December, although fat reserves in the well fed fawns were twice those of the nutritionally deprived fawns.

Concentration of dietary energy is the strongest determinant of bodyweight gain. Cows must not be overfed during the winter, or calving problems will increase. Rations should be designed for

Feed Value of Common Feeds

Feed Type	Dry Matter %	Available Energy MJME(= feed units)/ KgDM	Crude Protein %DM
Hay			
Meadow hay			
good - green leafy	86	9	15
poor - brown stemmed	86	7.5	12
Alfalfa hay			
green, leafy / brown, stocky	86	9.5 / 7.5	20 /12
Cereal hay - green / golden	86	8.5 / 7.5	9 / 6
Clover hay - leafy / stalky	87	9.5 / 7	16 / 8
Straw			
Barley straw			
leafy, some grain / golden	40	7	4.5
Oat straw - average	87	6.5	3
Silage / Haylage			
Pasture - leafy, direct cut	18	9.5	14
stalky, direct cut	22	8	12
leafy, wilted	28	9.5	14
stalky, wilted	32	8	12
Corn - whole plant	32	10	7
Green feed - leafy	17	12.5	12
early flowering	20	9	7
Italian Ryegrass	15	12	18
Grain			
Barley - good sample	87	12.5	10
pinched grain	87	11.5	11
Wheat - good sample	87	13	12
pinched grain	87	11.5	12
Oats - good sample	87	11.5	12
pinched grain	87	10	12
Corn - good sample	87	13.5	9
Peas	85	13	30
Lupins	85	13	30
Horticultural Crops			
Apples	18	11	3
Carrots	13	13	7
Cabbage	12	13	14
Cauliflower	12	13	14
Potatoes	23	12.5	10
Pumpkin	25	12.5	16
Squash	25	12.5	16
By-products			
Bran	87	9.6	14.5
Brewer's grains	35	10	25
Apple pomace (fresh)	25	8.44	6
Prepared feeds			
(note: check specification of individual products)			
High energy, fat fortified dairy rations	86	13.5	16
Grain based pellets	86	12.5	14
General all purpose stock pellets	86	10.5	12
Pasture			
Grass / white clover			
Spring	12	13	25
Summer	16	10	20
Autumn	15	11.5	18
Winter	15	11.5	18

Table 3 Feed Value of Common Feeds

maintenance only. Calves should at least maintain their weight over winter.

Rations for immature elk calves can be designed to ensure winter weight gains but need to be assessed from an overall economic status. Weaned elk fed a marginal diet during their first autumn and winter will demonstrate compensatory gain on pasture the following summer. However, despite compensatory gain on summer pasture, elk fed a marginal diet during their first winter can never catch up in bodyweight to elk fed a well balanced diet throughout their growth period.

Bulls need to enter the rut in good condition since bodyweight losses as high as 30% are typical. Failure to build up adequate energy reserves before rutting season begins can compromise reproductive performance. Bodyweight reserves need to be built back up after the rut, after which animals can be fed a maintenance diet until antler growth begins.

Feeds relatively high in energy content include the cereal grains: corn, oats, barley and mixed grain. Corn contains the most energy of any of the grains. Vegetable meals such as soybean, linseed and canola meal also contain high levels of energy but are also high in protein value. Oil seeds such as soybeans and canola are high in fat and therefore high in energy content.

Elk are very susceptible to rumen acidosis caused by overfeeding of grain. Grain feeding levels need to be changed slowly to allow the rumen microbes to adapt. Similarly, changes in feedstuffs need to be introduced gradually. Elk need to be closely monitored during winter to ensure they are still consuming adequate forage to maintain rumen function.

Calcium, Phosphorus and other Minerals

The metabolism of calcium and phosphorus is interrelated. These two minerals are considered simultaneously when designing or evaluating a ration. Both minerals are important constituents of bone. Calcium is also essential for nerve impulse transmission, muscle contraction, secretion of some hormones and enzyme activation. Phosphorus is an essential constituent of enzymes involved in fat, carbohydrate and energy metabolism.

The actual amount of calcium and phosphorus required by elk increases greatly during growth, lactation and antler growth. A minimum of 0.7% calcium and 0.4% phosphorus is suggested for elk rations. Of equal importance is to ensure a calcium : phosphorus ratio of at least 1.5:1. How wide a ratio is tolerated by elk has not been researched although other ruminant species can tolerate calcium : phosphorus ratios as high as 5:1.

Calcium levels are relatively high in hay, especially legumes, but phosphorus content is quite low. Grains are the opposite, being relatively high in phosphorus and low in calcium. Rations comprised of hay/haylage and natural grains and oilseeds or meals will not provide enough minerals and vitamins. Hay analysis is required to decide which type of mineral/vitamin supplement is most suitable. Access to browse (twigs and leaves) improves natural mineral intake on pastures.

Commercially prepared mineral-vitamin supplements are extremely variable in both nutrient content and price. Follow these general guidelines:

- Choose either a 2 : 1 (2x as much calcium as phosphorus), or 1 : 1 (equal calcium and phosphorus levels) depending upon forage analysis and grain feeding levels. Avoid low calcium, high phosphorus minerals as they tend to be both unpalatable and expensive.
- Next check the levels of vitamins A, D and E and of selenium. All of these nutrients are expensive and will add to the cost. Selenium and vitamin E levels should be increased in late winter/early spring; feed vitamin levels decrease during storage and will be quite low by this time. Also vitamin E and selenium are important to prevent nutritional myopathy and to defray the stress imposed by handling. Some cases of deer unthriftiness have also responded to increased dietary vitamin E and selenium levels (Knox et al. 1987). Supplemental vitamin E feeding can decrease during pasturing; fresh grass is a good source of natural vitamins.
- Check trace mineral levels for copper, zinc and manganese. Elk, unlike sheep, require supplemental copper. A feed or mineral designed for sheep will not contain any added copper.
- Check the feedtag for salt levels (if present, will be listed as sodium or Na). Some minerals include salt, others don't. Additional salt will be necessary if no other source is provided, particularly in spring and summer. Elk will not voluntarily consume much salt in winter.
- Lastly, decide if you need to purchase a mineral (mineral-vitamin supplement) or a premix (mineral-vitamin supplement containing higher levels of vitamins and trace minerals). If you purchase a commercial concentrate (i.e. prepared ration that already had minerals and vitamins added), a mineral will be sufficient. If, however, you are supplementing with straight grains, a premix is preferable.

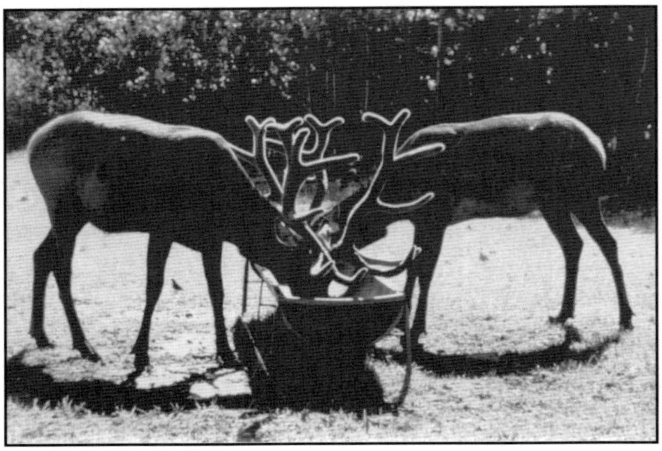

Fig. 1 Elk bulls at a well-constructed feeder. Remember to keep the feeder clean.

Probable Nutrient Requirements for Elk (dry matter basis) Whole Ration

Total / Whole Ration (DM)			GROWTH			GESTATION		LACTATION	
	Maintenance	Velvet	3 - 6 months	6 - 9 months	9 - 18 months	12 - 24 weeks	24 - 36 weeks	0 - 6 weeks	6 - 12 weeks
Feed Units (MJ of *ME*)	45	66	35	36	50	46	49	70	82
Crude Protein %	7 - 10	16	18 - 20	16 - 18	12 - 14	14 - 16	12 - 14	14 - 16	12 - 14
Fat % Minimum	3	3	3	3	3	3	3	3	3
ADF % Min - Max	25 - 45	25 - 45	16 - 35	20 - 40	20 - 45	20 - 45	20 - 45	20 - 40	20 - 40
Calcium %	0.35	1.40	0.60	0.55	0.50	0.50	0.60	0.70	0.60
Phosphorus %	0.25	0.50	0.30	0.30	0.30	0.40	0.40	0.40	0.40
Potassium %	0.65	1.0	0.65	0.65	0.65	0.65	0.65	1.0	1.0
Magnesium %	0.20	0.40	0.20	0.20	0.20	0.20	0.20	0.25	0.20
Salt %	0.15	0.20	0.20	0.20	0.20	0.20	0.20	0.20	0.20
Copper (ppm)	15	15	15	15	15	15	15	15	15
Molybdenum (ppm) [1]	0.30	0.30	0.30	0.30	0.30	0.30	0.30	0.30	0.30
Manganese (ppm)	40	40	40	40	40	40	40	40	40
Zinc (ppm)	50	50	50	50	50	50	50	50	50
Iron (ppm)	50	50	50	50	50	50	50	50	50
Iodine (ppm)	0.50	0.50	0.50	0.50	0.50	0.50	0.50	0.60	0.60
Cobalt (ppm)	0.15	0.25	0.15	0.15	0.15	0.15	0.15	0.20	0.20
Selenium (ppm) [2]	0.30	0.30	0.30	0.30	0.30	0.30	0.30	0.30	0.30
Vitamin A (IU/kg)	3000	5000	4000	4000	3000	5000	5000	5000	5000
Vitamin D (IU/kg)	600	1000	800	800	600	1000	1000	1000	1000
Vitamin E (IU/kg)	30	40	30	30	30	40	40	40	40

1. Molybdenum 0.3 to enhance copper utilization
2. Selenium availability is highly variable geographically - test your soils and feeds

ME = Metabolizable Energy

Table 4 Total/Whole Ration Nutrient Requirements for Elk, *see References overleaf.*

REFERENCES FOR NUTRIENT REQUIREMENTS

Drew, K.R., and Suttie, J.M., *Quality Meat from Farmed Deer*, Proceedings of the New Zealand Deer Farmers Association Annual Conference, No. 7, Technical Addresses, 1982.

Fennessy, P.F., Moore, G.H., Corson, I.D., *Energy Requirements of Red Deer.* Proceedings of the New Zealand Society of Animal Production, 41: 167-173, 1981.

International Minerals and Chemicals Technical Committee, *Trace Elements in Animal Nutrition*, IMC, 1970 Edition.

National Research Council
Nutrient Requirements of Dairy Cattle, National Academy Press, 1988.

Olson, John C., *Nutritional Comparisons: Mule Deer and Other Ruminants, Recent Advances in Wildlife Nutrition*, Colorado State University Technical Symposium, Fall 1969.

Varner, Larry W., *Cervid Nutrition and Supplementation*, Purina Mills, Inc., Gonzales, Texas, 1995.

Watkins, Bruce, *Nutrition of Rocky Mountain Elk, Part 1: Requirements for Energy, Fiber, and Protein*, Journal of the North American Elk Breeders Association, Spring, 1991.

DEVELOPING A FEEDING STRATEGY

To develop a feeding strategy for your elk, start by looking at your basic assets:

- livestock
- land, both fenced pasture and crop-growing land
- buildings and facilities
- available crops
- available human resources

Your livestock, your elk, are classic "mixed feeders". This means that they both graze and browse, selecting the most nutritious grasses, legumes, forbs, leaves, buds, fruits, and twigs of an endless variety of plants. The shape, structure and function of their teeth and digestive systems allow them to successfully use this great variety. Elk also have strong seasonal peaks and lows in appetite and nutritient needs, as shown in Table 2 on page 124. Fortunately, their appetites, growth rates and lactation all peak at the same time as pasture production peaks in temperate climates.

Elk, with careful management, will do quite well in a feedlot situation. However, the cost of growing pasture will almost always be substantially less than the cost of stored feed. Try it yourself - even calculating initial land purchase and fencing costs, spread over 20 or 25 years, plus annual maintenance and fertilizer, and assuming a conservative figure of five tonnes of dry matter (DM) consumed per acre, pasture should cost less than 2.2¢ /lb., or $50/tonne. Compared to the cost of other feed, to maximize pasture area and production during the 200 days possible is the best start to our feeding strategy. Pasture management is covered in greater detail in later in this chapter.

The next step in our feeding strategy is to establish a supplementary feeding program. Generally, the most available and suitable supplementary feeds for elk are hay, haylage, silage, corn, oats and barley. Hay, haylage and silage as the roughage component, is the most important part of the ration, and must be of good quality. Elk are selective eaters, and can easily separate out the rich leaves, leaving coarse stems behind. Choose hay that is fine stemmed, full of leaves, green in colour and sweet-smelling. Poorer quality may need to be chopped and mixed with better quality feed.

Baled haylage and silage can be exceptionally palatable and excellent feed for deer. Ensiling does increase the breakdown of true protein into non-protein nitrogen. Rations based on ensiled feeds should not contain any additional sources of non-protein nitrogen (urea or ammonia). Short-chopped silage made from grasses, cereals and legumes, piled and tarped, is probably the best quality and lowest cost supplemental feed for elk if it is tested and balanced with grain, minerals and vitamins to target requirements.

Large round bales, while convenient to handle and feed, often are poor choices for deer rations. To store properly, hay put up in large bales must be drier. Consequently, more leaves are lost during harvest, handling and storage than with small bales. Also, many round bales are stored outside with inadequate covering, resulting in molding and spoilage of significant amounts of hay. Since feed intake is already depressed during winter when hay feeding is the greatest, relying on poor quality, moldy roughage is detrimental to performance. Quality of large round bales may be improved by wrapping them with plastic to create haylage, or baleage, however, this is not as good a solution as piled silage. Dairy farmers have noticed that every time you change silos or cuttings within a silo, the cows go slightly off feed and the result is a drop in milk production. Using baleage from different bales that vary in dry matter and nutrient content is like feeding forage from a different silo everyday. In addition, quality is more difficult to assess when wrapped in plastic, and entire loads of bales are being delivered.

Moldy feeds of any kind must be avoided. Not only does mold reduce palatability, the mold spores ingested and inhaled will tax the animal's immune system, leaving it more prone to infections.

Purchasing either commercial rations and protein supplements or cereal grains and vegetable protein sources is usually a matter of individual preference. Commercially prepared feeds may be convenient and offer an "all-in-one" package. Commercial feeds can either be custom formulated (according to your specifications or recipe) or purchased as standard stock. Usually standard feeds will be less expensive although you must decide whether to purchase a beef, dairy, sheep or, in some cases, a specially prepared elk ration. Sheep feeds are unsuitable because they lack added copper. Dairy feeds often are more highly fortified with vitamins and trace minerals than beef feed. Dairy feeds also are formulated for a larger protein : energy ratio than beef feeds. The ratio of protein : energy will influence growth and fattening characteristics.

Commercial feeds can be formulated as pellets or as textured (sweet) feed. A completely pelleted ration prevents any sorting. Some elk have proven really adept at sorting through sweet feed, leaving the finer grained protein feeds and pelleted minerals at the bottom of the feeder. Pelleted rations vary in pellet size and density; it is important to find one that your elk will readily consume. Once you've found a feed company and product that your deer readily consume and perform well on, think carefully before changing!

A disadvantage of a commercially prepared feed is that it is impossible to know which feeds are included in the ration unless the company guarantees a fixed formula. Generally, feed formulas vary depending upon ingredient prices. Palatability, therefore, might change from batch to batch. On the plus side, commercially prepared feeds will be well formulated with precise control over ingredient inclusion rates and uniform mixing of ingredients; often more difficult to achieve on-farm.

Elk farmers who choose to feed straight grains or protein supplements gain flexibility in their feeding program at the expense of convenience. With two or three sources of feeds on-farm it is easy to custom mix batches to suit the particular needs of each management group. On the negative side, feeds mixed on-farm may not be as uniform as a commercially mixed feed. Natural grains and oilseed or protein meals need to be fortified with minerals and vitamins. Unfortunately, finer grained mineral and protein feeds often settle out of a ration and are left on the bottom of the feeder. Options include: adding molasses to bind the mineral; pelleting the mineral and protein feeds together; or purchasing the mineral in a pelleted rather than loose form. Offering minerals only on a self-fed free choice basis is not a good practice.

When we say "supplemental feeding", the winter season always comes to mind. We tend to think that our livestock can always get by on pasture from May to October, without giving much thought to the quality of that pasture. Early in the season, rapidly growing vegetation may be up to 80% moisture. An elk fills its rumen but doesn't get enough nutrients to meet its daily needs. Later in the season, elk may be grazing "belly-deep" grass which is so high in lignin and other indigestible fibres that it becomes unpalatable and nutritionally inadequate.

As was mentioned before, pastures, made up of a variety of grasses, forbs, legumes, shrubs and trees are always the least expensive source of elk feed. Good management can provide adequate nutrition from grazing and browsing for much of the year. However, when even our best efforts fall short of nutritional requirements, we need to provide supplemental feeds.

Our aim as producers of farmed elk must be to provide all of the nutrition our elk require in the feed we make available. These requirements must also be provided in a palatable and reasonably priced form. This is quite a balancing act, one that requires our regular and continued attention. Feed quality and requirements change substantially with the seasons, local conditions (including soil, water, terrain type and climate) and type and source of the feed. Feed analysis is an essential tool in balancing rations. If we buy all of our feed, we can often get away with requiring the suppliers to do all the analysis and balancing according to our instructions. If we produce our own feed, we may become quite skillful at assessing feed quality and adding the nutrients we are confident will be lacking. Accumulating this kind of knowledge requires a lot of feed analysis and consultation with neighbours in the early years of our management of a piece of land and a herd of elk.

The most important supplemental feed we provide is forage. We store and feed this as hay, haylage, baleage or silage. These names mostly differentiate stored forage on the basis of moisture content, ranging from 15% to 65% water content vs. dry matter. Where weather conditions allow adequate drying, hay is generally the least cost/least work alternative.

Because we tend to underestimate the value of hay, especially if we make it ourselves, and we generally feed it on the ground, wastage is one of the primary concerns. To minimize these wastage losses, we use various kinds of hay feeders. Photos 1, 2 and 3 show three different types.

Photo 1 Common hay feeder

Photo 1 shows one of the most common hay feeders, a "bale ring". This is a relatively poor choice for feeding elk, for two reasons. As the animals reach in and grab mouthfuls of hay, they tend to "dribble" bits out of the feeder and the edges of their mouths. These bits - often the most nutritious leaf material - fall to the ground around their feet, where they are trampled and soiled to the point of being unpalatable and consequently wasted. The second problem with this feeder arises when bulls start growing antlers. Unless they are tall enough to reach over the ring and the feeder is full enough to allow them to reach the hay, their antlers may be damaged by contact with the steel ring. If they injure their antlers, their velvet value and their voluntary feed intake both drop.

Photo 2 Better design for a hay feeder

Photo 2 shows an improved design for a hay feeder. The "tombstone" sides allow better access to the hay, with less potential for antler damage. However, wastage is just as bad as with the first design. In addition, the possibility exists that an animal will attempt to jump into the ring if hay is low, and may become wedged between the "tombstone" uprights.

Photo 3 Best design hay feeder

Photo 3 shows the best design. The hay is cradled in the rings, and therefore always sits at the bottom where the elk can reach it. Antlers are quite safe, as the bulls can tilt their heads and carefully avoid the steel frame. As the elk pull and dribble, most of the valuable bits drop into the bottom frame. This prevents most trampling and soiling wastage.

If forage is stored at a higher moisture content (18% +), it must be protected from spoilage. This is accomplished by restricting airflow with plastic wrap, concrete, or hard packing of piled forage. The limited amount of air (oxygen) left in the wrapped or sealed forage permits a limited fermentation process that actually increases palatability and quality for feeding elk. If the forage is chopped and piled or stored in large plastic tubes, it is typically stored at around 45 to 60 % moisture, and referred to as silage. If the forage is cut and conditioned, left to wilt for about 24 hours, then baled without chopping and wrapped with plastic, it is referred to as "haylage" or "baleage"

Haylage is typically stored at 20 to 45% moisture. Its palatability is excellent, but nutrient availability is less than with chopped silage. Both silage and haylage offer advantages and disadvantages compared to hay. The higher acceptable moisture contents allow forage to be safely stored much more quickly than with hay.

Photo 4 Low cost, effective silage feeding strategy

This is an obvious advantage when high humidity or frequent rainfall is a problem. Palatability and nutrient availability is superior in the softer and fermented higher moisture products. However, costs for storage equipment are usually higher for silage or haylage than for dry hay.

Silage for elk is typically chopped grass and legumes or cereal crops. The most cost effective and successful approach may to hire custom harvesters to cut, chop and store all one season's production in a few days and store it in large plastic tubes as shown in photo 4. As you can see, the plastic fence and steel apparatus keeps the elk off the sides of the tube. They feed at the face, and wastage is minimal on frozen ground. Silage may require considerable work and equipment to feed if stored in a large pile or pit. A loader is used to fill a forage wagon or TMR (total mixed ration) wagon, and the silage augured into feed bunks or onto the

Photo 6 Feeding on the ground, on fresh clean snow

ground. Feeding on the ground has associated hazards. Parasites and disease organisms are consumed with the feed. It is essential to feed in a different area each day, and avoid any heavily soiled areas. Try to put the feed on fresh grass or snow each day.

No matter what storage method is chosen, elk require good quality "YTRG" (young, tender, rapidly growing) forage. Chopping does not make coarse, poor quality forage into excellent feed. However, chopping does improve nutrient

Photo 5 Bale shredder

availability and voluntary feed intake.
A bale shredder, as shown in photo 5, will do an excellent job of preparing long hay or haylage. The chopped feed, like silage, can be blown into feed troughs or onto the ground. This is particularly useful when feeding alfalfa, as elk are very good at separating the leaves and leaving the stems behind. Chopping will encourage them to clean up most of the forage.

Photo 7 "Elk-Kube" TMR in feed trough

When feed testing identifies shortfalls in nutrition targets, supplements are fed in the form of grain or pellets. Grains provide energy and phosphorus, plus a variety of other nutrients that may not be present in sufficient quantities in forage.

Prepared TMR's, such as the "Elk - Kubes" shown in photo 7, are an easy but expensive way for new or small-scale producers to deliver a high quality balanced ration to their elk. If fed in open troughs, these cubes should be fed daily if any rain is expected. They may also be fed ad lib in creep feeders, but the producer loses the value of interaction with the animals.

This interaction between the farmer and his or her elk is an important part of good stockmanship and livestock marketing. In fact, other than provision of nutrients identified as being deficient in the forage, this "taming" ritual is the most valuable part of grain feeding. The grain is a treat for the elk, and they look forward to being fed. The person who feeds becomes a trusted individual, and further training allows the elk to accept the presence of new people as well. Nothing sells elk like the experience of going with the herdsman to the pasture, calling the elk and having them run up close and comfortable with the prospective buyer.

Oats are the best grain to use for supplemental or treat feeding. Their unique digestive characteristics make them safer than other grains. If the farmer is always careful to slowly introduce the elk to any new feed, and to increase the grain portion of their ration very slowly, that feeding should never cause problems such as grain overload. An example formula for a good supplemental energy, mineral and vitamin pellet is shown in the table below. This pellet is designed for farmed elk in the Prairies and Parkland of NW North America, to be fed in addition to fresh, green leafy legume/grass hay or pasture. Ideally, these pellets would contain 60% mixed grains (barley no more than half that), 30% peas, 2% flax and the remainder being added mineral mix. Add molasses at about 10 litres/tonne to enhance palatability.

Photo 9 Large creep feeder

SUPPLEMENTAL PELLET FORMULA

Crude Protein	16 %	Copper	250 mg/kg
Crude Fat	5 %	Cobalt	0.3 mg/kg
Crude Fibre	10 %	Zinc	800 mg/kg
Calcium	2 %	Iodine	6 mg/kg
Phosphorus	1.5 %	Iron	525 mg/kg
Sodium	0.6 %	Manganese	1040 mg/kg
Magnesium	0.4 %	Selenium	4 mg/kg
Potassium	1.0 %	Vitamin A	80,000 IU/kg
Sulphur	0.4 %	Vitamin D	8,000 IU/kg
Chloride	1.5 %	Vitamin E	400 IU/kg
		mg/kg = ppm	

Feed these pellets mixed with oats (50% oats / 50% pellets).

Feed bulls 4 - 6 lbs of the oats/pellets mix per head per day all year around, as long as they eat up all you give them.

Feed cows 2 - 5 lbs from early May to November, depending on pasture quality and availability, then drop slowly to 1 - 2 lbs / head / day in December to April.

Feed calves 4 - 8 lbs from August of their calf year through to December of the next year when they reach about eighteen months of age, then treat them as adults. Remember that these feed levels constitute luxury feeding. If the regular forage portion of their ration is as good as it should be, you needn't feed these levels.

Always make any changes in ration content or amount quite slowly and carefully, over a period of one to two weeks. In addition, keep a trace mineralized salt block available to your livestock all year around. Typically, elk will only use this salt lick extensively in the summer.

Each different area or farm and each different feeding program may need slight adjustments made to the formula described above in order to balance the levels of nutrients in your pasture or stored feed. Begin this process by sampling your soil, looking both at macronutrients like NPK, and micronutrients as well. Consult your local agronomist, and farmer neighbours as well for advice at this stage, remembering that elk are not exactly the same as cattle or sheep in their

Photo 8 Rugged and efficient feed trough made of steel pipe and conveyor belting

requirements. The next step is to test your feed - including your growing pastures and all of your stored or supplemental feeds. If you use a great variety, test different batches regularly and balance your supplement accordingly. Calcium, phosphorus and selenium, in particular, must be balanced for each area and diet. Copper should be increased if no browse is available.

Always make sure that adequate feeding space is provided for each of your elk. Social pressures will exclude lower-ranking elk from feeding areas if space is inadequate. They all tend to sleep at the same time, and feed at the same time, so allow at least 2' (60 cm) of feed bunk or trough space per elk in that feeding area.

Supplemental feeding is a regular part of farmed elk management. Done right, it makes all the difference in producing quiet, excellent quality elk livestock and elk products. Be sure to work with all your local sources of information and knowledge, and don't be afraid to ask questions or make suggestions. Elk are new livestock - we're all still learning!

Elk calves begin to utilize forages at about 20 to 40 days of age. At this stage, they will often eat soil, apparently to begin the growth of rumen flora (digestive bacteria). Give them fresh topsoil (preferably forest topsoil) in a pile near where their mothers are fed or watered. This will help avoid the situation where they pick up disease organisms from contaminated soil. At about 80 to 90 days of age, calves will begin to eat grains if they have access to these with their mothers. Spread oats plus supplement pellets out in a long line in feed troughs or on clean grass, in a different spot each day. By weaning time at 100 days, the calves will adapt more easily to a diet of forage and grain without milk.

Formulating the most appropriate rations involves many choices and decisions. No single approach will be appropriate for every herd. As an owner, you need to find a feed company and nutritionist you trust and feel comfortable with. It is important that you as an owner communicate your desired goals to your nutritionist and continue to seek their advice in fine tuning rations to meet your herd's needs. Nutrition is a vital cornerstone to healthy, productive deer. As a team you, your veterinarian and nutritionist can help to optimize the health, productive and reproductive status of your herd.

Body Condition Scoring

A helpful technique for monitoring how well your feeding program is working is Body Condition Scoring (BCS).

BCS involves the establishment of a numerical range using a 5-point scoring system where an extremely thin cow scores a "1" and an obese cow scores a

"5". Scoring is done both visually and by feeling the amount of flesh covering the ribs, rump and tail areas.

Do not consider frame size, state of lactation or gestation and health when assigning a condition score. These factors are considered when interpreting condition scores, but not in determining a score.

In initial scoring, the "hands-on" approach will be most accurate, especially during the time when the animals have a heavy winter coat. With practice one should be able to make mental comparisons using only visual clues. Body areas to evaluate:

Ribs
Spine
Pelvic area
Brisket
Muscles of hindquarters & shoulders
Tail head

BCS is only intended to be one tool of several used to evaluate and monitor your herd. It should assist in monitoring your herd health and nutritional programs.

An example of a possible use would be to have females score between 2.5 and 3.5 at calving and 3.0 to 4.0 for breeding.

1.0 BCS: Spine - very prominent above the ribs; each vertebrae distinct. Transverse processes - visible and prominent. Hooks - very sharp; no flesh covering. All Ribs very visible - no flesh on palpitation. Thurl - sharp angle. Tail head region - concave with no flesh. Withers - very sharp.
2.0 BCS: Spine individual processes are evident. Spine is evident above the ribs. Transverse process visible. All ribs are visible. Hooks prominent. Thurl is sunken. Tail head is sunken with only minor flesh covering. Withers is prominent.
3.0 BCS: Spine smooth ridge - the individual processes are not evident. Transverse process is

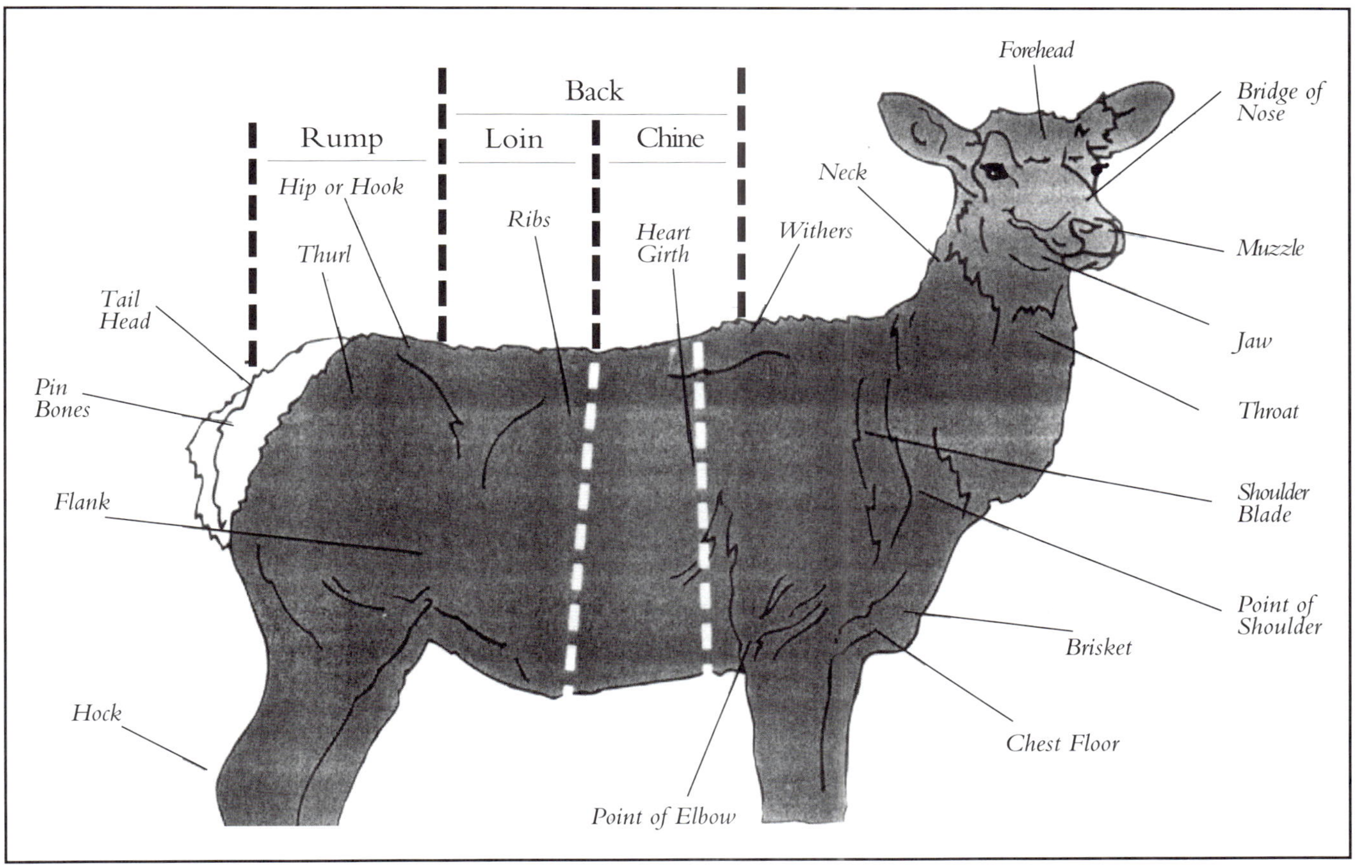

Fig. 2 Terminology of Elk Anatomy for BCS. Thanks to Dr. Glen Zebarth for the preceeding information on BCS.

visible. Hooks - smooth. Last two ribs barely visible. Thurl - very slight depression. Tail head region is flat and smooth. Withers - smooth V.

4.0 BCS: Spine is flat - no processes visible. Transverse process is barely visible and barely palpable. Hooks - rounded with fat. Ribs - none evident. Thurl - flat. Tail head rounded with tail starting to form a dip in fat. Withers - rounded.

5.0 BCS: Spine buried in fat. Hooks buried in fat. Thurl is rounded. Tail head is buried in fat. Withers very rounded. Brisket is wide and deep with fat. Fat build-up under abdomen.

As previous calculations show, pasture growth is the most important and most inexpensive feed available to elk farmers. Not only that, but our farmed elk are seasonally tuned to achieve their maximum growth rates just when our pastures are prepared to supply a flush of the required nutrients. With good management and the right mix of species, the pasture season may extend for more than half the year, and provide two-thirds of annual feed requirements.

There are two major grazing strategies; rotational grazing or continuous grazing also known as set stocking. Set stocking involves running an estimated ideal number of animals over a known area for long periods of time. The selective feeding nature of elk allows them to continually crop and graze preferential species and will in time eliminate the most desired species thus reducing the overall production of pasture. This pattern produces an uneven forage intake throughout the season and can often mismatch animal requirements and available feed. Maximum production and achieving target weights can be very difficult under this regime.

Rotational grazing is the principle of offering a larger number of mouths better quality feed in a smaller area for a very short space of time before moving on to another block. It requires much greater control through fencing and pasture subdivision and more time input. Rotational grazing management promotes good pasture utilization and productivity and advances legume persistence and even pasture growth pattern. As the spring and summer surpluses occur individual pastures can be shut out of the rotation for a period, the surplus growth harvested for hay and silage and then the highly palatable regrowth used for lactation or young stock production. In any rotation the principle is to quickly eat the available green leafy material, move the stock on and rest the pasture for recovery.

In continual grazing, plants are not allowed to grow and recover and root reserves are depleted over time.

Young elk particularly like to eat the succulent and highest nutritive value parts of plants first and will stop eating if the pasture becomes dirty from rain/mud or soiled from over manuring and high stock concentrations, even though it may appear there is ample feed remaining.

If pasture becomes too long, however, dead and unpalatable material builds up in the base and rapidly reduces quality on offer. If forced to graze too hard, intake is restricted and again animal production is limited.

The art of fine tuning a rotational grazing system is best learned by experience and consultation with local experts, to perfect the techniques and species composition most suited to your area, climate, herd make-up and management style.

PREFERRED SPECIES

Selection of pasture species must balance considerations of preference and palatability, total production estimates, and seasonal production characteristics. P.J. Fargey, from the University of Alberta, conducted a field trial to establish what species elk preferentially graze. (Table 6)

Comparisons between results observed in this study and those from literature provide some interesting contrasts. Wheat grasses are often mentioned as preferred species because they frequently occur in elk diets. However, results from this study suggest that when elk are given a choice, they will rarely select wheat grasses. Furthermore, cicer milkvetch (Astragalus cicer) is often considered to be an unpalatable species for elk, however, in summer and winter trials of this study it was frequently selected. Caution should also be taken when extrapolating from domestic animal preferences to wild species. In a comparable study with domestic cattle, researchers found that sainfoin was avoided which is in dramatic contrast to elk which used this species extensively in every season.

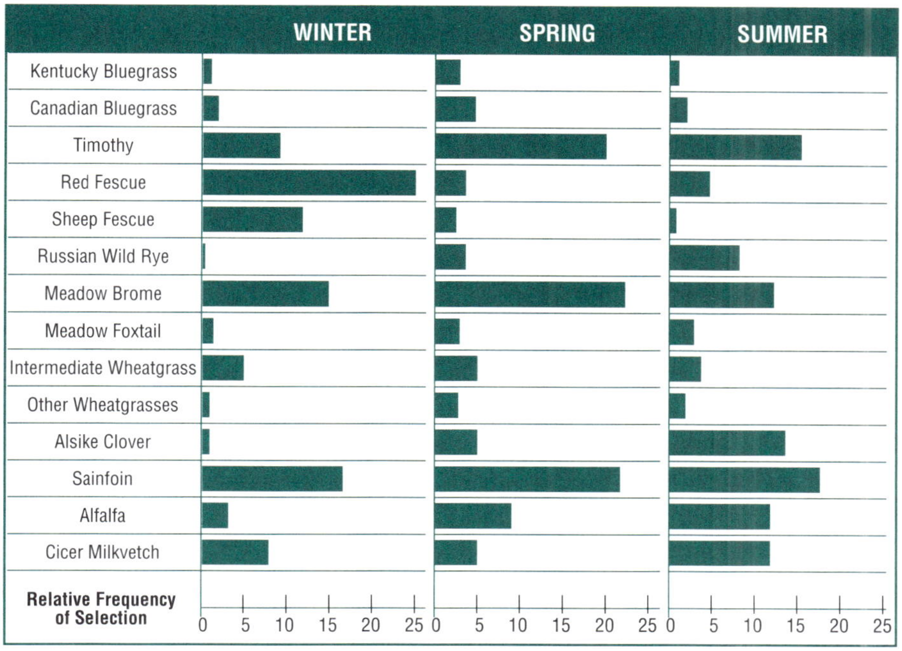

Table 6 Elk Selection of 18 Species of Grass and Legumes

For the elk farming scenario, the question remains, "Is knowledge of forage preference important?" Forage selection is clearly important in free-ranging situations where animals choose from a wide variety of species, however, in relatively small tame pastures the primary consideration is how well animals grow or maintain themselves. Many studies have shown that feeding preferences do not always relate to voluntary intake. Consequently, an elk farmer making species seeding decisions on the basis of foraging preferences would not necessarily be optimizing animal productivity. Data on animal production parameters, like average daily gain, are what is really required.

Despite the fact that the relationship between diet preferences and animal production is likely imperfect, some general recommendations can be made using the results from this study. An Alberta elk farmer is primarily concerned with providing pasture during the spring and summer seasons and will likely be supplementally feeding during the winter months. Spring pasture requirements could be met by grass species like smooth or meadow brome grass and timothy which green up early to provide nutritious spring pasture. Sainfoin is a highly nutritious legume that begins growing early in the spring and is worth considering as a spring pasture species. Summer pasture should have a mix of legumes and grasses. The selection of a legume species is less important than the adoption of a grazing management strategy which would prevent the legumes from being overgrazed. For the same reason, the grass(es) used should ideally be bunch grasses to minimize their competition with the legumes.

The recommendations outlined here can be used successfully as general guidelines. Considerable scope remains for refining these recommendations through further research. These studies should, and no doubt will, focus on plant species combinations and pasture management strategies that maximize animal production.

MAXIMIZING PASTURE PRODUCTION

The goal in pasture utilization is to provide, in the most efficient and inexpensive manner, highly palatable forage for the longest possible grazing season. Remember, pasture is the least expensive option for feeding elk.

Soil testing and analysis, with responsive addition of nutrients (fertilizers and minerals) to achieve balanced target levels in your soil, is the mandatory first step. The right mix of fertilizers will depend on the species mix you choose to grow, and the production goals you aim for. Grasses need large amounts of nitrogen. Legumes need relatively large amounts of potassium. Both need phosphorus and several other nutrients. Some soils will have adequate levels of many nutrients but most will need nitrogen for grass growth. Pastures with over 30% legumes will have a reduced need for nitrogen fertilizer. Excess nitrogen in a grass/legume mix will favour grass growth at the expense of legumes. Nitrogen can be

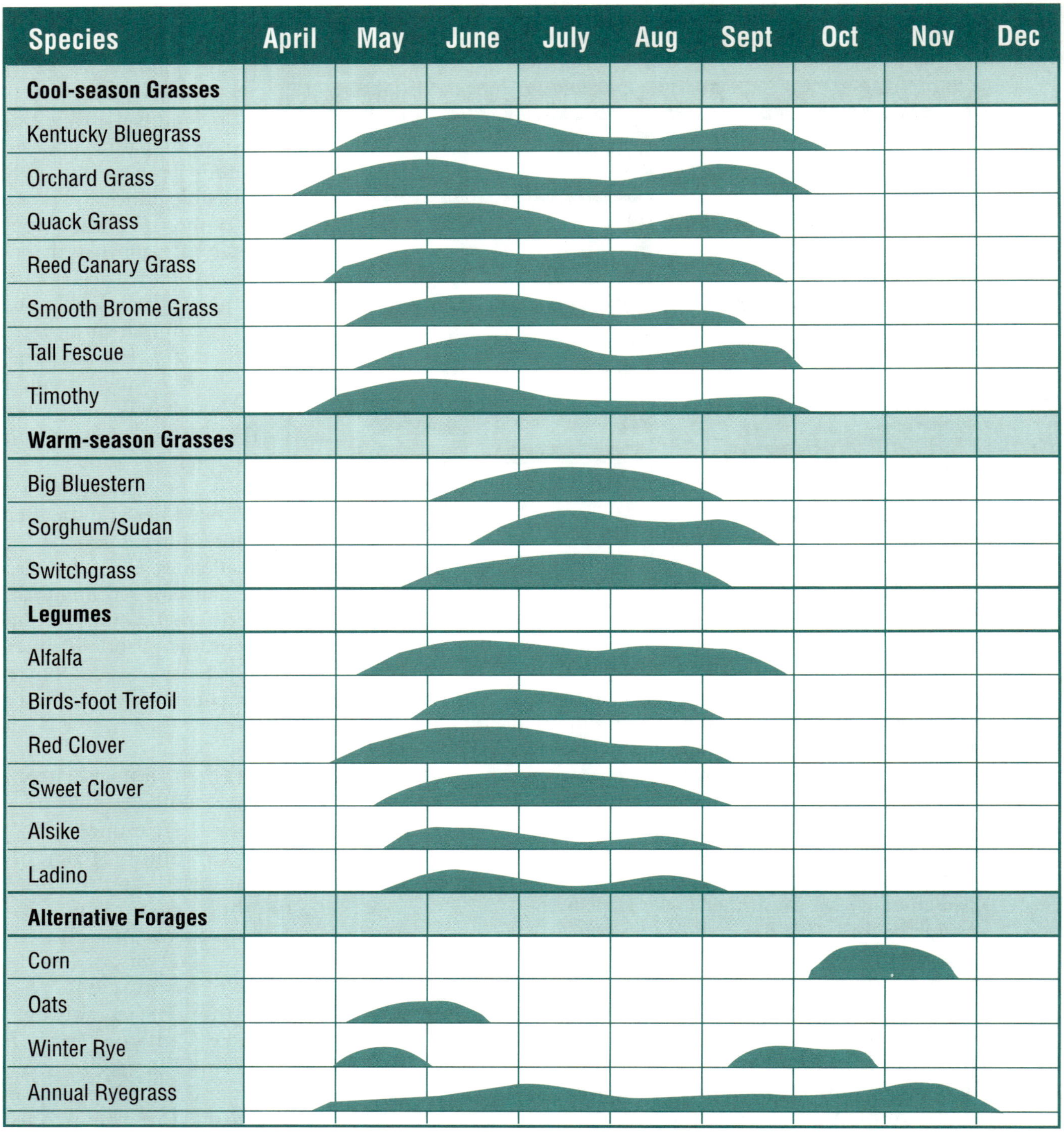

Table 7 Seasonal Growth Pattern for Forages

TARGET SOIL NUTRIENT LEVELS FOR OPTIMAL ELK PASTURE

Phosphorus	150 - 200 lbs/acre (30 - 40 ppm)	Zinc	2 - 5 ppm
Potassium	2 - 5% base saturation (at least 200 lbs/acre or 100 ppm	Molybdenum	0.25 - 0.5 ppm
Magnesium	10 - 20% base saturation (200 lbs/acre or 100 ppm)	Aluminum	less than 1,000 ppm
Calcium	60 - 70% base saturation	Sulphur	20 - 30 ppm
Sodium	0.5 - 1% base saturation	Iron	100 ppm
pH	6.0 - 7.0	Copper	5 - 8 ppm
Boron	0.6 - 1 ppm	Cobalt	0.10 ppm
Manganese	30 ppm	Selenium	0.5 - 1.5 ppm

Thanks to Wesley McQuown, Crop Consultant and to Brookside Laboratoires, Inc. for some of the above soil nutrient levels information.

Forage Species Characteristics and Tolerance to Soil Problems

Crop	Special Characteristics	Potential Limitations	Hardiness[1]	Salinity Tolerance[2]	Alkalinity Tolerance[3] (high pH)	Acid Tolerance[4] (low pH)	Flood Tolerance	Drought Tolerance	Normal Longevity in years[5]	Uses[6]
Alfalfa (Flemish)	rapid regrowth	hardiness	F - G	X	X				M	H,P,Pr
Alfalfa (Standard)	hardy	regrowth	G - E	X	X				M	H,P,Pr
Alfalfa (Creeping Rooted)	drought tolerant	regrowth	G - E	X	X			X	M - L	H,P,Pr
Alfalfa (Siberian)	very hardy	regrowth	E	X	X			X	L	H,P,Pr
Bird's-foot Trefoil	bloat-free	hardiness	F		X	X	X		S - M	P
Cicer Milk-vetch	bloat-free	establishment	G	X	X			X	M - L	P
Alsike Clover	acid & flood tolerant	not drought tolerant	G			X	X		S	H,P
Red Clover	acid tolerant	not drought tolerant	G			X			S	H,P
Sweet Clover	plowdown	biennial	G	X	X			X	S	H
Sainfoin	bloat-free	hardiness	F		X				S - M	H,P
Orchard Grass	rapid regrowth	hardiness	F - G			X			M	H,P
Meadow Brome	rapid regrowth	seed size	E	X					L	H,P
Timothy	peat soils	drought tolerant	G - E			X	X		M	H,P,Pr
Creeping Red Fescue	strong creeper		E	X		X	X		L	P
Tall Fescue		hardiness	F	X	X	X	X		M	P
Reed Canary Grass	peat soils	alkaloids (use new varieties)	E	X		X	X		L	H,P
Meadow Foxtail	peat soils/early growth	fluffy seed/early heading	E			X	X		L	H,P
Crested Wheatgrass	early growth	regrowth/palatability	E	X	X			X	L	H,P
Intermediate Wheatgrass			G	X				X	M	H,P
Streambank Wheatgrass	reclamation		E	X				X	L	P
Altai Wildrye	fall pasture	establishment	E	X				X	L	P
Russian Wildrye	fall pasture	establishment	E	X				X	L	P
Dahurian Wildrye	easily established	short-lived	G	X					S	H,P

1. Hardiness rated as Fair (F), Good (G) or Excellent (E).
2. Salinity - For the purpose of this chart, forages are considered to be tolerant if they are adapted to soils of 4 -6 mS/cm (mmhos) salinity in the top three to four feet of soil under average moisture conditions
3. Alkalinity (pH above 7.0) - For purpose of this chart, forages are considered to be tolerant if they are adapted to soils with a pH of 8.5 or more.
4. Acidity (pH below 7.0) - For purpose of this chart, legumes are considered to be acid tolerant if they are adapted to soils in the range of pH 5.5 - 6.0. Grasses are considered to be tolerant if they are adapted to soils with pH lower than 5.0.
5. Longevity rate as Short (S) 2 to 3 years, Medium (M) 3 to 5 years, and Long (L) 5+ years. Varieties will greatly affect longevity of this species.
6. Uses - Hay (H), Pasture (P), Processing (Pr).

Table 8

applied at any time during the growing season that growth is desired. Spring application will usually be necessary only if harvest of surplus growth is planned. Otherwise, the elk will probably not keep up to the growth in June. Table 6 re: preferences, Table 7 re: seasonal growth patterns, Table 8 re: forage species characteristics, plus more specific information regarding the production ability of various plants in your area according to your soil and climate will guide you in your selection.

Some "Best Picks" for Elk Pasture

1. Alfalfa, the "Queen of Forages", will be the preferred legume in most areas where elk are raised. Alfalfa is the most productive legume in most situations, yielding well over five tons per acre if well managed. Good management includes rotational grazing. Alfalfa should be allowed to grow to a height of eight to ten inches before grazing down to three inches, then allowed to recover for 30 to 40 days before grazing again. One recovery period must cover the 30 to 40 day period before the first killing frost of the autumn, to allow root reserves to build up for spring growth. Residue after frost may be grazed in autumn or winter.

Rapidly growing alfalfa in a pure stand may exhibit an excessive protein level. The by-products of this high protein diet are excreted in urine, possibly resulting in "burns" to the penis sheath or vulva. To avoid this, alfalfa should be grown in a mixed stand with grasses such as timothy or meadow brome, or with forbs such as chicory, and allowed to grow to eight to ten inches before grazing. Aim for an alfalfa content of 30 - 40% in the stand. Best varieties vary with location and soils, but consider Alfagraze, Rangelander, Stampede, or Algonquin.

2. Timothy and orchard grass both are highly palatable when kept grazed to a height of three to ten inches, and both begin growth early in the spring. These grasses will both tolerate earlier and more frequent grazing than alfalfa, and therefore might do better when planted with clovers or in a pure stand with applications of nitrogen fertilizer. Best varieties: OKAY orchard grass, local preference for timothy.

3. Reed canary grass, specifically one of the new low alkaloid varieties such as Palaton or Bellevue, will outproduce most other grasses on both dry and wet soils, responds well to a legume mix or to nitrogen fertilizers, and is highly palatable if grazed from three to ten inches.

4. Meadow Brome Grass is a proven standard for prairie and parkland pastures. Its longevity, palatability and rapid regrowth after harvesting makes it a highly preferred variety, particularly in a mix with alfalfa. Talk to your local specialist when choosing the right variety for your area.

5. Forage Chicory, such as Puna, is a deep tap-rooted plant, highly productive and palatable, grows well in an alfalfa mix, provides higher levels of minerals and is more drought-resistant due to its deep rooted nature.

PASTURE SEED MIX FOR ELK

Designed for the Parkland area of North Central North America, for medium textured soils and average precipitation. Innoculated and blended.

For medium to well drained soils:

(by weight)
20% Orchardgrass
15% Meadow Bromegrass
5% Creeping Red Fescue
20% Able or AC Caribou Alfalfa
15% Rangelander or Spredor 2 Alfalfa
10% Sainfoin
10% Red Clover
5% Puna Chicory if available

Seed at 15 to 20 lbs per acre, fertilize with Nitrogen, Phosphorus, Potash and micronutrients as adsived by a local soil specialist. Typical fertilizer requirements would be 30 lbs per acre nitrogen, 45 lbs per acre phosphorus, 15 lbs per acre potash, 15 lbs per acre copper sulfate.

For wet or poorly drained soils:

(by weight)
20% Orchardgrass
40% Belelvue or Palaton Reed Canary grass
15% Alsike Clover or Magnum 3 Alfalfa
20% Red Clover
5% Sainfoin

Seed at 10 to 15 lbs per acre, fertilize as above.

EXTENDING THE PASTURE SEASON

Winter feeding is one of the biggest contributors to the expense of elk farm management. Several strategies may be used to reduce this cost, including annual crops, swath grazing and forage stockpiling.

Annual Crops are planted every year, either in autumn for a crop such as winter rye, or in spring for a crop such as annual ryegrass or oats. Winter rye is the cereal crop we are familiar with. Planted in August, it provides good growth until frost slows it down and snow buries it. As soon as the sun touches it in the spring, it begins growing again, and continues vigorous growth as long as grazing is heavy enough to prevent heading out.

Annual ryegrass and oats are both highly palatable. Annual rye will outproduce oats or any other cereal by a wide margin when regularly grazed. All three of these "annuals" may survive another winter if frost is light, but the key to best production is high nitrogen and heavy grazing to prevent formation of seed heads. Annual ryegrass and oats work particularly well for producing forage in heavily manured and trampled wintering areas. Best varieties of annual ryegrass are the "Italian" types, particlarly Lemtal or Maris Ledger.

Swath Grazing involves the late planting of a crop such as oats, then swathing just before frost, and allowing livestock to eat the swaths. Cold weather slows spoilage, and the concentrations of feed in the swath assists stock in accessing adequate nutrition under snow cover.

Forage Stockpiling refers to the practise of allowing forage crops to grow in late summer and grazing them standing in late autumn, winter, and early spring. Depending on local climate, a great variety of forages may work well, ranging from sweet-stemmed forage sorghums and bermuda grass in the South and Southeast, bluestems and gamagrass in the Southern Plains, or red fescue, meadow brome or orchard grass in the northern Prairies and Parkland. Generally, the highest quality forages stockpiled for winter and spring grazing are the more immature ones grown under higher fertility. These crops are cut or grazed in July or early August, then left untouched until needed. Low growing plants that can stay under snow cover for longer periods of time and have a finer leaf tend to hold their quality well, (eg.) meadow brome, orchard grass, or creeping red fescue. Tall growing plants tend to suffer more from exposure, but there may be an advantage to having these species in the stand to draw the attention of the grazing animals. The density of the lower growing, higher quality feed below must be high enough to justify the energy expended in digging through the snow.

Grain corn or forage corn may be excellent choices for grazing in early winter, particularly for bulls in their post-rut weight gain period.

GRAZING MANAGEMENT

Pasture and grazing management is an art, best learned with good advice and earned experience. The best techniques are unique to each farm, herd and manager. Here are some bits of advice to start:

- Plan to fence enough pasture to keep elk densities at from two adults per acre to two acres per adult, depending on soil and climate.
- Fence in a diversity of land. Include cultivated land, native prairie, plus bush areas. Each delivers a different complement of nutrients, all valuable to the elk.
- To provide good quality pasture or stored feed, think "YTRG" - Young, Tender, Rapidly Growing.
- When Rotational Grazing, move the stock while there is still foliage left, usually at an average height of three inches. This will allow more rapid regrowth.
- Graze with larger herds, higher stock density, to improve animal distribution in the pasture, to achieve more even utilization of forage, and to reduce selective grazing of preferred species.
- Put calving groups into pastures with good amounts of cover, either as brush or small patches of trees, or stockpiled tall grass, to provide hiding cover for the calves. Have a productive pasture next to this, cut the forage in that productive pasture in early June, then let the cows and calves move themselves, after at least 2/3 of the calves are born.

SUMMARY

ELK FEEDING AND NUTRITION

- Elk are highly influenced by seasonality in both nutrition and reproduction.
- Production curves, target liveweights and seasonality influences on each class and sex are quite different, but well known. To feed to these requirements elk should be managed, up to a point, in their classes; young stock; adult breeding or velvetting bulls; pregnant and lactating cows.

- The biggest feed demands are for cows during lactation, almost double that of early pregnancy and winter. Bulls have high energy requirements in winter and because of weight losses during the rut and very little body fat cover are particularly vulnerable to underfeeding stress.
- Growth from birth to weaning is critical, i.e. lactation feeding and good pasture management are vital. If pasture is poor quality, supplementation is indicated with grains or high quality legume hay.
- Good pasture management, including rotational grazing, will provide a continuous supply of young, tender, rapidly growing forage sufficient to meet all feed requirements from mid-spring through to winter.
- Liveweight plateaus occur at 15 months and 27 months of age. Elk must exceed minimum threshold weights for mating performance or target venison weights.
- Maximum growth in young stock occurs in spring, summer and autumn. From mid February intake increases at 10% per week. Feeding must be increased to take advantage of the natural ability to grow rapidly.
- Develop target weights as a check on management, and practice regular weighing.
- Balance winter rations based on energy and protein needs. Do not over feed protein as there is little bonus and additional protein is costly.
- Many feeds are suitable for elk. In selecting choices know the dry matter, energy content and crude protein contents so that an informed decision on cost effectiveness can be made. Cost per amount of energy available is a more realistic value than cost per tonne.
- Always use quality feed. Elk are selective eaters and may leave poor quality feed behind. This should not be misinterpreted as adequate feeding.
- Do not over feed grains or formulated rations. These are expensive, can have a depressing effect on intake and production at high levels, and there is always a risk of acidosis (grain poisoning) or other problems.
- Good feeding particularly in winter encourages contact between farmer and stock and adds considerably to ease of handling and ultimately good production. This feature can be used very positively for a sound base in profitable elk farming.

How Do Your Elk Compare?

Production Parameters of Farmed Elk - Based on Surveys of Alberta Producers 1990 - 91

Average					**Averages**	**Minimum**	**Maximum**
Age of Dam (mother)	**Two years (first calves)**		**Adult Cows**		4.7		16
Weight of Dam (lbs)	483	471	531	534	526	352	660
Calving Rate (%)	81	81	96	96	93		
Weaning Rate (%)	73	73	91	91	88		
Calf Gender (sex)	**Male**	**Female**	**Male**	**Female**	**all**		
Calving Date	11-Jun	11-Jun	06-Jun	06-Jun	07-Jun	21-Apr	10-Sep
Birthweight (lbs)	42	40	38.5	36	38.9	25	49
Weight @ 100 days (lbs)	195	180	210	190	197	125	277
Growth Rate (lb/day)	1.52	1.43	1.69	1.52	1.58	0.86	2.35
Weight @ 200 days (adj. lbs)	279	251	286	269	275	197	375
Weight Gain (lbs/day)	0.84	0.75	1.06	0.95	0.95	0.44	1.61
Date Exposed to Bull	11-Aug	19-Aug	28-Jul	03-Aug	05-Aug	01-Jan	15-Sep
End of Breeding	17-Nov	22-Nov	27-Nov	26-Nov	24-Nov	14-Oct	31-Dec
No. days Exposed	97.9	94.4	121.4	115.3	111.4	48	365
Bull/Cow Ratio	18.2	18.2	22.3	22.6	21	4	47
Age of Bull	4.2	3.6	4.3	4.4	4.3	2	10
Fenced Acres per Elk	2.9	3.4	3.1	2.9	2.9	0.7	15.5

Frequency of multiple births: one in 271 calvings

Antler Production

Age	**1994 Alberta Average wt (lbs)**		**Target wt (lbs) for improvement**
Spiker	2		
Two year olds	8	2 year	10
Three year olds	11	3 year	14
Four year olds	14	4 year	18
Five years & older	18	5 year	22
		6 year +	26

Chapter 1 ***Wapiti***

INTRODUCTION

North American Elk farmers are just beginning to develop the business of consistent supply of quality meat products for the food and hospitality trade. Elk meat may be sold as "elk" or as "wapiti". There is a distinct advantage to selling as "elk" or "wapiti" rather than venison, since the supply of elk is more limited, and the name "venison" is commonly associated either with the variable quality of hunter-killed wildlife or with imported red deer venison. Most of our current knowledge of venison qualities is based on information from New Zealand regarding red deer, but most of this information is applicable also to farmed elk. Wapiti meat has one distinct advantage over New Zealand red deer venison or Cervena in its texture. Red deer venison, and particularly the New Zealand product, has a much higher iron content than North American Wapiti. This may be desirable from a nutritional standpoint, but that high iron/hemoglobin content tends to make venison taste and feel like liver. Wapiti tends to be more like beef.

Meat from farm-raised elk easily exceeds the traditional limitations of the feral or wild shot product. Quality control specifications regarding age of the animal, potential yield, and product specification can be selected on-farm with health status and "fit to purpose" guaranteed. The animals can then be provided for processing on a "just in time" basis for a fresh chilled supply of truly exquisite product to chefs and upmarket retailers.

Farm raised wapiti is "new style" food - light, lean and tender. Its nature and quality is consistent, with a mild but distinct flavour. Wapiti fits in with the new food trends emerging in North America, including relaxed eating styles and ethnic eating experiences. People are willing to experiment with new products and wapiti is ideally positioned to override the nutrition and health concerns that have led to a decline in traditional red meat consumption.

Suppliers of farmed elk meat are fortunate that their product naturally fits the optimum nutritional profile of currently recommended diets. But education is needed to change the perception of hunted venison as a tough, dry, gamey, and inconsistent product. Appreciation for farm raised wapiti is something most consumers need education and coaxing to develop.

Successful elk farming doesn't quit at the farm gate. Support and promotion of the product and its versatility, quality and excellence, are as much a part of elk raising as quality feeding and handling. It is important, however, to distinguish the farming process from the product and its marketing.

NUTRITIONAL PROFILE

Nutrient comparisons (Table 1) emphasize the obvious features, i.e. low in fat, cholesterol and calories. Protein levels are comparable to other meats. Venison is uniquely rich in minerals, particularly iron and phosphorus which in turn account for its rich dark colouring. This nutritional bonus can be a disadvantage in retail marketing where consumers are used to bright red coloured meat, contrasted by (often trimmed) white fat. Farm raised elk is one of the few naturally reared meats with no growth hormones or steroids used since elk are naturally effective producers of meat, and artificial manipulations negatively affect velvet growth and reproductive performance.

Producers must ensure that, if the market emphasis is leanness and tenderness, the animals put forward for slaughter must qualify. They should be young (ideally less than 29 months of age) and selected for rapid growth rate and superior muscle conformation. Younger animals display the most desirable carcass composition. Mature elk can be appreciably fatter than live weight guidelines recommend. Farmers with that prior knowledge can stream this type of animal into the processed chain for value added products such as sausage, salami, smoked hams and processed meats where some fat is an advantage.

Comparative Nutritive Value*
Based on 100 gram Portions

	Calories	Fat (gm)	Cholesterol (mg)	Protein (gm)
Venison, loin cut	**159**	**3.3**	**66**	**25**
Beef Bottom Round, lean	214	9.76	92	31
Ground Beef	265	18.4	85	24
Pork Shoulder Cut, lean	219	10.64	101	29
Lamb Leg Roast, lean	178	7.62	83	25
Veal Cutlet	213	10.35	125	26
Chicken Breast	159	3.42	83	31
Turkey (light meat)	154	3.42	68	29
Salmon (pink)	138	5.75	39	20
Scallops (breaded)	215	11.0	77	17

Mineral Compositions
mg per 100 g Fresh Meat

	Phosphorus	Sodium	Potassium	Iron	Calcium
Venison loin	**250**	**52**	**320**	**3.82**	**9.1**
Beef loin steer	208	39	399	2.30	3.0
heifer	210	40	417	2.20	2.9
Lamb loin wether	168	71	304	1.48	11.7
ram	167	77	314	1.40	13.7
Pork	200	76	370	.90	8.0
Chicken (white meat)	210	72	330	.50	10
Veal fillet	260	110	360	1.20	8.0

Table 1 Nutritional Profile of Venison (raw)

MARKETING POSITION

While its nutritional advantages allow it potential in a health conscious, "heart smart" dietary role, most farm raised wapiti suppliers are currently marketing at top range prices to the hospitality trade. This strategy involves professional chefs and food distributors promoting the product and educating consumers until the volume of supply can satisfy the demand of retail outlets where use and preparation methods are more variable.

Because of its market positioning and relative scarcity, buyers of venison expect a consistently high standard of supply and product that meets rigid specifications in terms of :

1) muscle cut and size, (repeatability),
2) tenderness (and by implication moisture content),
3) quality of preparation and presentation (chilled/fresh and of high hygiene - no spoilage or waste).

These standards aim to set the farm raised product as a superior product completely separate from the traditional consumer perception of wild shot venison.

The product must meet the highest standards of:

- on-farm production (on- farm food safety and quality assurance programs);
- transportation of both livestock and processed product (accredited transport operator standards);
- slaughter & processing (HACCP accreditation of slaughter and processing plants)

The marketing and quality control programs currently in place for New Zealand-produced venison provide the bench mark for farm-raised wapiti in North America. Premiums are available for local product because of its identification, freshness and relatively uncomplicated supply route. Competitive pricing against New Zealand products is a controversial issue for North American producers, many of whom have established direct links from farm, through local processing plant to end user, and find that their returns have been eroded by a wholesaler-

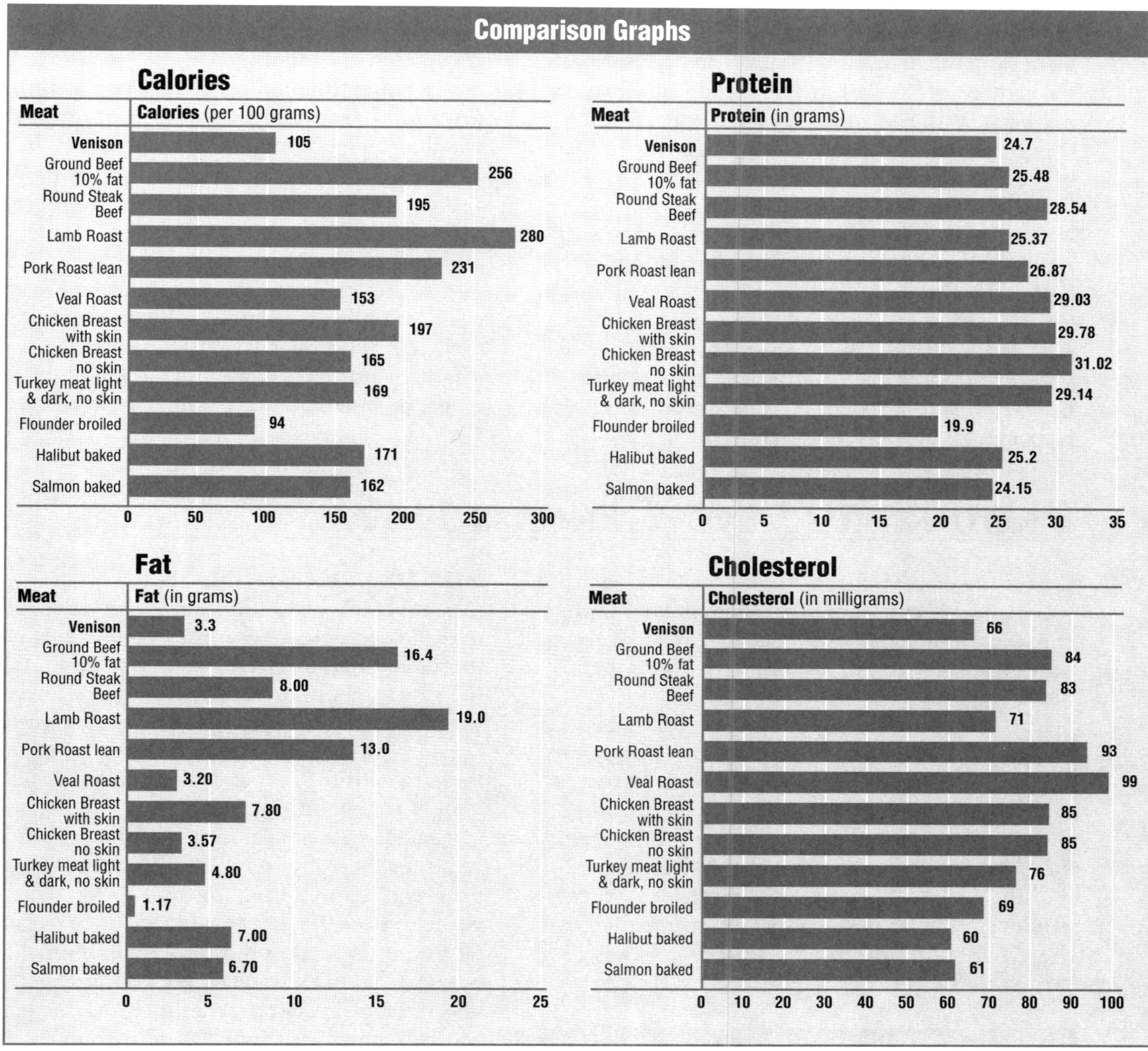

distributor operation based on a more consistent supply of imported venison. The current small size of the North American industry creates difficulties in formulating an integrated marketing plan, but its smallness is also an advantage in developing tight and profitable relationships individually with a host of outlets not available to competing importers.

Suppliers and processors together can concentrate on value rather than volume by reducing the variability in seasonal supply, adding value through knowledge of local and ethnic eating trends and producing variations to suit. They can capitalize on education and communication by retaining full association with the product and slowly developing new markets for increased volumes.

As demand builds, competition based on quality and consistency should increase and enhance producer returns, rather than competition based on price which has a negative effect on producer returns.

Therefore, the market position from the outset

must be at the top end, and elk farming systems and products must continue to be able to support that position.

WORLD MARKETING

Traditional demands for venison and game meats are centred on West European and Scandinavian markets. With the advent of large scale elk and deer farming, feral venison is being in part replaced by higher quality, farmed origin product. However, the market is rapidly growing, with greater availability of products and the taste and nutritional attributes of farmed elk and venison being promoted.

New markets are rapidly emerging with S.E. Asia and North America considered to be those with most potential. Appropriate pre and post mortem health inspection at dedicated slaughter plants allow health requirements and meat inspection standards to be met universally, while allowing for kosher venison and halal slaughter.

It is vital that the distinction between farmed elk and wild shot venison is preserved, to add to the associated perceptions of enhanced quality and consistency in standards.

In emerging markets quality production and top end market positioning are the targets - age of slaughter, quality hygiene standards and the ability to supply with consistency are important. With this approach elk farmers world wide, irrespective of their stage of industry development, enjoy a growing and appreciative demand for their high quality and healthful meat products in the enviable position of undersupply. In the world marketing of agricultural production, this aspect alone is a unique feature of elk farming and a significant aspect of sustained profitability.

PRODUCTION AND CARCASS DATA

Carcass quality, yield and return are dependent on weight and age of slaughter. Growth rates and seasonal influences define specific carcass weight limits for age, and are modified by genetics and management.

In males, peak slaughter weight for age is reached at 15 and 27 months and these represent ideal end points for premium venison. Variation within a group and the role of quality feeding through the seasonal changes allow these target weights to be extended through 11 months - 20 months of age and from 27 - 34 months, with producers able to target a specific weight and use the variability to extend the season of supply for a greater period.

Breed Type					
	Red	Super Red	Elk/ Red	Elk	Fallow
Live - weight (average) Age 12-15 months	104kg	121	145kg	200	48
Carcass weight	59.2kg	69.2	84	112	28
Dressing %	57%	57.5%	58	56	59
Range	52-59	53-59	56-59	51-58	56-60
Live - weight 24-27 months	140	165	190	300	64
Carcass weight	80	95	110	170	38

Table 2 Dressing % : Ratio of Liveweight : hot carcass weight
To estimate optimal slaughter weights it is necessary to know accurate liveweight : carcass weight ratios as represented in this Table.

YIELDS, TERMINOLOGY

Elk and deer yield significantly high ratios of meat to bone and fat and, compared with beef, yield substantially more from the higher value parts of the carcass - loin or saddle, and hind quarters.

Body confirmation is virtually the same for all breeds although red/elk hybrids show appreciably higher yields for loin and hindquarters at an earlier weight for age (Table 3).

An analysis of lean, fat, and bone yields also indicates there are significant productive advantages from elk/red hybrids compared to red stags at over twice the age. The red stags' tendency to lay down fat in the rib section indicates the potential for over-fatness in even relatively young stags. Two and three year old prime animals may have a rewarding use for shoulder and trim in the further processed venison market. Elk and hybrids are always leaner at the same age, or weight for age. Table 4 demonstrates these trends.

Fallow deer show similar yields to that of red deer at 13 - 15 months, but can increase fat % yields from 7.2% to 11% by 21 months of age which can be a significant grading problem in animals older than two years of age.

Female deer produce excellent venison even at advanced age if their body condition is good. Venison quality tests, when based on physical testing of tenderness and juiciness and colour, and supported by consumer perception, show there is an appreciable decline in tenderness in stags older than three years. There is also a significant darkening of stag venison from older venison.

Hinds show no change from 15 months to 13 years. However, once the mature body weight threshold (95 - 100 kg) has been reached, will markedly increase in fat content and GR above 53 - 55 kg carcass weight. This may be significant in an annual cull of aged non-pregnant animals.

Average Stag Carcass Composition			
	Red Deer		Elk x Red Crossbreed
	15 month	26 month	11 month
Liveweight (kg)	104	130	116
Hot carcass weight (kg)	58.2	74.1	68.4
Cold carcass weight (kg)	56.4	72.3	66.6
Dressing %	56	57	59
Typical Carcass Yields	**kg (%CW)**	**kg (%CW)**	**kg (%CW)**
Saddle	8.4 (14.5)	10.9 (15)	11.8 (18)
Hind legs	21.7 (38.5)	28.1 (39)	26.6 (40
Shoulder	11.5 (21)	13.7 (19)	12.9 (20)
Neck	8.5 (15)	11.6 (16)	9.3 (14)
Ribs	5.7 (10)	8.0 (11)	5.5 (8)
G.R. (tissue depth 11/12th ribs) -fatness indicator	6 mm	10 mm	4.7 mm

Table 3

There are early indications that under Canadian climate and feed conditions that favour production of temperate deer, yields may be as much as 78% lean : bone.

Lean, Fat & Bone Content of Primal Cuts (Tissue as % Primal Cuts)						
Primal Cuts	Lean		Fat		Bone	
	Red	Elk/Red	Red	Elk/Red	Red	Elk/Red
Saddle	69	74	7.3	3.9	24	22
Shoulder	74	76	6.2	4.6	16	16
Hind Leg	77	79	5.2	3.5	21	21
Neck	67	70	6.8	5.7	26	24
Ribs	68	75	12.3	5.5.	20	19
Total Yield	72.7	76	7.0	4.7	20	19

Table 4 After Drew 1989

GRADING SYSTEMS

While not yet widely applied in North America, carcass grading is well established in other countries, and the USDA has recently proposed standards for venison.

13 to 15 Month Fallow Spiker	
Live Weight	45 Kg
Carcass Weight	26.5 Kg
DP%	59%
% Carcass Wt.	
Saddle	15.7
Hind legs	42.5
Shoulder	16.9
Neck	12.1
Ribs	12.8

Table 5

Carcasses are generally assessed at intervals of 5 kg in weight. Once a level of carcass fatness is determined as a limit, the GRD or tissue depth measurement, (i.e. total tissue depth measured between the 11th and 12th rib, 16 cm from the midline) is taken at a stable, easily located point that accurately reflects both chemical, and dissectable fat content.

Grading should also include an assessment of carcass damage, and while this is commonly slight in deer, it does serve as an indicator of management difficulties either on-farm, in transport or immediately prior to slaughter.

A basic grading system that is already a standard in Australia and New Zealand defines the sex of the animal, and a hot carcass weight range which in turn is related to a tissue depth measurement at the GR site.

Usually a tolerance of 2 mm of tissue depth is defined in the grading which gives a "trimmer" class, where the surplus fat can be removed with little difficulty and without increasing the fat content of ribs and shoulder trim too much. Beyond that carcasses are graded as overfat, and there can be a heavy discount in price.

Bruising, which is often caused by poor handling in the period immediately prior to slaughter either on-farm, or commonly during transport or pre-slaughter handling, is usually scored on the basis of how many primal cuts are affected. Bruises can be deep seated and require extensive trimming to remove.

Their most common cause is overcrowding in pens or transport and poorly trimmed hard antler stubs.

DEFINITION

A venison carcass is the body of a slaughtered cervidae animal after:

- bleeding
- skinning
- removal of all the internal digestive, respiratory, excretory, reproductive and circulatory organs
- minimum trimming as required by the meat inspection service for the carcass to be passed fit for human consumption

and removal of:

- head between the skull (occipital bone) and the first cervical vertebra
- feet between the knee joint (carpus and metacarpus) and the hock joint (tarsus and metatarsus)
- tail so that it is no longer than four coccygeal vertebrae
- thick skirt by separating the connective tissues as close as possible from the bodies of the lumbar vertebrae
- kidneys, kidney knobs and the pelvic channel fat
- udder, or udder and cod fat
- the carcass must be weighed Hot Standard Carcass Weight (HSCW)

VENISON CUTS: TERMINOLOGY

Primal Cut	Sale Cuts	
Hinds (39% Carcass Weight)	1) Bone-In Leg	Whole leg - standard carcass removed between 5th/6th lumbar vertebra *shank can be removed at the stifle joint
	2) Boneless Leg (Whole)	Shank removed and tunnel and seam boned, retaining all bones and cartilage
	3) Boneless Leg Cuts 4 cuts	As for beef - silverside, topside and thick flank removed along natural seams, rump removed from ileum and sacrum bone
	4) Denver Leg 8 cuts	Leg cuts are further divided into individual muscle groups. The silverskin and sinews are completely removed. Thick flank is removed
Saddle (15% Carcass Weight)	1) Bone-in Saddle	7,8,9 ribs retained. Whole loin and tenderloins retained. Trimmed to specification at the junction of eye muscle and ribs - 10-12 cm from the midline
	2) Bone-in Rack	Split saddle, cut from a side by removal of the loin between 3rd/4th and 11th/12th ribs - 50-70 mm of rib retained, cleaned and trimmed. Loin trimmed to silverskin (striploins)
	3) Boneless Saddle 4 pieces	Backstraps and tenderloins removed from backbone and trimmed. These premium muscle cuts can be further broken down by deskinning and desinewing (Denver) and sold as Denver Saddle, striploin and tenderloins
Fores (27.5% Carcass Weight)	1) Bone in Shoulder	Removed by following the natural contour of shoulder blade and seam of underlying muscle
	2a) Boneless Shoulder	All bones removed plus basal neck meat
	2b) Boneless Blade	Boned out shoulderless neck muscle
	3) Shepherd Steaks, Oyster Cut	Individual muscle groups (3) removed, desinewed and deskinned
Boneless A Neck (14.5% Carcass Weight)	Boneless Neck	Removed from bone whole - often desinewed and prepared in strips or coarsely diced (25 mm x 25 mm)
Boneless B	1) Rib, Flank and all primal cut trimmings	Diced or minced

Inventory of Ten Yearling Red Deer Males Processed 1/31/97

Total Hanging Weight **1,235.00 lbs**

Cut	Weight (lbs)	% of hanging weight	Total Price	Price per lb.
Tenderloin	23.05	1.87%	362.50	15.70
Loin Chops	7.78	0.63%	111.90	14.40
Loin	65.38	5.29%	854.60	13.10
8 - Rib	20.78	1.68%	197.40	9.50
Sirloin	26.98	2.18%	238.80	8.85
Top Round	54.91	4.45%	431.00	7.85
5 - Rib	11.51	0.93%	90.35	7.85
Medallions	21.63	1.75%	169.80	7.85
Sirloin Tips	51.54	4.17%	386.55	7.50
Denver Leg	21.59	1.75%	161.90	7.50
Bottom Round Steak	42.15	3.41%	261.33	6.20
Bottom Round Roast	6.19	0.50%	38.38	6.20
Shoulder Roast	16.46	1.33%	80.65	4.90
Chuck Tender	34.72	2.81%	170.12	4.90
Spare Rib	15.00	1.21%	73.50	4.90
Stew Meat	76.86	6.22%	326.65	4.25
Trim	219.00	17.73%	711.75	3.25
Neck	82.19	6.66%	213.69	2.60
Heart	11.95	0.97%	31.07	2.60
Liver	22.57	1.83%	45.14	2.00
Osoboco	103.00	8.34%	169.95	1.65
Total Finished Weight	**935.24**	**75.73%**	**5,137.03**	
			Average Price per lb of finished weight	5.48
			Average Price per lb of hanging weight	4.15

VENISON

TRIMMING IS LIMITED TO REMOVAL OF:

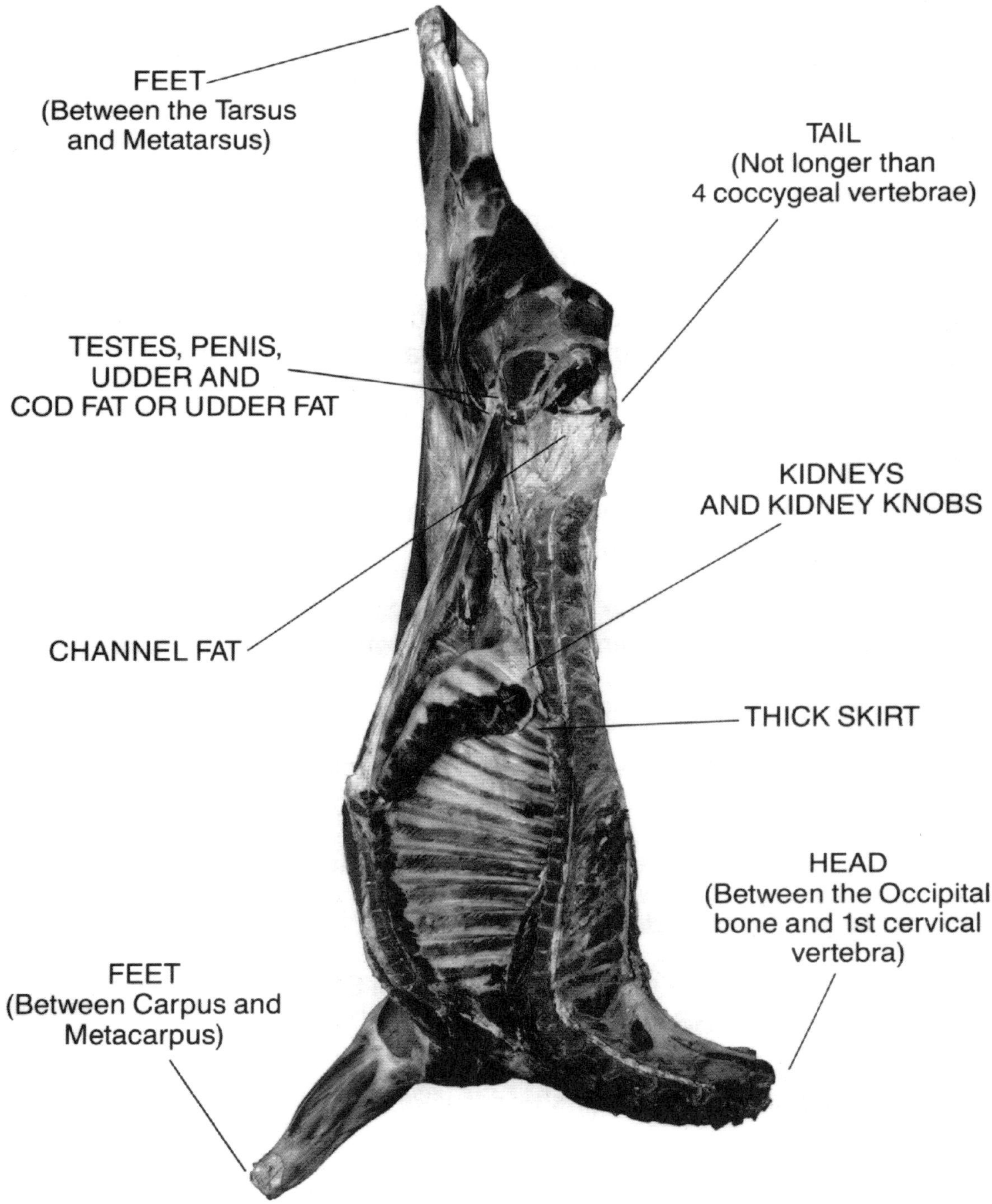

- Where the penis-testes are required to be removed as a speciality item – the removal is 10mm either side of the mid-line of the ventral pelvic "pubis."
- Where the tail is required to be removed as a speciality item – the removal is at the junction between the sacral and coccygeal vertebrae. The ischiatic glands must not be damaged.

Table 6 Presentation of Venison Carcass

VENISON CUTS

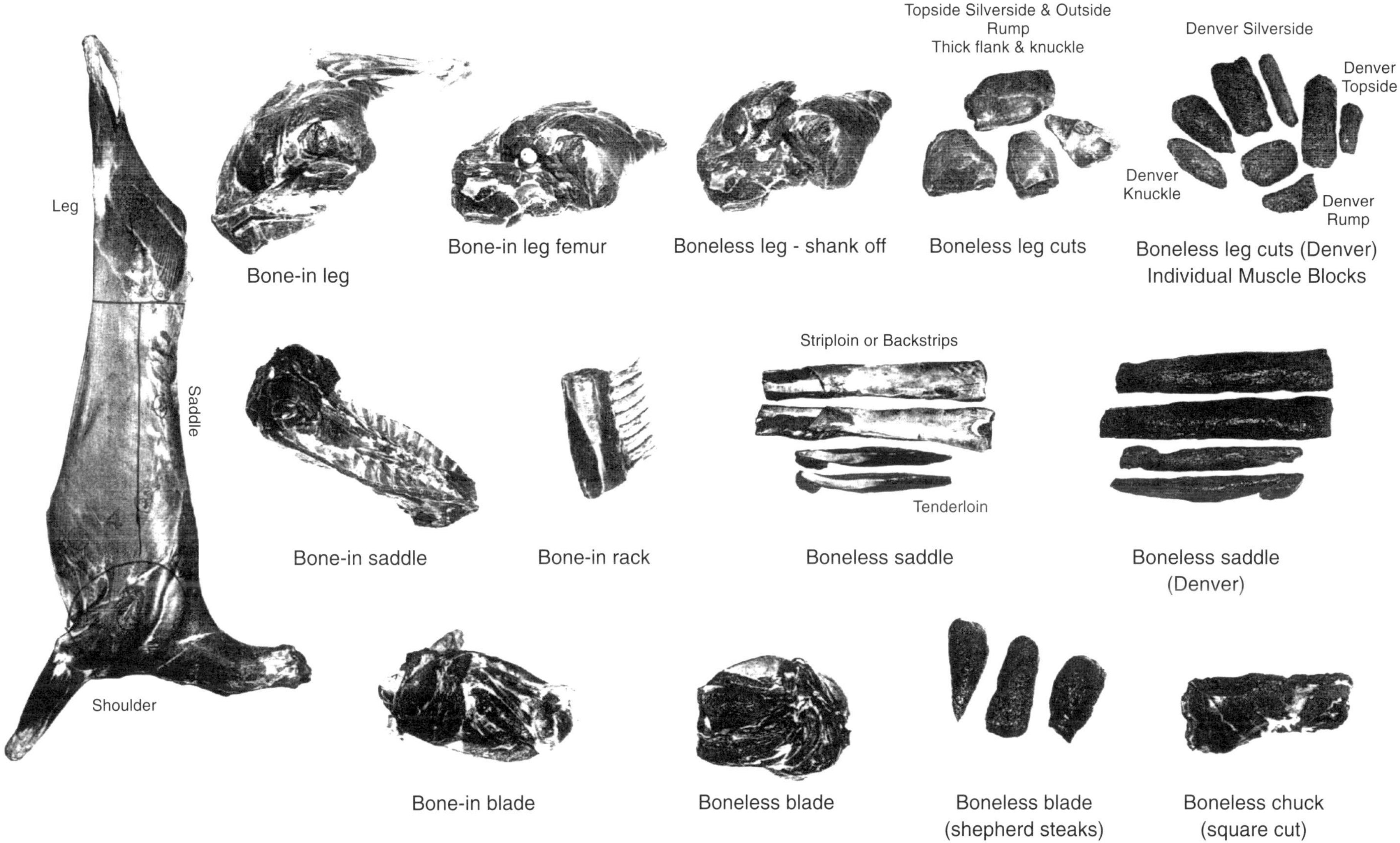

This page presents the usual venison carcass terminology. Often, for retail and restaurant marketing the individual cuts are boned out off the whole carcass, rather than in split carcass form. The terminology can be either the same as for beef and preparation for local and home use done accordingly, but the farm raised venison carcass does have its own terminology and distinct cuts. However, in relating to butchery a good knowledge of both terminologies is beneficial. These are represented in detail in Table 2.

According to the NZ system, GR limits for fatness in red deer are:

Grade	Carcass Weight Range	GR (mm)
AP3	<50 kg	<10
AP2	50-70 kg	10-12
AP1	>70 kg	<14
AT	(Trimmer grades)	+ 2-3/each class
AF	(Fat)	>+4/each class

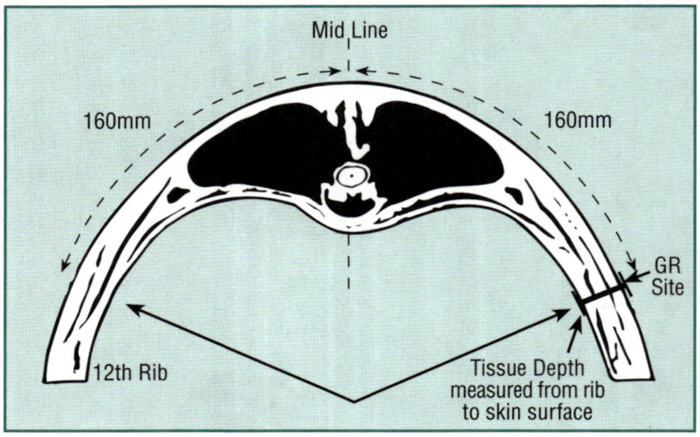

Fig. 1 Standard site of tissue depth measurement (GRD) in deer. GRD best represents the fat content of that animal.

Individual buyers are often targeting a very specific customer with strict cut specifications. They may use a price variation in 5 kg ranges within the broad fatness/weight ranges to attract the product they need.

Typically AT grades are priced at 20% less than prime and AF grades may be 50% less.

Very large or very light carcasses often fall outside the established range of market needs and may also be penalized with lower per kilogram returns.

Some buyers may also distinguish between hinds and stags based on the difference in weight. It takes an equivalent amount of work to dress a smaller carcass for less weight yield. Premiums for fallow deer have been established in certain markets, but these or any premiums for species type are difficult to maintain in the face of volume supply from New Zealand and other sources.

Research work has shown there is some value to be added in grading for age and fatness. Stags up to 36 months old show no change in tenderness, colour and texture, but toughness and strength of flavour increase beyond that. Hinds show highly acceptable quality at all ages, although they are prone to over-fatness above 55 - 58 kg carcass weight.

Adult stags are noticeably to grossly fat from the time of velvet harvest until the rut when behavioural characteristics and rapid changes in body composition make them unsuitable for quality venison production. Ideal slaughter time is from December to March after recovery post rut and they may be more appropriate for use in further processed products.

Fig. 2 shows the relationship of fat content to carcass weight and age. At the same age elk and hybrids which are slower maturing are always leaner and will yield consistently better than red deer.

Trials on castration in red and hybrid stags have shown few advantages, except that of temperament during the rut. Growth rates are appreciably slower with loss of velvet antler production, and fat content may reach unacceptable levels even at an early age.

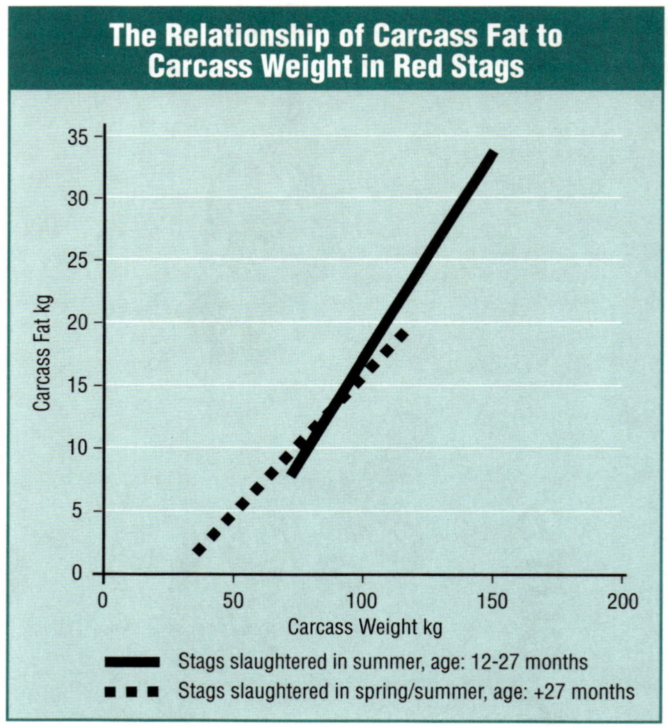

Fig. 2

WAPITI MEAT PROCESSING & QUALITY

As elk meat is primarily destined for the fresh or chilled market, processing procedures play an important role in maintaining quality. Reduction of contamination should be a priority. Elk can be processed under dry, non-washed conditions extremely well. The principal enemy is elk hair which is brittle, especially during coat change seasons, and difficult to control during depelting. Inverted dressing systems, where the carcass is prepared from the fore legs and brisket first and the skin mechanically rolled down over the loin and hindquarters, keeps the high value parts of the carcass virtually sterile and all knife work restricted to trim areas. Potential slaughter animals should be kept in mud and pond free pastures prior to slaughter, or washed thoroughly as a final resort.

Bruising and dark-cutting both substantially devalue a carcass, and both are the result of poor management and handling, prior to slaughter. Slaughter animals must be grouped with similar size, age and gender peers long before slaughter, and kept in those groups right up to entry to the kill box. Proper transport construction and handling will also reduce bruising.

Dark-cutting is a consequence of stress, resulting in an undesirable deep reddish-black meat colour. Ideally, our slaughter animals would be slaughtered on-farm, but failing that, we must handle them with appropriate care and caution.

Tenderness is the major consumer acceptance criteria and is most affected in post slaughter venison management by cooling and chilling. Low voltage electrical stimulation of carcasses is essential for elk being frozen or rapidly chilled (pH falls to 6.0 within 2 hours after stimulation and the meat can then be rapidly chilled for carcass breakdown or frozen without the risk of cold shortening). Meat can also be tenderized by holding it for extended periods above freezing point with ideal temperatures 10°C for 24 hours, 4°C for 24 - 48 hours post slaughter, however, some processing regulations may prevent such controlled aging.

Elk, as a quality high value food item, warrants particular attention in processing and handling. Quality production and processing, on and off farm, will ensure that farm raised elk meat production remains both profitable and in a situation of growing demand.

Chapter 1 The Nature of Antler

The annual growth, and harvest of velvet antler offers deer farmers an attractive product source and return, but carries with it a number of management and welfare responsibilities that cannot be ignored. Antler harvest forms the basis of traditional deer farming industries in China, Korea and Russia. There is considerable market potential in these countries and in western culture, with recognition that the pharmacological and bioactivity properties of prepared antler are

real and diverse in many medicinal and tonic applications.

Antlers are unique to the deer family, and a focus of man's interaction and appreciation of them. From a farmed viewpoint, an understanding of the growth, form and function of antlers allows for successful management, either for harvest, or should the choice be taken, to allow the herd sires to grow hard antler and realize the full expression of a breeding and selection program.

GROWTH AND THE SEASONAL CYCLE

Following the rut and winter recuperation, levels of the male hormone testosterone fall to negligible levels. Under these conditions the annual cycle of casting, antler growth and hardening (calcification and mineralization) begin. Testosterone levels remain low through spring and summer as velvet antler grows rapidly. After the summer midpoint (the solstice), testosterone levels begin to rise and signal the hardening and mineralization process. The velvet skin covering the antler is shed, and the antlers become coloured by display activity and territory marking. The annual pattern is illustrated in Fig. 1. They emerge as the physical weapons of dominance and the visual indicators of breeding potential. Changes in antler structure match the changes in behaviour and appearance that accompany the ability to breed successfully in a narrowly defined breeding season. To correctly understand the various stages of these events the terminology needs to be explained (Fig. 2).

PEDICLE

Antlers grow from pedicles, permanent living bone projections of the skull outgrown from the frontal bones. They begin to grow at three to seven months of age, with the speed determined by juvenile maturity and a threshold body weight. Around one year of age, pedicles generally grow the first antler, a single spike, although with advanced genetics and good feeding multibranched spikes are increasingly common. The pedicle is distinguished from the velvet antler itself by a zone of transition in hair style and colour. It is not as clear in yearlings. Care must be taken not to damage the pedicle through accident or when removing antler as permanent damaging

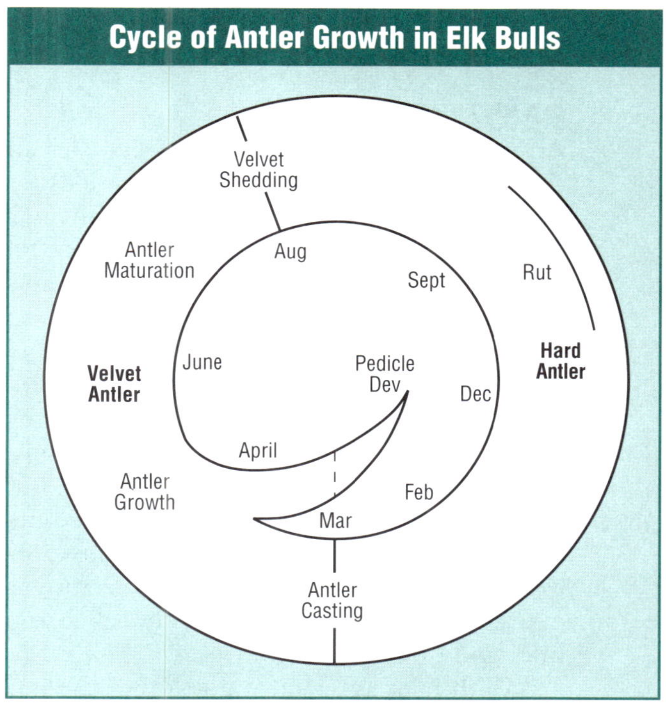

Fig. 1

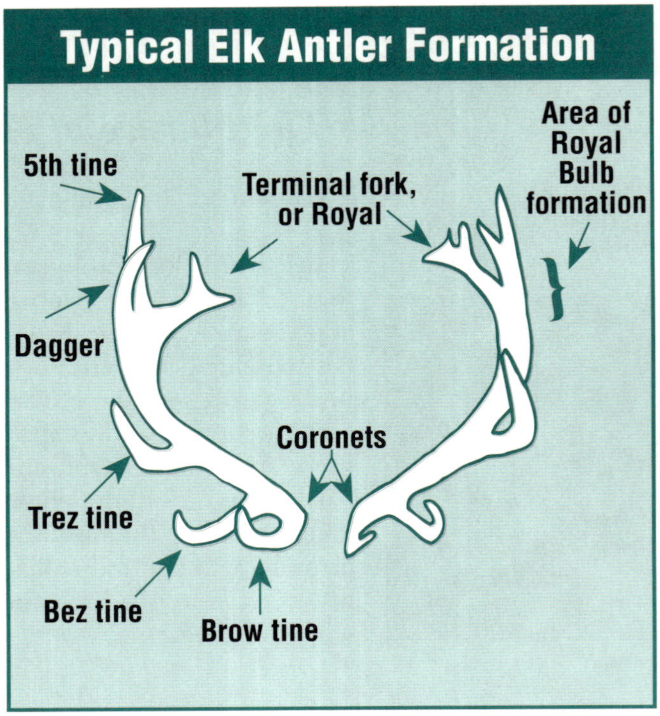

Fig. 2

effects on antler growth and style will result.

The timing of pedicle growth and antler growth is dependent on nutrition, and genetics. On-farm well fed animals reach a threshold body weight early and begin pedicle and antler growth in the middle of the first winter. Speed of development is varied both in time and rate. Weather, particularly sunshine, also has an influence.

Strong pedicle diameter at a young age may be an indicator of good antler potential, but not always.

CASTING

Annually (March - April - May) the hard antler, or the "buttons" from velvetted stags, are cast. This is under hormonal control directed by changing daylight length. A thin junction between the pedicle and antler resorbs the bone elements, and shortly before casting the pedicle will enlarge and show a swollen blackened skin at the base of the antler. Casting of both sides typically occurs within 3 - 12 hours of each other and there may be slight bleeding. The wound rapidly heals and velvet growth begins.

Occasionally, in previously velvetted bulls a button may not drop and may become embedded in the new growth. This should be removed with care, so that ongoing growth is not deformed or infection does not set in.

With increasing age a portion of the pedicle is lost at each casting and in older animals new antler growth will start at skull level.

CORONET

In two year and older bulls the coronet is the growing point from which the antler develops, and like the pedicle is extremely susceptible to mechanical damage affecting growth and form. When cutting antler, (soft or hard) the cut should be made 8 - 12 mm above the coronet/pedicle junction as any damage remains permanent.

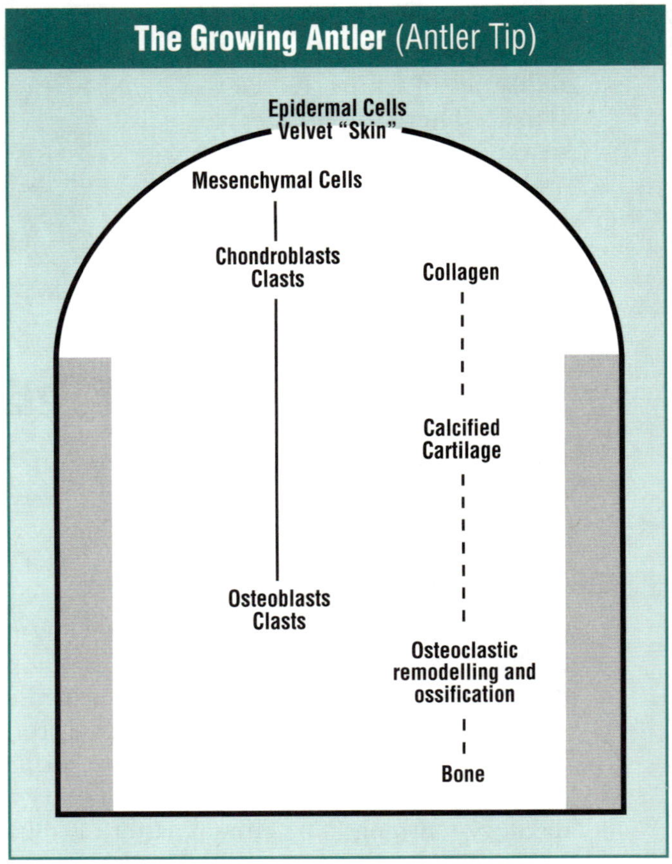

Fig. 3 Pattern of cell types and differentiation of the growing antler (adapted from Banks 1974 in Banks and Newbrey 1983)

VELVET ANTLER

Velvet Antler is an actively growing cartilagineous type tissue, which is supplied with a complex vascular and nervous system (Fig. 3). The skin surface covering the antler is termed velvet and is variable in colour. It is often very dark; black or dark brown in red deer and a distinctive light orange brown in some elk strains, although greys and blacks are common. The skin is covered by short hairs growing at right angles to the surface and the whole antler often has a greasy feel from sebaceous gland secretions.

Antler grows from the tips of the various tines and main beam. It has a corresponding gradation of tissue types moving down the antler from an undifferentiated cell growing mass, changing to a cartilage matrix, a honeycombed cartilage and finally to calcified cartilage and early bone formation at the base.

Oriental Systems for Classifying the Sections of the Velvet Antler

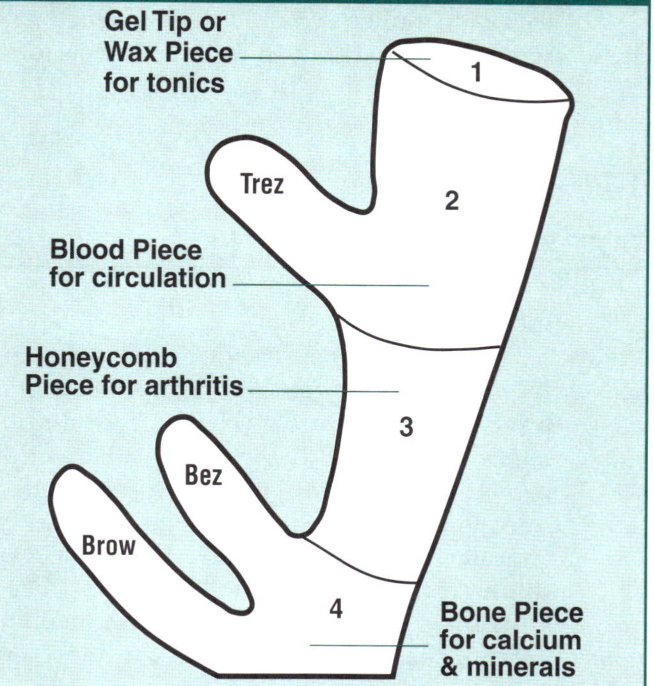

Fig. 4

Calcium, Phosphorus and Ash contents of a single velvet antler cut at the A grade stage classified according to Oriental description of the velvet antler sections

Velvet Antler Section	% of Dried Velvet Antler Weight	Composition		
		Ca	P	Ash
1	2.2	3.9	24	22
2	30	4.6	16	16
3 - main beam	24	3.5	21	21
3 - brow, bez	26	5.7	26	24
4	18	5.5.	20	19
Entire	100	4.7	20	19

Table 1 Dried velvet antler components in relation to quality and oriental antler description

Such differentiation is shown in the traditional Oriental antler grading system - a system that reflects both price and pharmacological properties (Fig. 4 and Table 1).

Antler Growth Patterns

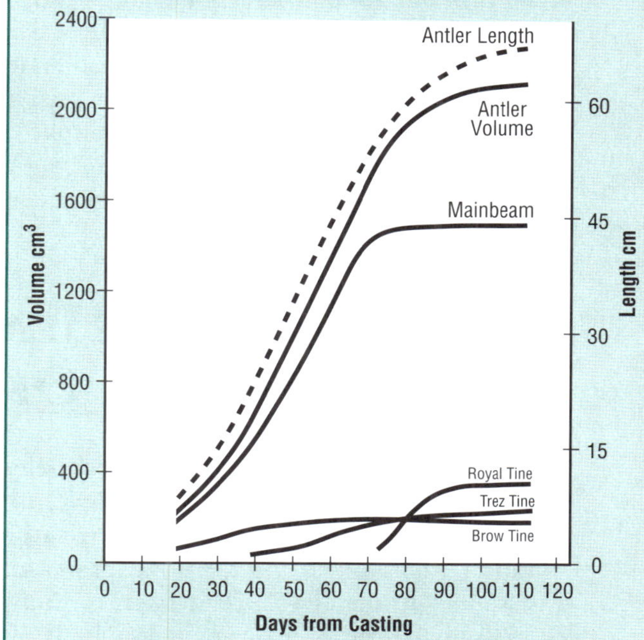

TINES

Tine growth follows a distinct pattern. They start as buds developed from the main beam and grow at differential rates to the faster growth of the main beam. The bez tine can be variable in size and position on the antler, and can be absent either on one or both sides.

This may not occur every year and it is important to have knowledge of the time of casting and the stage of development to properly determine whether an unusual bez tine is growing, or whether the trez tine is emerging (Fig. 4 and Table 2).

Velvet Growth Events	
1. Pedicle swelling	7 days
2. Casting of old antler	0 days
3. Brow bud formation	14-16 days
4. Bez bud forms (if present)	30 days
5. Trez bud formation	48 days
6. Royal bulb formation	60 days
7. Max bulbing	68-75 days
8. Complete growth	115-130 days

Table 2 gives a broad guide to the timing of antler growth

ROYAL BULB

The top of the growing beam is used as the major assessment of whether antler is ready to harvest or not. Typically it thickens above the trez tine and will bulb depending on age and the extent that royal tines will branch off. An average growth period to this stage is 63 - 65 days after casting but is extremely variable and an individual feature. The farmer should not sacrifice quality for weight and as the task is to determine when the antler has reached maximum growth prior to the onset of rapid calcification at the base, he must observe the whole antler.

Ideal harvest time is when the tip of the royal bulb just begins to flatten, or in larger antlers a small crease (0.25 - 0.5 cm) will appear in the bulb area. The bulb should be at about 80% of the height from the brow to the trez tine, above the trez. If, however, the brow or bez tines show signs of turning up and pointing, calcification has begun and the antler should be harvested irrespective of bulb development.

As a general rule, younger animals must be harvested earlier and only thick beamed antlers left to grow to the limits of royal bulb development.

Using EVA for Health - the Traditional Oriental Approach

Traditional Oriental medicine has been keeping Asians well for thousands of years. Fundamental to its success is the practice of preventive, rather than curative, techniques, and the concept of balance. Elk Velvet Antler, EVA, is an essential component of many balanced preparations. The basis of this approach to maintenance of health is the philosophy of Yin and Yang, cosmic forces which are said to control all natural phenomena and life processes. Both forces are dependent on each other, and the ideal state within the body and within the universe is to have the two forces in relative balance and harmony. Traditional Oriental medicine is used to keep the two in balance, preventing the ill health that results whenever imbalance occurs.

EVA is so highly valued because the deer is the animal with the greatest Yang energy, recognized as a good luck animal and associated closely with health and longevity. Yin and Yang have their own domains within the human body. Half the vital organs pertain to the Yin and the other half are part of Yang. Yang energy is the force that carries blood throughout the body. In Chinese medicine, EVA is the main medicine for increasing blood volume and its components as well as blood flow. When blood is increased more energy will be generated and the body will be maintained in better health. This enhanced blood activity aids in promotion of health and the prevention of illness through righting the Yin and Yang balance. Often a doctor is only paid when the patient is well, not sick - a practice which probably would not be greeted with enthusiasm by western medical professionals.

The Traditional Oriental doctor does not simply look at the patient's symptoms when prescribing a treatment. He also takes into full consideration many other things such as the patient's constitution, medical history, age, possible side effects and even the patients physical appearance and personality. EVA is not prescribed to everyone, and it is not always prescribed in the same form or mixture. People with plenty of Yang energy, characterized by their hot body, aggressive and positive nature, do not need to take EVA regularly. On the other hand, EVA is perfect for people with more of the Yin forces - laid back, negative or timid nature, cool body people.

A book called Shin Nong Bon Cho Kyung, written 1800 years ago, suggests, "Deer velvet tastes sweet and its property is warm. It is used for reinforcing vital energy, strengthening memory and will, generating teeth, curing persistent vaginal blood discharges, and treating fevers and epilepsy. Deer antler cures sores, carbuncles (boils), expels pathogens as well as retained blood in the uterus. It is also used for treating consumptive disease and illness caused by over-exertion, lumbago, excessive loss of weight, repairing the body, reinforcing vital energy, curing infertility, stopping pain and preventing miscarriage. Prolonged consumption would keep the body light and extend longevity."

Today, 1800 years later, velvet is still being used to prevent and right many ills in Asia. Westerners have only recently begun to make use of this ancient remedy and tonic. Most have been unwilling to take it seriously because of the notion that it is an aphrodisiac. Velvet is used to remedy sexual disorders in men such as impotence, premature ejaculation and watery semen. According to modern research, velvet antler has shown gonadotropic activity. This means that it can stimulate growth and increase the weight of both the seminal vesicles and prostate. But few users claim any direct aphrodisiac effects, rather they suggest that as their health improves with EVA, their libido increases correspondingly.

Positive effects have been reported when EVA is used to increase the blood, remedy kidney deficiency and gastro-intestinal disorders, to improve low blood pressure and eliminate anaemia. It is also used successfully to promote rapid healing by aiding the regeneration of damaged tissue, to treat menstrual disorders and menopause in women, and to improve weak bones and sinews. In Asia, it is mostly consumed in spring and autumn and is often prescribed with a mixture of other herbal medicines, commonly served as a soup. EVA is usually divided into four sections, classified according to the different composition of each section. Each section has different uses. The two upper sections are used as preventative medicines and tonics for children and young people while middle portion is used in the treatment of arthritis and osteomyelitis. The lower part is regarded as being of particular benefit to older people subject to calcium deficiency and osteoporosis.

Dr. Lee Sang-In of Kyung Hee University in South Korea says velvet is also used to treat infertility in woman and promote skin growth where there has been long lasting infection. Liver problems and high cholesterol levels can also be alleviated with velvet, he says.

EVA can boost hormonal activity, increase oxygen uptake to the brain, liver and kidneys, decrease the rate of muscle fatigue, and promote muscular growth. It can be very useful for both physical and mental activity as well as promoting healing. But EVA should not be casually used by everyone. It should not be used or should only be used cautiously in people with prostate problems, heart conditions and lupus. EVA is a powerful substance, and consequently we must be aware that improper or excessive consumption may not be beneficial. The best advice is to try EVA in small amounts for several months, and notice the effects. Think of the balance associated the calming Yin force and the energizing Yang. Consider what your individual nature is, what you need to improve, and consider how EVA has helped so many other people. Talk to people who understand natural health products, and enjoy the effects of improved health!

RESULTS OF WESTERN MEDICAL RESEARCH REGARDING THE EFFECTS OF SUPPLEMENTATION WITH EVA

Writing in the North American Elk magazine, Frederick Obey summarized the results of modern research on velvet antler. This research has been conducted in eight general areas:

Hormonal Effects Velvet antler has been shown to boost testosterone levels. (This hormone promotes male characteristics.) It has also been shown to boost estrogen levels. (This hormone promotes female characteristics.) The estrogen hormone most affected by velvet antler is estradiol. Estradiol is a precursor to testosterone.

Blood Building Effects Tests with preparations of velvet antler have shown it to increase production of both red blood cells and white blood cells. These facts are linked with velvet antler's ability to increase oxygen uptake to the brain, liver and kidneys.

Protection Against Stress Velvet antler helps the body to maintain homeostasis against heat, cold and electric shock. This has been linked to velvet antler's ability to decrease mast cell degranulation.

Protection from Liver Damage In a controlled test, velvet antler helped to protect lab animals from liver damage by carbon tetrachloride.

Stimulation of Growth Velvet antler is an extremely rich, fast growing tissue that contains many growth factors. Elk can grow up to 50 pounds of new bone in approximately two months. Due to its fast rate of growth, deer antler is being looked at as a model for studies on osteoporosis, arteriosclerosis, as a possible graft for healing fractures and as a model for cancer studies.

Aging Retardation Recent studies in Japan show that velvet reduced signs normally associated with senility. This is probably due to its hormonal effects.

Recovery from Traumatic Injury Antler, as an extremely fast growing tissue, is comprised of many cell types. These include: fibroblasts, chondroblasts, chondrocytes and others. All of these are required for healthy growing bones and tissue. Velvet antler is high in calcium and phosphates which aid in the healing of bones. Research has shown that velvet antler helps to heal neural (nerve) tissue. This mechanism can be explained by velvet antler's ability to enhance glycosis to nerve tissue. Velvet antler's relationship to recovery from traumatic injury could possibly explain its effectiveness against arthritis.

Reduction of Blood Pressure Blood pressure reduction is due to velvet antler's ability to increase dilation of the peripheral blood vessels.

Velvet antler is also widely known as an aphrodisiac. Its reputation as an aphrodisiac originated through ancient medicine and folklore from the East. According to modern research, velvet antler has shown gonadotropic activity. This means that it can stimulate growth and increase the weight of both the seminal vesicles and prostate.

Velvet antler can boost hormonal activity, increase oxygen uptake to the brain, liver and kidneys, decrease the rate of muscle fatigue, and promote muscular growth. It is extremely useful for both physical and mental activity as well as promoting healing.

Velvet should not be used or should only be used cautiously in people with prostate problems, heart conditions and lupus.

VELVET DEVELOPMENT

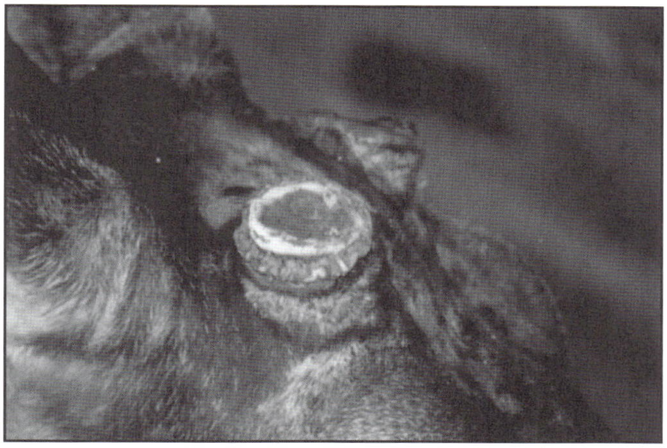

1. Just prior to casting, the pedicle enlarges, showing a ring of swollen blackened skin at the base of the antler button.

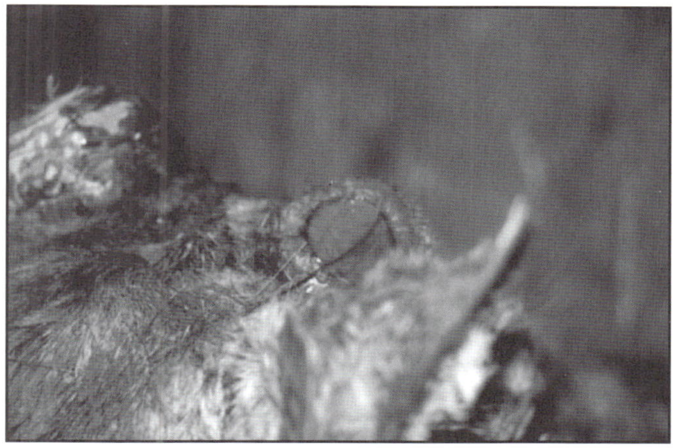

2. Just after casting, some slight bleeding may occur.

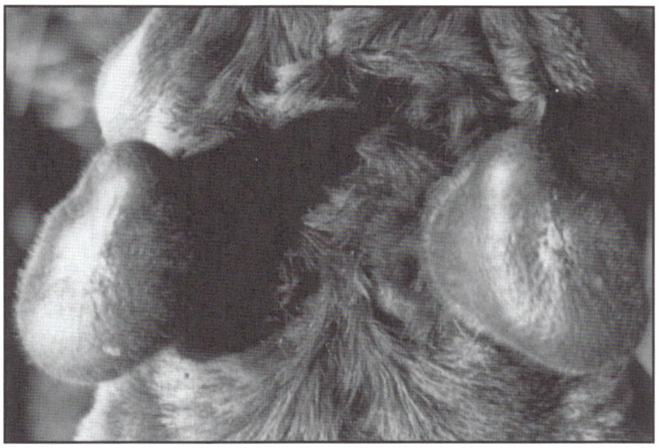

3. Brow bud formation: 14 -16 days after casting.

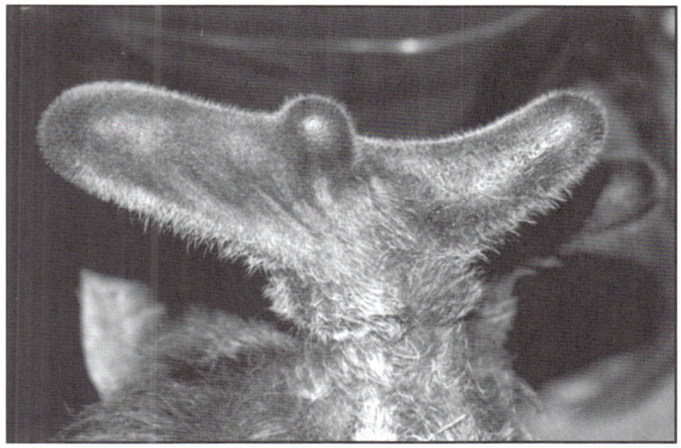

4. Bez bud formation: approximately 30 days of growth.

5. About 32 days growth. Strong brow and bez development.

6. Well-developed 3 year old at about 70 days. Ready to cut.

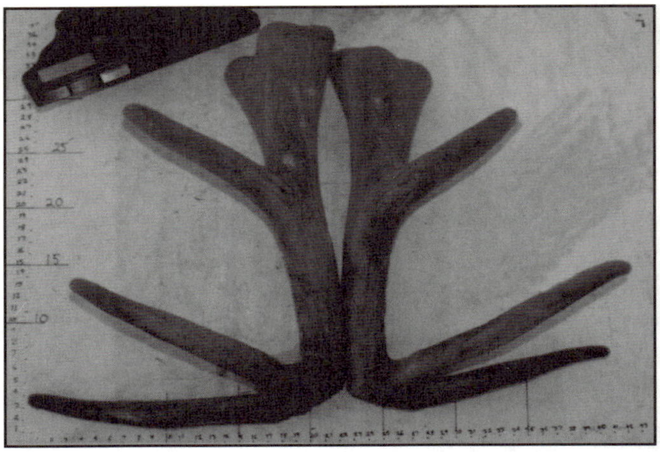

7. *A good balanced elk "head" with exceptional "tops". Cut about 70 days.*

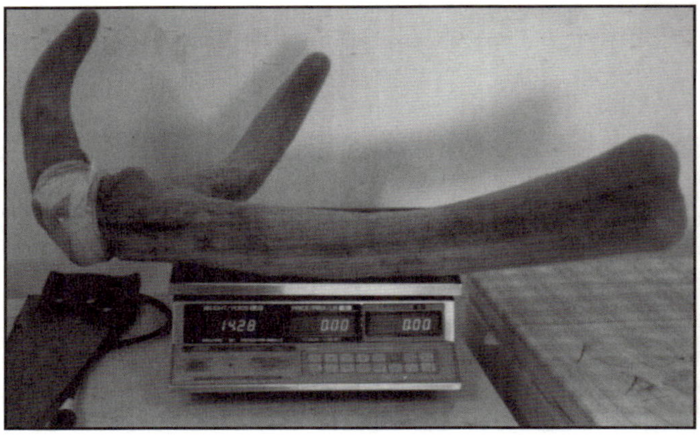

8. *Elk - 6 year - 64 days, cut right. Note: trez missing*

9. *3 year Elk, 70 - 75 days, close to overgrown for this age.*

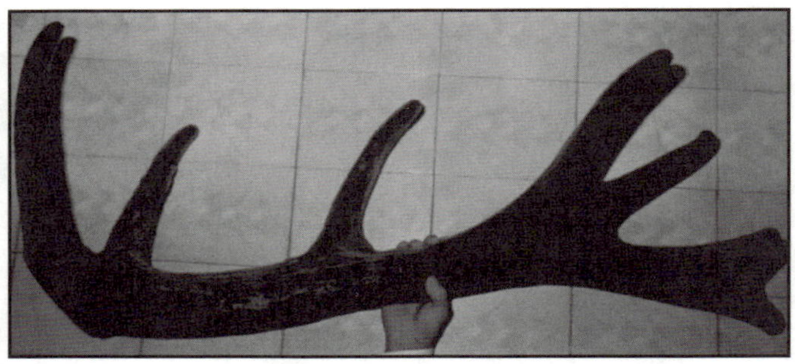

10. *Russian Antler probably cut at at least 90 days, but note the brow & bez are still straight and furry.*

Management for Optimum Growth

The following broad features of antler growth are important for production and are significant for age, breed type and individuals.

- Older bulls cast significantly earlier than younger bulls with up to six weeks difference to two year olds, but individual variation is very high.
- The pattern and timing of tine initiation and growth is broadly similar over all ages.
- Velvet antler weight increases with age until at peak production (elk 8 - 10 years old) and then stabilizes or gradually declines.
- A general relationship exists between liveweight and velvet antler weight within an age group and a blood line.
- Across breed types, elk yield more velvet per unit liveweight than red deer and take longer to grow antler while still producing premium grades.
- Above the trez tine the main beam will grow rapidly relative to the development and differentiation of the royal bulb (up to 2 cm per day) and appreciable gains in weight can be made by closely watching the growth patterns (up to 0.5 kg in 7 days).
- Once the royal bulb flattens, differentiation into royal tine buds is rapid as is calcification of the basal tines. Severe penalties exist for overgrown product in spite of the apparent greater size and weight.

• Older bulls, with greater beam and earlier growth and casting, can be grown out to a greater extent. Young animals should be cut earlier rather than later to preserve quality.

• The whole antler must be assessed in determining cutting stage as brow and bez tines can calcify rapidly as indicated by pointing and upward curvature of the tips.

It can be clearly seen that once the main bulb has creased there is rapid calcification of the lower tines, with little addition in weight in relation to the large drop in quality. The growth in antler length from stage 1 - 2 is significant in weight terms with little loss in quality.

It pays then to observe bulls on a daily basis once the beam has developed from above the trez tine. Older animals, with thick beamed antlers can be left to grow out more than young animals or thin antlers.

Production Trends in Farmed Deer Antler Growth

	Breed Type			
	Red	F2-25% Elk	F1-50% Elk/Red	Elk
Velvet antler yield (kg, adjusted to 65 days, average)				
3 years old	1.9	2.5	3.1	5.5
6 years old	2.9	3.8	5.5	7.5
Casting Dates (expect variation +/- 3-9 days)				
3 years old	11 April	10 April	8 April	5 April
6 years old	28 March	4 March	16 March	14 March
Days until harvest (individual variation +/- 5-10 days)				
3 year olds	56	59	64	69
6 year olds	69	69	75	75
Liveweight (kg, average)				
3 year olds	150	205	255	320
6 year olds	195	270	330	440
Relative antler yield (Red deer at 1.00, unadjusted)				
3 year olds	1.00	1.3	1.6	2.9
6 year olds	1.00	1.3	1.9	2.6
Relative antler yield (Red deer at 1.00, adjusted*)				
3 year olds	1.00	1.02	1.08	1.63
6 year olds	1.00	1.01	1.28	1.52

* Antler yield adjusted for relative liveweight and metabolic efficiency. Research has shown that, in cold climates, an increase in LW of 100% results in an increased food intake of only 68%, a fact which allows us to adjust for both LW and efficiency.

Table 3 Production Trends in Farmed Deer Antler Growth

SELECTION AND FEEDING

NUTRITION

Somewhere between 35 - 40% of the variability in velvet antler is of genetic origin and therefore heritable. The animals must be fed well to express this genetic potential, but there is little evidence to suggest that feeding and genetics influence the market perception of velvet quality. Evidence from nutritional trials confirms only that underfeeding in late winter and during actual antler growth impacts on velvet production. A high protein diet during velvet growth has little effect on velvet antler weight, but may improve hard antler weight.

A variety of complex formulas and different diet additives including herbs, browse, kelps, lichens and vitamin, mineral and protein additives have been tried with little repeatability from year to year or consistent achievement. Individual producers may strongly defend their use, but the evidence would suggest otherwise.

Feeding a high quality, high energy, diet (adlibitum) towards the end of the rut in early winter, to recover lost body condition, seems to be the most positive step to take to improve velvet yields. Under North American winter conditions this type of feeding will be maintained pre and post casting as an aid to production until spring grass growth is underway.

Pre-rut body weight and condition have an influence on velvet antler production. This reinforces the importance of quality feeding.

Average Composition of Hard Antlers

Mineral		Organic	
Calcium %	19.8	Organic matter %	48.0
Phosphorus %	10.1	Protein %	36.3
Magnesium %	1.1	Energy (Kcal/gram)	2.1
Sodium %	0.5		
Potassium%	0.1		
Iron, ppm	55.0		
Manganese, ppm	6.6		
Zinc, ppm	116.00		
Copper, ppm	48.0		

Table 4

Rank on yearling liveweight	Yearling liveweight (kg)	Cumulative (2 to 5 years) velvet antler production (kg)
Top 1/6	119.6	10.06
1/3	107.5	9.14
1/3	101.6	8.88
Bottom 1/6	94.1	8.57
Overall mean (/S.D.)	105.3 (9.57)	9.11 (1.60)

Table title: **Cumulative velvet antler production from 2 to 5 years of age for a group of 90 red stags categorized according to their yearling liveweights**

Table 5a Cumulative velvet antler production from 2 to 5 years of age for a group of 90 red stags categorized according to their yearling liveweights

Rank on 2 year old velvet antler	2 year old velvet weight (kg)	Cumulative (2 to 5 years) velvet production (kg)
Top 1/6	2.18	11.37*
Next 1/3	1.55	9.35
Next 1/3	1.29	8.47
Bottom 1/6	1.02	7.69
Overall mean (/S.D.)	1.48 (0.43)	9.11 (1.60)

Table title: **Cumulative velvet antler production from 2 to 5 years of age for a group of 90 red stags categorized according to their 2 year old velvet antler production**

* Gain in selection on antler weight alone

Table 5b Cumulative velvet antler production from 2 to 5 years of age for a group of 90 red stags categorized according to their 2 year old velvet antler production

Breeding option	Velvet antler production (kg) as 4 year old			
	Breeding Animals		Genetic progress/ generation	Yield of progeny
	Stags	Red deer hinds		
Selection within strain				
Average stags	2.5	2.5[2]	0.0	2.50
Top 3% of stags	3.6[1]	2.5	0.22	2.72
Hybridization between strains[3]				
A. Average Stags	3.6	2.5	0.55	3.05
Top 3% of stags	5.2	2.5	0.87	3.37
B. Average Stags	5.2	2.5	1.35	3.85
Top 3% of stags	7.5	2.5	1.81	4.31

Table title: **Comparison of various breeding options to improve velvet antler production (comparisons as 4 year olds)**

Table 6 Comparison of various breeding options to improve velvet antler production (comparisons as 4 year olds)

SELECTION

Rapid improvements in velvet antler production can be achieved by selecting sires for velvet production and liveweight. Antler weight at two and three years of age is positively correlated to both liveweight at 15 months of age and subsequent antler production. Liveweight at 15 months has a positive correlation to future velvet production, as do two and three year old cuts.

But if the objective is solely to improve antler weight, then selection on antler weight alone will make faster progress. (Tables 5a & 5b)

It is important to record casting dates so an accurate velvet growth, corrected for age, can be estimated. Casting dates are easily recorded during late winter feeding by simple observation. Selection for the top 5% of young stags ensures that progress is rapid.

Breeding and selection within a strain is one option. There are even more dramatic gains to be made by crossbreeding between divergent strains of East European deer (A) or elk (B) as shown in Table 6.

The illustration A could represent a European stag x base hind situation. B is typical of an Elk cross Red potential.

In 1990 - 1991, Bruce Friedel and Robert Hudson surveyed the Alberta velvet industry to establish baseline performance data on elk velvet production. They also estimated which velvet production parameters are more useful in predicting velvet antler performance, and their practical applications in the genetic selection process.

The average age of bulls in the survey was 3.69 years, with the oldest bull recorded being 15 years old. Mean pre-rut weight was 766 lbs. A total of 1,026 mixed-age bulls were velvetted, producing 10,750 lbs, averaging 10.5 lbs/bull. Velvet production was affected more by pre-rut body weight than bull age (Fig. 5) and increased 3 lb between two and three years of age, 2.5 lbs between years three and four, and 1.6 lbs between ages four and five.

Older bulls cast their antler buttons earlier than younger bulls. Weight within individual age classes also had a slight influence on antler casting dates and antler casting date had a marked effect on velvetting date across all age classes.

Producers in the survey velvetted their bulls between royal and sur-royal initiation (indentation of the fourth and fifth points before antler calcification occurs) after 59, 69, 67, and 71 days of

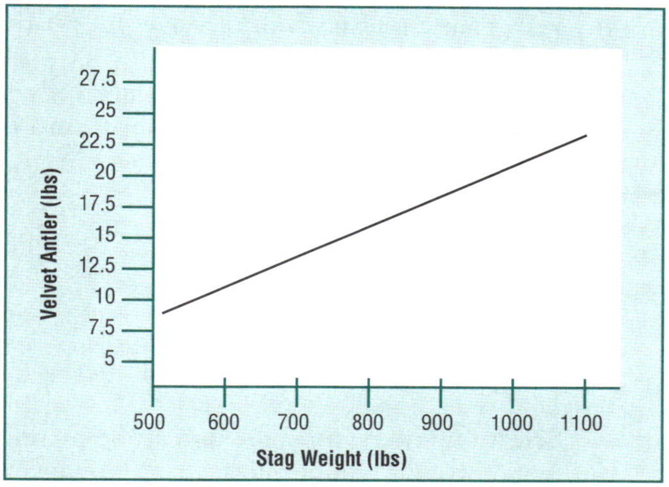

Fig. 5 The effect of stag weight on velvet antler production

growth in two, three, four, and five year old bulls, respectively. An increase of three days between antler casting dates and velvetting dates resulted in an additional 1 lb of velvet antler.

Comparisons between velvet growth can be made on a metabolic weight basis, with elk (79 g W-0.75 of bull) being more efficient velvet producers per unit of live weight than the smaller red deer (50 g W-0.75 of stag) in New Zealand. Velvet production efficiency peaked at age seven, which has important implications for managing the bull herd.

Antler velvet production was related more strongly to past velvet yield (Fig. 6), than either age, body weight or antler casting dates. Within age groups, pre-rut body weights were related to velvet antler weights, which indicates an effect of frame size on velvet antler growth in elk, similar to red deer in New Zealand. Improvements in velvet antler production in elk are possible if producers select bulls on the basis of velvet antler yield adjusted for age and live weight.

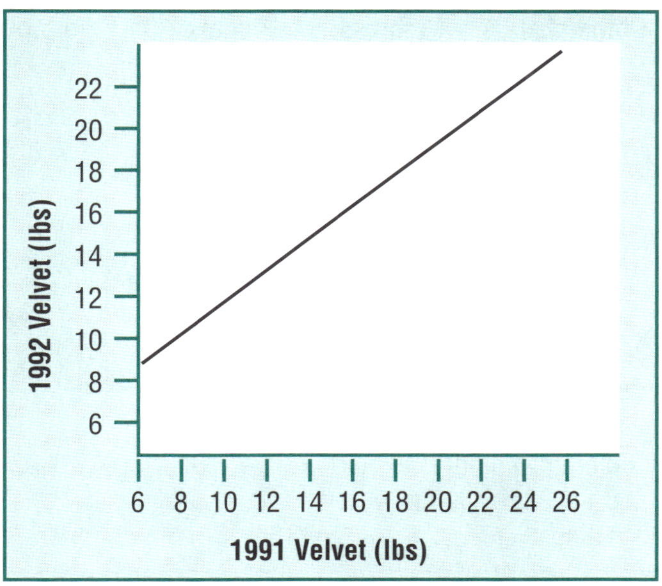

Fig. 6 The effect of past velvet antler performance on velvet production

SELECTING FOR IMPROVED VELVET PRODUCTION

About 35 percent of the variability in velvet antler weight is of genetic origin, which makes it a moderately heritable trait. While the relationship between body weight and velvet antler weight is well established, individual variation of antler weights within an age group are a better prediction of subsequent velvet growth. This is also the best selection index if the sole objective is to improve velvet production. The selection process requires that producers know the mean +/- standard deviation of velvet production within an age group, after which the expected rate of increase can be calculated using an equation for estimating gains in velvet antler production:

$$\text{Genetic gain per year} = \frac{\text{Heritability x Selection differential}}{\text{Generation interval}}$$

Phenotypic deviations of velvet production between average velvetting bulls and selected breeding bulls of similar ages become the actual selection differentials. Generation interval reflects the average age of parents when progeny are born and can be estimated by adding average cow age (five years in Alberta) to bull age and dividing by two. Genetic gains in velvet antler production are estimated in Table 7. Producers that use older breeding bulls and have complete velvet records can calculate genetic gain with more accuracy. If a producer wants to use a six year old breeding bull and has velvet records from age two he can calculate heritability by:

$$\text{Heritability (5 yrs)} = \frac{\text{number of yrs x .35 (heritability 1 yr)}}{\text{1 +[number of years -1] x repeatability (9.05)}}$$

Example for a six year old breeding bull:

$$\text{Heritability (5 yrs)} = \frac{5 \text{ x } .35}{1 + [5 - 1] \text{ x } 9.05} = \frac{1.75}{4.62} = .33$$

Recalculate an average selection differential for those five years (say that six year old bull cut 7.5 lb above average each year). The expected genetic gain is:

$$\text{Genetic gain from your six year old bull (lbs/yr)} = \frac{.38 \times [(7.5 + 0)/2]}{6.0}$$

Within Alberta's elk industry there appears to be a higher selection differential among older aged bulls, which suggests that there is more variability in velvet antler production as bulls age. Most elk producers use older breeding bulls with proven velvet antler performance, but as generation intervals increase the expected rate of genetic gain slows. If we consider the impact that a selective breeding program has on both male and female progeny, the use of highly selected younger-aged bulls is more dramatic. As selection pressure on a particular trait increases, the variability or 'selection differential' usually decreases, resulting in slower genetic progress.

Annual rates of genetic gain may seem insignificant to many producers, but the implications of selecting better velvet producing two year old bulls for breeding would improve velvet growth in progeny born 10 years hence by 2.6 lb per year (0.26 lb/yr x 10 yr = 2.6 lb) Elk bulls born in 2003 producing 2.6 lb per year more velvet during their productive life span (aged two to 11), could yield a farmer an additional 26 lb of velvet antler! That's a lot of velvet for relatively little effort.

Velvet selection programs are only as good as the records that are kept, and their interpretation.

Essential to these records are:

- Days of growth - harvest date - casting date.
- Corrected antler weight (2 yrs old) ★ see Table 14.
- Antler weight (mature).
- Grade - beam thickness and tines.

Selection criteria must be high as the present market trends will penalize small thin velvet, overgrown velvet, and poor quality. It takes some years to establish a quality mature herd, given the age that bulls reach peak production, but the potential rewards and satisfaction are many if this pathway of elk farming is chosen. However, no matter how well the animals respond to nutrition and selection, if management of yarding and harvest are poor, then the effort is partially wasted. There should also be selection pressure for handling and temperament. Quiet bulls, well managed, should never break or damage antler and with division into appropriate group sizes can be

Expected Improvements in Velvet Antler Production from Following a Progeny Recording and Selection Program

Stag Age	Alberta Av. Velvet (lb)	NAEBA Winners	Selection Differential [1]	Genetic Progress [2]	Yield Progeny (lbs) [3]	Percent Improvement [4]
2 yr olds	7.8	13.9	6.1	1.22	9.02	0.16
3 yr olds	11.1	22.0	10.9	2.19	13.26	0.20
4 yr olds	14.2	30.5	16.4	3.27	17.42	0.23
5 yr olds	17.5	34.6	17.1	3.43	20.89	0.20
6 yr olds	19.9	37.5	17.7	3.53	23.38	0.18
7 yr olds	22.1	42.4	20.4	4.07	26.12	0.18

1. Selection Differential = breeding stag's velvet yield (NAEBA winner) - average stag's velvet
2. Genetic Progress = ([sel. differential stag + sel. differential hind] /2) x hert. = ([6.1 + 0] /2) x .4 = 1.22 lbs.
3. Yield Progeny = av. stags + genetic progress = 7.8 + 1.22 = 9.02 lbs. velvet
4. Percent Improvement = genetic progress /average stags = 1.22 /7.8 = .16 or 16%

Table 7

handled in the most simple facilities.

The conversion factors to a 55 day or 65 day standard are given in Table 8. A 40 day antler is about half (0.51) the weight of a 65 day antler; therefore to convert the antler weight, you simply divide the actual velvet yield by the appropriate day factor from the table. This table is designed for red deer, and may not be accurate for elk.

Actual growth period- days from casting to harvest	Adjustment factor for standardized growth period of	
	55 days	65 days
35	0.51	0.41
40	0.63	0.51
45	0.76	0.61
50	0.88	0.71
55	1.00	0.80
60	1.12	0.90
65	1.24	1.00
70	1.37	1.10

Table 8 Adjustment factors for converting two year old velvet antler yields to a standardized period of growth (days from casting to harvest)

MANAGEMENT OF VELVET ANTLER HARVEST

- Organize groups on casting dates, maximum 20 - 25 animals.
- Familiarize bulls with yards during winter and casting, not when they are in full velvet.
- Establish a routine; same vet, same handlers, same time.
- Welfare considerations, drug safety and hygiene are paramount.

Based on grouping of casting date, velvet herds should not exceed 20 - 25 animals under most situations.

No amount of selection, breeding and recording can compensate for poor management, harvest and storage techniques, which result in damage to velvet, or spoilage during storage prior to processing. The following summary highlights management factors important in preserving animal welfare, and optimum velvet antler returns for producers.

ANIMAL HANDLING

Bulls, particularly two and three year olds should be thoroughly familiar with the yards and squeezes or restraints long before harvesting. With casting dates being recorded, it is appropriate to group animals in small mobs according to casting date to allow easy assessment of the stages of velvet antler development before removal. Therefore, it is necessary to yard only those animals ready for harvest over a 10 - 14 day period. Yard familiarization can be achieved during early spring as they cast. Older animals will naturally thereby be grouped separately from younger and later developing bulls.

Overcrowding in yards is the primary source of antler damage, particularly in young animals due to rearing and boxing with the front feet, or with bulls being under crowd pressure, lowering their heads in large milling groups and suffering breakage. A spacious pre-yard entry pen allows ample room for quick drafting into smaller groups of three to five animals. Bulls should not be left on their own for extended periods unless tranquillized as they easily become stressed.

Raceways and entrances to squeezes should have sufficient width at head height to avoid contact of antlers with walls or entrances, and squeezes should be well positioned in yards so that little effort is needed to encourage the bull to enter. Short curving raceways from a central drafting arena utilize the bulls' natural desire to move around corners. Subdued lighting in these pressure areas seems to assist. Yards should be thoroughly checked prior to the season for broken gates, latches, projections, and broken fence wires. Aggressive, nervous, or flighty animals should be culled as they can initiate further disruption in an established group.

Elk respond well to a simple established routine. Conduct velvet antler harvest at a similar time each day with one or two handlers the bulls are familiar with, and who have good rapport with the overseeing veterinarian. The task becomes non-stressful and efficient and large numbers can be handled under such a routine.

Velvetting in hot, unsettled weather or the heat of the day should be avoided, as should velvetting

Rick Alsager's "Forest of Money"

if rapid weather changes such as thunderstorms are forecast. Bulls are also likely to become upset by high winds. Visitors or outside disturbances during velvetting should be discouraged and, above all, the procedure must be conducted with respect and regard for animal welfare. There are important legal and ethical requirements to be observed in antler removal techniques and producers have a responsibility to observe these at all times.

Post-harvest care of bulls demands similar thought and care. Shade and ample fresh drinking water are required particularly during summer. Bulls recovering from sedation should be encouraged to leave the yards as soon as is practical after tourniquet removal. They must be checked frequently for signs of recumbency, bloat or distress. Tourniquets need be applied for only 15 - 20 minutes post-harvest, but antlers should be checked for excessive bleeding after tourniquet removal. Fresh air and clean conditions promote rapid healing of the cut surfaces. Normally no post-operative antibiotics, fly repellents or dressings are required. Damage to coronets by leaving tourniquets on or poor cutting techniques can be permanent and costly.

DO:
Understand the growth and development patterns of the antler.
Group bulls based on age and casting dates.
Use small groups of familiar animals.
Establish a time, yarding and handling routine.
Make welfare and veterinarian involvement a priority.
Plan times and events for minimal handling and waiting.
Check yards and facilities for safety and hygiene.
Provide post harvest shade, water and observation.
Respect the food safety and quality of the product.
Store antler correctly.
Have knowledge of drug action - and animal and human safety with drugs.
Keep accurate records.
DON'T:
Mix breed types.
Be in a hurry.
Overcrowd elk in yards.
Remove antler in unsuitable weathers (excessive heat, storms).
Encourage visitors, or change the routine.

ANTLER HARVEST

WELFARE CONSIDERATIONS

Elk farmers assume a deep responsibility in the harvesting of velvet antler and it is significant that Canada has produced its own Code of Practice for farming of deer as an Industry Gold Standard and is among the first deer farming countries to do so.

There are two issues. First, for safe farming there is no place for hard antler animals in routine elk management because of the potential for aggression and injury to animals and their handlers, particularly during the rut.

Second, the correct removal of antler has been shown to be less stressful than common practices like ear tagging or vaccination. It also occurs at a time when the elk are at their most tractable and responsive to handling. That the product is in demand and significant in pharmacology in Oriental culture is both an incentive and a bonus of this farm management operation. Velvet antler is a valuable food product and maintaining unequalled hygiene and welfare standards is important to international marketing and acceptance by the public.

There is considerable public opposition to the concept of velvet antler harvesting, supported by speculation, sensationalism and an uninformed view of velvet antlers' properties and uses. As farmers we take the responsibility to not only ensure that velvet removal is the most sophisticated and correct operation on farm, but to understand and promote the product carefully and respect it and the welfare of the animals equally.

FARMER RESPONSIBILITIES

Selecting the right stage for cutting

Once the antler has reached about 55 days growth, the form and growth of the antler should be checked every two days at least, and preferably every 24 hours. The best time is in early morning with sunlight behind the animals as the hairs on the velvet skin form a halo effect and the flattening or division of the royal bulb can be seen clearly. If animals are grouped in smaller groups according to casting dates, then there is no need to yard and handle groups at different stages of growth every time.

In deciding when to cut, the most important indicators are your records from years before. Each bull is slightly different re: density, growth style, and pattern of calcification. Some must be harvested at a certain stage, others at a different point in development. The number of days since casting is a very useful guide, plus the shape of the royal bulb with its 4th and 5th tines, and the shape and character of the brow tine. The royal bulb should be just starting to "dimple" and divide in young bulls. In older bulls, the division might be as much as several inches, if you know that bull does not calcify early. The brow tine tips should be straight, round, soft and velvet covered. When the tip begins to turn up, becomes pointed and the velvet covering flattens, it is time to cut. Get to know your bulls.

Timing veterinary visits

It can often be prohibitively expensive to involve a veterinarian visit for just one animal. Often on small units it is better to time a visit for an average stage of growth and harvest all animals within five to six days of each other in development. This is effective use of veterinary time, reduces the amount of yarding and sorting required and produces a quality product. The veterinary visit should be planned in advance and if possible, for larger groups, at a similar time each day. This not only encourages the routine but reduces the time spent with animals waiting in yards, or scrambled activity when the vet arrives.

The farmer should quietly yard and sort his animals prior to the vet arriving, leaving 15 - 20 minutes maximum as a time to settle.

Bulls not ready should be returned to the pasture as soon as sorting is complete, and the selected bulls left two or three to a pen in the largest areas available for them.

It is important that bulls in velvet are never overcrowded in pens, or rushed as damage is inevitable under these conditions.

Most experienced velvetters favour early morning or late evening as the optimum time, avoiding the heat of the day.

Summary of Velvet Removal Process

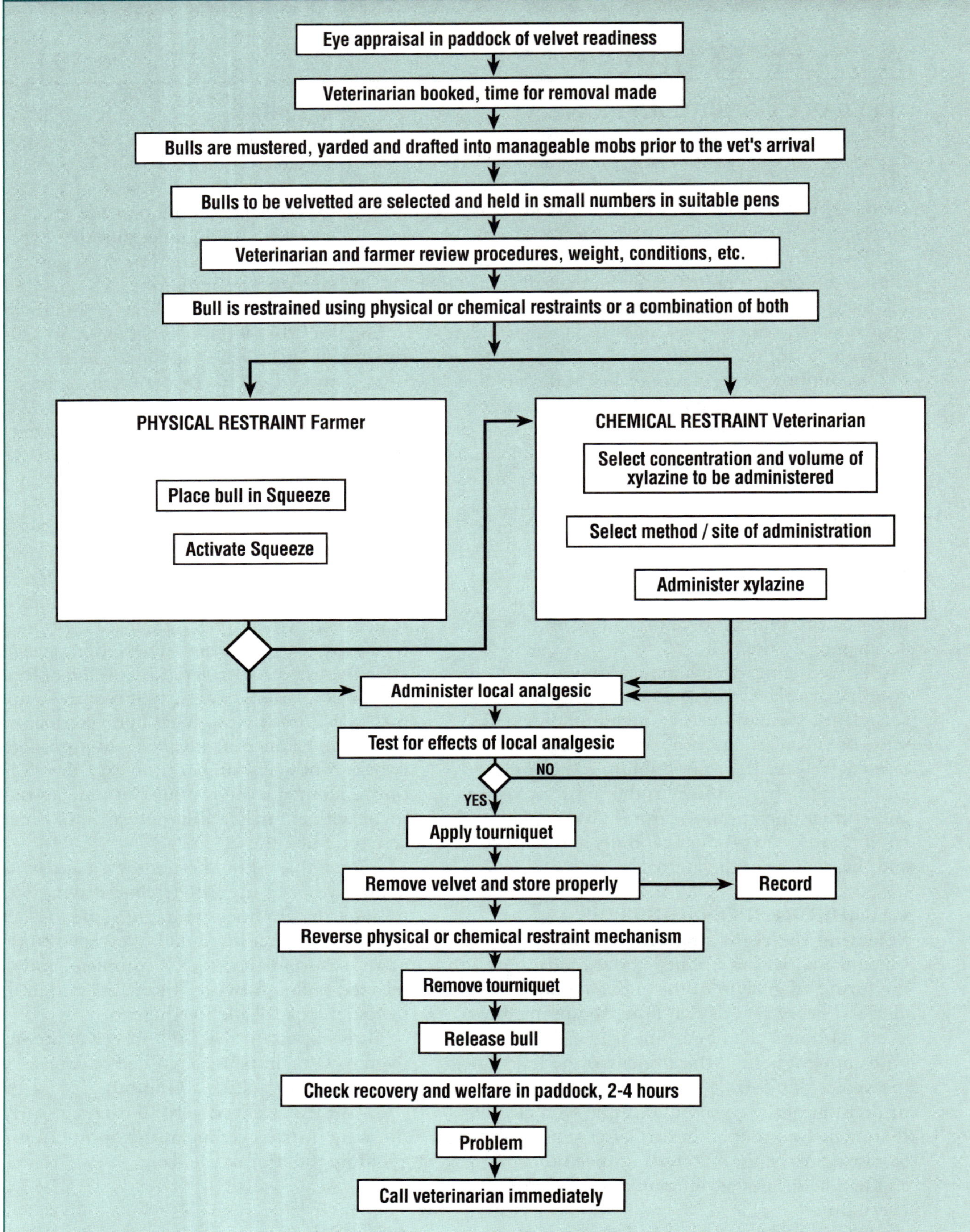

Equipment

- Physical restraint and local analgesia. The farmer must ensure that the squeeze to be used is operating correctly, i.e. will hold bull with sufficient pressure, yet not cause discomfort, and has adequate facility for immediate head restraint. The squeeze area should be well lit so that site injection and removal procedures are easily seen and achieved.
- Disinfectant and container (for saw).
- Solid backed medium toothed (meat) saw that is sharp, functional, and clean.
- Tourniquet material; thick rubber tubing, strong elastic.
- Cloth for blindfold.
- Needles and syringes.
- Plastic film, tape and plastic bags for antler.
- Scales to weigh antlers.
- Velvet storage area. A meat freezer preferably, insect free, darkened and with adequate shelves, or rods to store the velvet until freezing which should be within one hour of removal.
- Record books - Bull I.D., casting date, velvet record tags, harvest date, drugs used, procedure, handling management notes, velvet weight, grade and description.
- For the use of chemical restraints, the farmer should supply a towel to place under the animal's head to prevent dirt and debris from invading the eyes and nostrils, and may like to have his own pole syringe and tranquillizing equipment.
- A wall chart showing immediate treatment and emergency numbers for drug procedures and accidents.
- Farmers must supply a secure disposal unit for needles, used syringes and drug bottles, and a lockable security cabinet for any drugs and equipment used on the farm for this procedure.

HARVEST

Hygiene is paramount, both for equipment and the harvested antler. Veterinary advice on the safe storage of sedative and anaesthetic drugs, including a strict procedure on coping with a drug emergency, should be known to all those involved in harvest, and these procedures reviewed at the start of each season.

Saws, tourniquets, recording materials, tags and storage facilities should all be on hand at the commencement of harvest. For antler removal a medium toothed saw is advised and it should be washed and disinfected between uses. Cutting should be parallel to the coronet, 5 - 10 mm above it and started from the outside of the antler towards the centre. If the animal is only lightly sedated, or held in a crush, then local anaesthetic administered via ring block at the base of the pedicle, must be given with time for the full effective pain block to develop - over 4 - 5 min. The head must be firmly restrained to prevent any sudden movement of the animal which may tear the velvet skin. The antler should also be fully supported during cutting. Sedated bulls can occasionally react to sudden movement or loud noise and can easily damage themselves or antler with involuntary action. If bulls are lying on pen floors, these should be clean to prevent contamination of the cut antler, or exposed eyes and mouth. If bulls have been recumbent for some time, the uppermost antler can become rapidly drained of blood and the head should be tilted and antler held down for some time prior to removal to ensure the matrix is blood filled. The entire operation should only take the time required for optimal care.

Tranquillizing, anaesthetizing and reviving bulls is a skilled operation which can use some potentially dangerous drugs and requires a good knowledge of anatomy and veterinary techniques, particularly when injecting. Advice and instruction is mandatory.

HANDLING AND STORAGE

After removal, velvet antler should be held for cooling in a clean, shaded insect-proof area, reclined cut end uppermost on a slope of 15 - 20° to allow contained blood to be evenly distributed through the antler. Velvet is also commonly hung upside down on racks until cool and then frozen in an inclined (15 - 30°) or upright position. If this occurs for too long blood can pool in the upper region and create processing problems. Turn antler as it freezes - at least every half hour for three hours. Great care must be taken to avoid deforming the bulb or tine shape during freezing to ensure the antler is presented in true form for grading.

Frozen velvet antler should be stored in airtight plastic bags and handled subsequently with the due care that a valuable edible medicinal product demands.

Attention to detail from all points of animal selection, grouping and handling and harvest techniques with an established simple routine ensure that on-farm breeding and feeding programmes are rewarded. Ultimately a knowledge of the changing market requirements for style and quality must be known each season. Bulls must be assessed during the 55 - 80 days of velvet antler growth after casting by close observation of antler shape, bulb and tine development to allow the correct decision to be made re: time to harvest. Maximum weight, lack of calcification and highest contained blood content are thus optimized.

POST VELVETTING RECOVERY

Farm responsibility does not end with the removal of tourniquets and release from the shed. Post velvetting, bulls should be let into nearby pasture with clean fresh drinking water and shade available and observed frequently (hourly until grazing and sleeping patterns are resumed). Under chemical restraint there is some chance that reversion may occur so there is a risk of bloating or going down in difficult conditions. There should be no water holes, creeks or steep ground in a recovery area. With good bleeding control (haemostatic powders, etc.) there will be few bleeding episodes, but excessive exertion should be avoided, (i.e. moving animals long distances, etc.). Bulls should not be confined in outside yards under conditions of dust and potential heat stress for any amount of time.

Tourniquets must be removed before bulls are released, and should not be retained for longer than 30 minutes. They can be reapplied if any excessive bleeding is a concern.

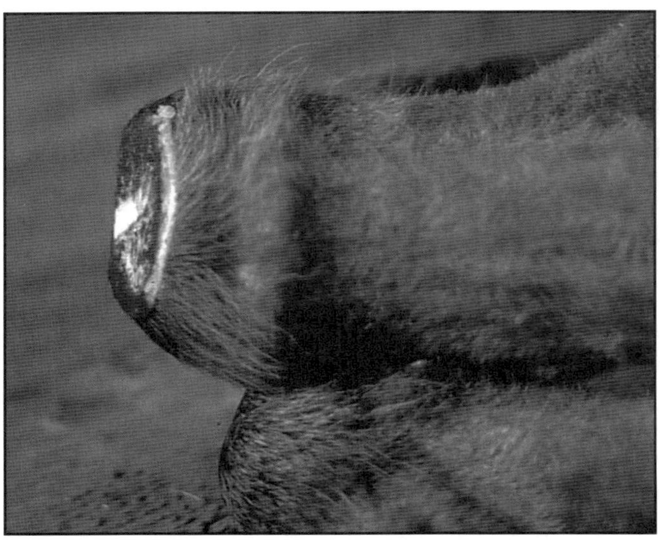

A terrible mistake - cut below the coronet

VETERINARY PROCEDURES
DRUGS

Dosage and choice of drug for immobilizing is the choice of the veterinarian and under his direct control, but the farmer's knowledge of liveweight is important. Individual animal variation is wide and a good knowledge of the animal's behaviour and past history is important.

Quiet stock, darkened conditions, and a lack of disturbance after injection aid effective tranquillizing. Noise, sudden light, or activity can cause a fright response in animals that appear sedated. A quiet approach from behind the animal, gentle covering of the eyes and firm support of the head, all aid comfort and correct procedure.

Injection should aim for the heavy musculature in the neck, preferably anterior to avoid risk to the primal venison cuts of the body. Using a pole syringe or close contact injection is very effective but does require good familiarization routines with the animals. Injection in the rump may not be effective because a buildup of fat reserves slows drug spread and effect. New needles should be used on each animal.

Xylazine alone is not an effective pain killer and 2% lidocaine is required, either as a ring block around the base of the pedicle, or by direct infusion at the nerve sites. See Figs. 8a & b. (adapted from Murray Woobbury and Jerry Haigh, in the Can. Vet. J. Volume 37, August 1996) Studies have shown the ring block to be more effective. Fig. 8b shows needle entry sites for injection. The ring block is actually a series of overlapping lines of lidocaine. Insert a 4 cm, 20 g needle up to the hub, parallel to the base of the pedicle where the skin curves up from the flat skull. Inject about 2 cm per site, slowly, as the needle is withdrawn, and repeat to totally encircle the pedicle. A total of 10 to 15 ml will be used. Injection is difficult between the pedicles, but this area is of lesser importance than the rest of the circle. Wait at least four minutes for a complete analgesic effect.

Electronic Analgesic (EA) may also be used for pain control. EA works by blocking pain signals before they reach the brain and by releasing naturally occurring biochemicals such as endorphins and enkaphalins to raise the threshold of pain.

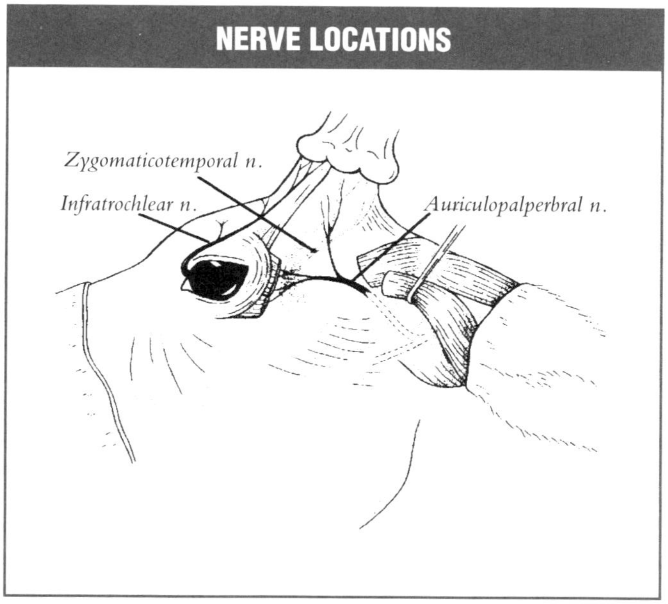

Fig. 8a Lateral view of the head of an elk showing the innervation of the antler pedicle

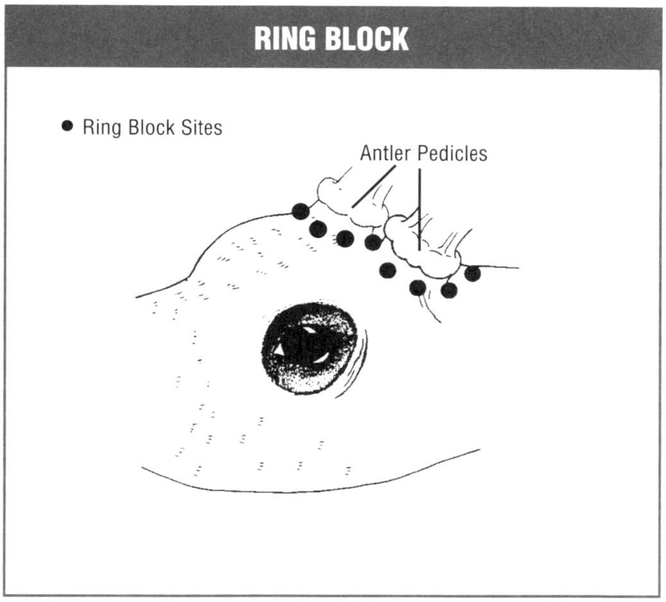

Fig. 8b Injection sites for local anesthesia of the antler pedicle, using the ring block method

EA is effected with the use of a specially designed device which induces an electrical current between two points on either side of the pedicles. This device must be used very carefully and only after adequate training in its use.

Tourniquets are applied in a figure 8 loop at the pedicle level below the coronet. They should be of material that has pliability and some width to comfortably compress blood vessels without crushing or cutting into underlying tissue. Tourniquets must be clean, and washed between uses.

Antlers should be cut from the outside towards the centre although this is not always easy. A small surface cut is made first, the level of analgesia checked by lack of reaction from the bull, and the antler cut through smoothly and cleanly from this position. The saw must be sharp and disinfected and the antler supported at all times. Cutting from the centre out can leave the potential for the antler to fall at the last minute, or the animal's head to move as the weight is relieved, and risks damage to the pedicle and tearing of the velvet skin. If the animal's head moves at any stage, cease cutting until control is regained.

Post operatively, veterinarians may select a variety of haemostatic treatments or antibiotic powders and insect repellents depending on local conditions. These all aid the arrest of bleeding, but are not critical to the operative success.

Tourniquets should be removed in no longer than 30 minutes and the animals released from the shed into fresh clean conditions. If additional drugs to xylazine have been used veterinary procedures will require additional reversal agents. This is a skilled technique and a direct veterinary responsibility.

The following post treatment abnormal signs of recovery indicate that veterinary advice must be sought and the event recorded:

- prolonged lack of alertness
- continued wide based stance
- continued unstable gait
- laboured breathing
- bloat
- excessive bleeding

By developing a strong veterinary-client relationship and by familiarization of the animals to a routine and strict observance of welfare and hygiene codes, antler removal is a safe, humane and clinical farm procedure and a normal part of farmed elk management. Most jurisdictions are requiring a "velvet antler removal accreditation" course be completed by all individuals who will actually remove velvet antler. This will insure that the highest standards of animal welfare and product quality are maintained.

STORAGE, GRADING AND MARKETING OF VELVET ANTLER

STORAGE

Antler must be stored frozen, and allowed to freeze without deformity of the bulb or tines. After harvest, and once cooled while inclining cut end uppermost, the blood should be evenly spread through the antler tissue. Antler should be cleaned, tagged, date recorded, weighed and measured for grading, then frozen as quickly as possible and sealed in plastic bags to avoid moisture loss and freezer burn. Antler must be turned as it freezes - at least once every half hour for at least three hours. Frozen antler must be kept absolutely clean and handled carefully. Never bump or drop it - cracks will form which make it useless for slicing after it has dried.

GRADING

The NZ Game Industry Board has developed a comprehensive grading system that is recognized internationally, based on objective measurement of length, diameter of beam above and below the trez tine, and diameter of the royal bulb, plus presence or absence of bez tines and abnormal growth features. The system reflects quality of cutting and recognizes calcification and overgrown product. Measurements of thickness are the most important and the minimum diameter, taken above the trez tine, represents the minimum slice size that can be made in the final market with dried product, effectively setting the market price for the stick.

As a general rule the highest quality antler within a breed type (remembering that elk antler is seen as higher quality than red) arises from the older bulls who cast and grow velvet earlier and for a longer period. This usually derives the A, B and C grades. Younger antler is thinner, prone to more calcification and shorter and lighter. Cut correctly it is still as high a quality but its beam measurements only allow C, D and E gradings.

As discussed under selection, two year old velvet antler weight forms the basis for choosing future herd sires and improved antler performance. Antler style and form as indicated by grade should also be considered in selection criteria.

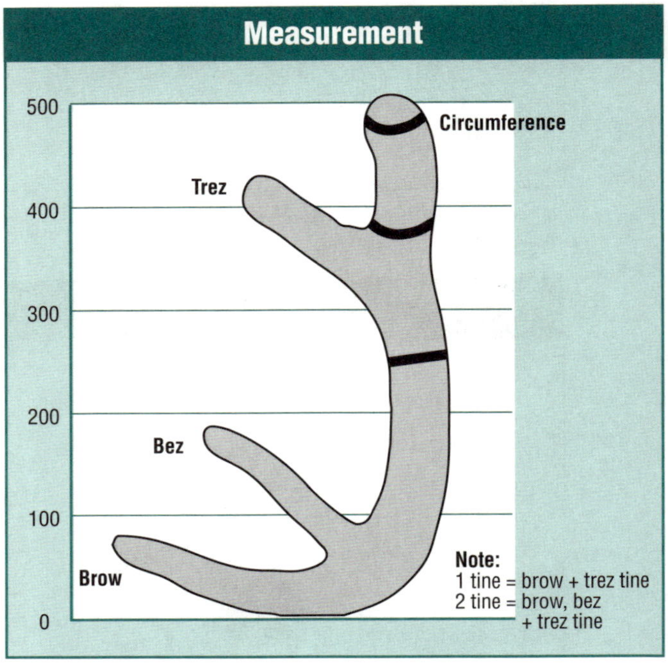

Fig. 9 shows a typical antler and measurement sites for a grading basis. The circumference is the lesser of the measurements taken around the clear beam immediately above the trez tine and around the mid point of the beam below the trez tine. Measurement is taken with the velvet fur compressed.

MARKETING VELVET ANTLER

In the early stages of industry development, marketing velvet is hindered by lack of volume and an uncoordinated approach. Many producers have established direct links to the growing Canadian Asian population through advertising in community newspapers and via direct inquiry. A number of processing plants have been established in Canada and these operators will buy on grade and contract delivery. Irrespective of the product end point, quality is reflected in price as the international market is increasingly competitive. New initiatives include capsulated velvet antler pills for local use and as these products gain more acceptance this market should increase rapidly.

Most of the industry associations are active in developing co-operative sales with pooling of product with defined standards and the industry should rapidly develop from there. Processed products are generally sold in dried whole stick form, but the drying procedure is a learned and closely protected operation best left to the experts who have invested in the industry.

A good relationship with a processor is very useful to have. Good communication on the quality and presentation of the farmer's product, particularly as it relates to contained blood content, calcification and cutting time is essential as it is only by this that combined objectives of producer and processor can be achieved.

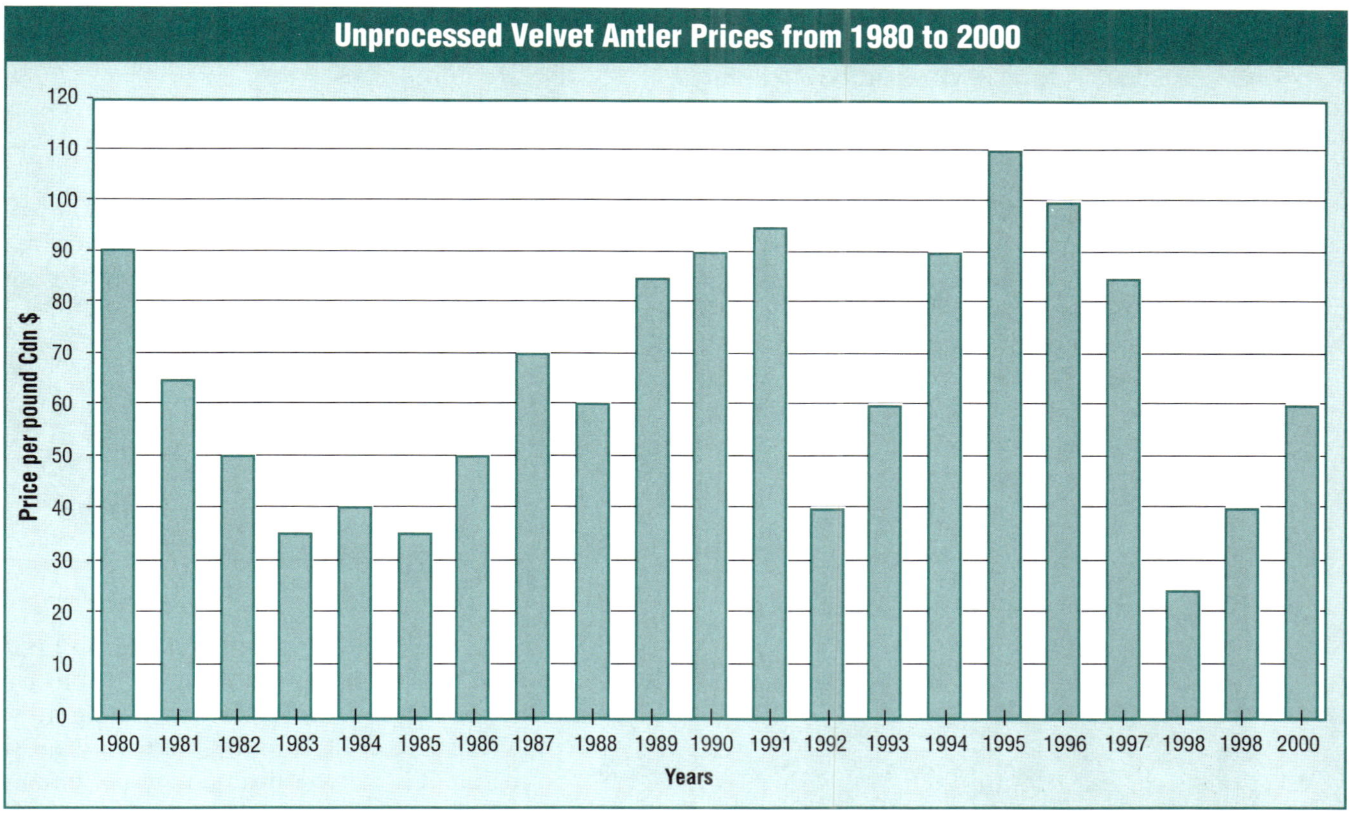

Average of prices paid in Canada for top quality, unprocessed ("green" frozen) red deer and elk velvet antler of various sizes (does not include spikes or regrowth).

Table 9 This chart illustrates informal survey of reputable North American Industry sources.

Judging Velvet Antler

Velvet antler competitions have become some of the most interesting, exciting and financially rewarding aspects of the elk farming industry. They are great social events, they allow breeders and buyers to compare the relative genetic merit of various breeding programs and they are a tremendous marketing tool. Winning a class at a Velvet Competition immediately adds thousands of dollars of value to your bull. With this kind of value involved, there is a lot of pressure on the judges. What kind of Criteria do they use to guide their evaluations?

The North American Elk Breeders Association, with input from hundreds of people from all around the world, has developed a system for judging elk velvet antler. It is called the "Certified Weight Index". This name reflects the great importance of antler Weight, while also considers the fact that antlers are compared to and rated against others in their class, and that the weight and indexed position is certified by a team of NAEBA approved judges.

Look at the CWI scoresheet on the next page. This is the actual scoresheet from the Five-year-old class at the International Finals, held in Fargo, North Dakota in August 1997. When antlers are entered, the rules require that a photograph is supplied to identify the bull, and some proof of age of the bull is also required. Each antler of the pair is then weighed and measured. As

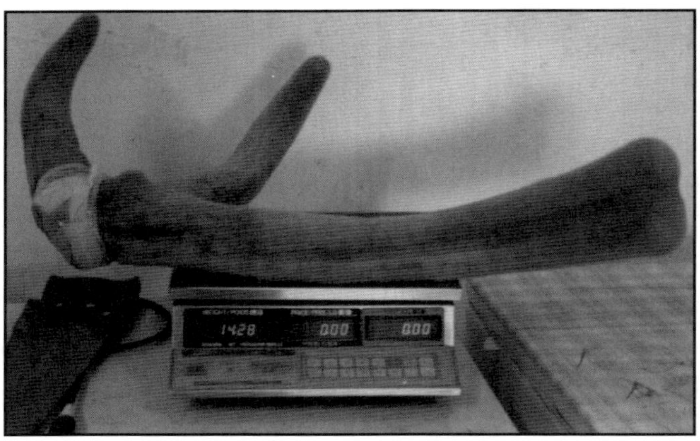

1. Weight - the most important criterion

you can see from the scoresheet, Weight is the most important criterion. The weight of each antler is added to the other, and the total for the pair is multiplied by two to arrive at the number of points. For the first bull, a total weight of 35.36 pounds times two gives over 70 points of a total CWI of almost 124. Beam circumference is measured next, and the total of the two beam measurements, in centimeters, are totaled and multiplied by 0.2 to arrive at a point score.

Uniformity is the next measure.

2. Judging Uniformity

This criterion, plus Symmetry, are included because we are not just producing velvet antler.

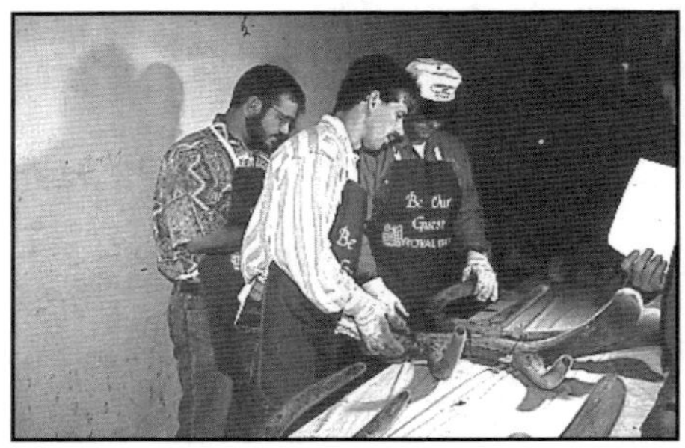

3. Are they Symmetrical?

We are also producing trophy bulls, and bulls with aesthetic value, so we want the antlers to have a desirable shape, with a smooth flowing line along the three points and the top, and we want the pair to match. These two criteria may be lumped together as "Style", worth a total of ten points.

The next consideration is Balance. The top part of a velvet antler may be worth up to ten times more than the bottom part. So, the judges look for large bulbed or webbed tops, with relatively small brow and bez tynes. Large beam diameter, especially if the diameter increases continuously from base to top, also contributes to a balance score approaching the possible total of ten.

NAEBA's Certified Weight Index (CWI)

Velvet Competition: Fargo, Class 5

Rank	Set #	Age Years	Bull Name/I.D.	Antler (Rt lbs)	Antler (Lt lbs)	Antler weight	Beam (Rt cm)	Beam (Lt cm)	Beam .2pt/cm	Uniformity 5pts	Symmetry 5pts	Balance 10 pts	Calcification 25 pts	Damage 5 pts	CWI (points)	Owner
1	5083	5	STARBUCK	18.41	16.95	35.36	22.5	22.5	9.00	4.29	4.29	8.66	22.05	4.91	123.92	Legacy Deer Group
2	5096	5	PRINCE OF PRIDE	15.71	16.69	32.40	23.2	23.6	9.36	4.60	4.49	7.93	19.92	4.98	116.07	Elk Hills Ranch
3	5003	5	595	17.48	17.08	34.56	23.0	23.0	9.20	4.20	4.41	7.07	16.41	4.93	115.34	Clearstone Breeding Farms
4	5095	5	OAK POINT RAYS	16.59	15.11	31.70	23.2	22.3	9.10	4.29	4.45	7.44	18.70	4.94	112.32	Oak Point Elk Farm
5	5005	5	11YW	16.65	16.15	32.80	23.4	21.3	8.94	3.86	4.37	7.11	16.35	4.94	111.17	Pierre Lemire
6	5156	5	ATLAS	16.34	14.74	31.08	23.5	23.0	9.30	4.43	4.30	7.70	17.80	4.69	110.37	Tynes Will Tell
7	5189	5	38 SL0K00313	13.98	16.00	29.98	21.6	22.9	8.90	4.36	4.54	7.85	19.69	4.98	110.27	Trad Elk Ranch
8	5187	5	B10SVCB0108	14.61	14.98	29.59	22.2	22.4	8.92	4.48	4.56	7.41	18.83	4.90	108.27	Bergen Game Farm
9	5072	5	FLASHBACK	14.87	15.93	90.80	21.0	20.0	8.20	4.34	4.39	7.40	17.46	4.73	108.11	Royal Rainbow Elk
10	5094	5	MINN-KOTA PRIDE	14.96	14.29	29.25	24.0	23.5	9.50	4.40	4.61	6.75	18.89	4.98	107.63	Flooded Creek
11	5052	5	C0170001	17.36	16.04	33.40	21.0	20.3	8.26	2.39	2.13	5.81	19.31	2.85	107.55	Plateau Valley Elk Ranch
12	5008	5	35	13.66	14.26	27.92	21.0	21.5	8.50	4.39	4.49	8.30	19.84	4.35	105.70	Elk View Farm
13	5054	5		13.90	14.07	27.97	20.4	19.9	8.06	4.48	4.69	7.64	19.31	4.64	104.76	Elk Echo Ranch
14	5152	5	AJ	13.58	13.82	27.40	22.0	22.0	8.80	4.29	4.15	8.75	19.65	4.29	104.73	Wapiti Hills Inc.
15	5158	5	BULL 16	14.21	15.24	29.45	21.5	21.0	8.50	3.71	3.91	6.66	16.98	4.61	103.28	Hanke Elk Farm
16	5055	5	Y736	13.82	13.52	27.34	20.4	19.8	8.04	4.06	4.28	7.13	20.13	4.91	103.22	Elk Echo Ranch
17	5147	5	47	9.07	9.89	18.96	20.0	20.5	8.10	3.86	4.20	5.05	17.25	4.85	81.25	Coleman Farms
18	5150	5	#1 BLUE	8.99	10.04	19.03	18.5	19.0	7.50	3.95	4.03	4.48	18.14	4.81	80.96	Johanson Elk
19	5148	5	42 GREEN	8.96	9.72	18.68	21.0	20.0	8.20	3.26	3.66	4.19	19.10	4.85	80.62	Coleman Farms

4. Huge webbed antler

5. Huge slingshot antler

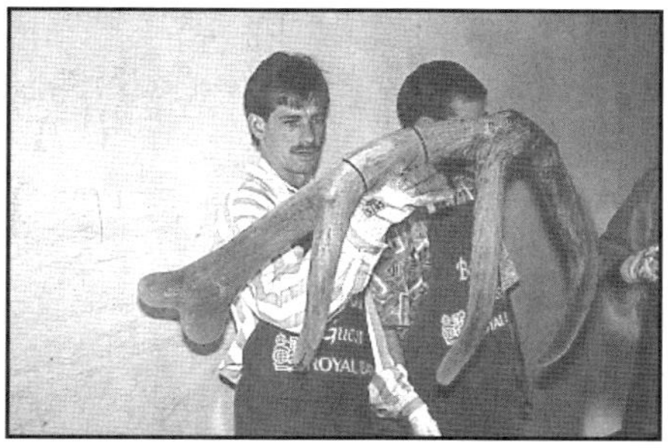

6. Assessing Balance

Calcification is the most difficult and subjective aspect of velvet antler judging, and also one of the most valuable.

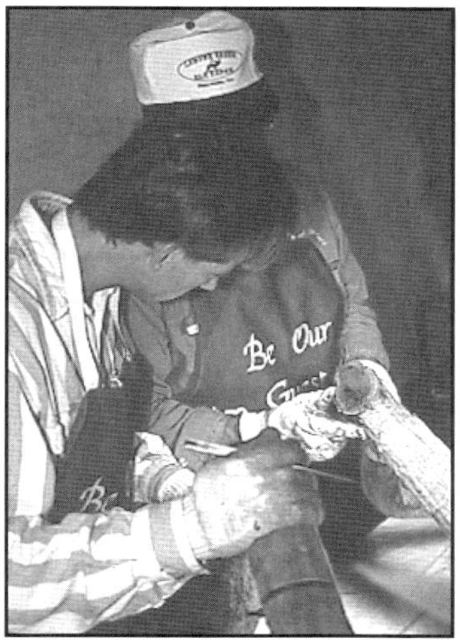

7. & 8. Calcification - the toughest measure of worth

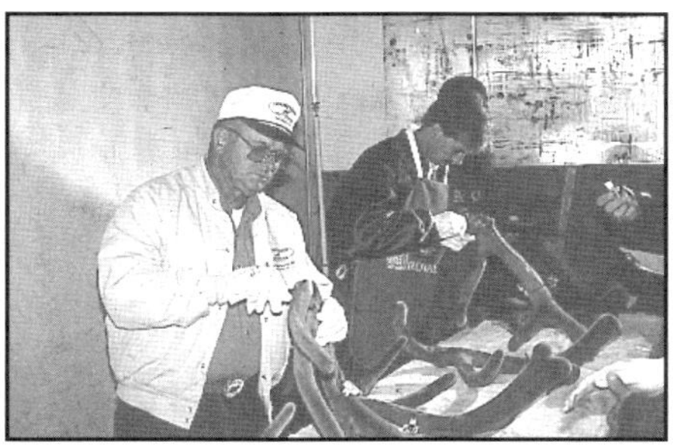

A total of 25 points are available in this category. That total is rarely awarded, partly because any producer is prepared to sacrifice a couple of points for Calcification in exchange for a couple of pounds of weight. The balance between weight and quality is the biggest struggle in the antler business. How do the judges assess the degree of calcification? They look at several criteria:

- appearance of the top - how far is it grown, how large and advanced is the bulbing, webbing or tyne development;
- the roundness of the tip of the brow tyne, compared to the roundness of the second (bez) and third (trez) tynes;
- appearance and density of the base - colour of the middle and outer ring, all relative to the point at which the cut was made (how close to the pedicle);
- diameter and shape of the beam - the larger the beam, the longer you can grow out the top, plus consider development of grooving, and circular or oval shape;
- all of the above are considered with the age of the bull in mind. Younger bulls can not usually be grown out to the same extent as older bulls.

Damage is the final criteria. This may come in many forms, none of which we want to see. A total of 5 points are available and all five points are awarded to an undamaged antler. In 1998, judges could deduct all five points plus penalize a further five points, a total of ten possible negative points.

With all these criteria considered, the judges compile the results with a laptop computer.

9. Adding them up

The crowds assemble, the antlers are displayed, the results announced, and the winners start marketing semen and offspring.

Beyond the immediate marketing advantages, Velvet Antler Competitions are a critical part of our industry's plans for genetic improvement. If we keep all our livestock at home, and never compare the results of our breeding and management to that of other producers - How will we ever move ahead?

ANTLER COMPETITION INSIGHTS

by Bruce Friedel

Many people have debated the merits of antler competitions and the emphasis that elk producers place on them. Winners often tout their achievements in their marketing campaigns. Those less fortunate often criticize competitions as a quest for "fortune and glory" among the "good old boys." Amid the verbal barrage, newcomers look on with skepticism, wondering how they, as rookies, are going to sort through the arguments in their quest for improved antler production. One wise old industry sage wondered if he was shopping at a feed store, a pharmacy, or an AI stud farm when he attended antler competitions.

BELIEVE IN THE VENUE!

First, to get the most from attending or competing in local antler competitions, you should believe in the process. In the absence of EBVs (Estimated Breeding Values) for the elk industry, producers are left to do their own homework in selecting improved antler genetics. Our modern antler competitions, which adopted individual velvet age classes in 1995 and individual hard antler age classes in 1999, raised the bar for within-age group comparisons.

As to the grumbling about the feeding programs or the cheating, two recent articles may point elk farmers in the right direction. Pearse (2000) reported that Invermay made an 80% improvement in velvet production during a 15-year time span (from the mid 70s until the early 90s). Their strategy was simply to mate the "best" males to their "best" females. After EBV analysis on the last 20 plus years of data, little permanent genetic gain was found in those first 15 years! Instead, almost all of the 80% improvement was due to management, feeding and environmental effects. By reviewing hind progeny performance and purchasing semen or young sires showing truly superior antler genetics, Invermay has shown recent positive progress in selecting for improved velvet production.

Undoubtedly, the majority of antler improvement in our industry in the last ten years has been due to improvement in management and nutrition! Producers who find themselves placing below the top ten each year should first address their management, nutrition, and genetic selection programs before casting criticism at others concerning at which feed store they shop!

In Alberta, producers have the benefit of data analysis from Alberta Agriculture's inventory reporting. Thorleifson (1999) reported that the 2-year-old Alberta velvet average was 7.8 lbs. in 1994 and only 7.63 lbs. in 1998, suggesting little genetic gain or management improvement. A recent study by Wang, et al (1999), which used Alberta's historical antler records, calculated a heritability of .27 +/- .03 and a repeatability of between .31 +/- .03 and .34 +/- .03, using two different animal models. For those who do not understand quantitative genetics, that means 27% of the variation in antler yield was due to genetic differences. The difference between heritability and repeatability, .31 - .27 = .04 or .34 - .27 = .07, means that 4 to 7% of the variation in velvet production could be explained by permanent environmental effects. The remaining 66 to 69% of the variation in antler yield is due to non-additive, non-genetic, temporary environmental effects (i.e., feed supply, nutrition and management). Science states that antler yield will respond to selection, and antler competitions are an effective venue for within-age group comparisons.

THE SECRETS OF COMPETING

Past arguments that antler competitions are for the "good old boys" should be a thing of the past. With 8 velvet and 12 hard antler classes at each of the 7 preliminary competitions and the international final, there will be 680 trophies, ribbons and certificates available for this summer's competition winners. Previously, the industry was lucky to even draw that many entries! With so many potential prizes up for grabs, even the faint of heart should be encouraged to enjoy the fun and camaraderie of their local antler competitions. Here are a few tips for those newcomers:

Live by the old 4-H motto of "Learn to do by doing." Not everyone can be a winner, but there's a lot to be said about the process of competing and the insight it brings. Each spring as North America's elk farmers are feeding their "top-secret, jet-propelled, antler-growing elixirs" to their bulls, the antler frenzy begins. Buttons start dropping, casting dates get recorded, velvet starts growing, and ... the telephone starts ringing. While everybody eagerly anticipates the fruits of their own labor, elk farmers can never quench their thirst to know how everybody else's bulls are doing. "Has King been cut yet? What did Korean Gold weigh? Is Starbuck as good as last year?"

In 1999, after hearing countless rumors of 2-year-olds cutting upwards of 20 lbs., we all began to get discouraged. Who wants to bring a 12-pound 2-year-old to a competition and get embarrassed by someone who has an "18 pounder," right? Wrong! Remember to put things into perspective. In 1998 there were over 1,100 two-year-old bulls cut in Alberta. There were only two "15 pounders" and five "14 pounders" in the entire province! Ten 2-year-olds that cut in the 12 pound or less range placed in the top 2% of the province's 2-year-old velvet yields. The best "12 pounder" (11.94 lbs.) placed fourth! That's pretty esteemed company, if you ask any quantitative geneticist.

THE VELVETING DECISION

Now that you've made a decision to take your best to the "county fair," it is time to get ready to compete. The first and most important decision is not when you should cut, but rather ... should you cut? With 12 hard antler classes in each competition, producers really need to assess whether they are looking at a good velveting bull or a better representation of a hard antler bull. Once you get the hang of it, it's an easy decision to make.

Is the velvet short and stocky with good beam measurement which gets progressively larger as you move higher up the beam? (Figure 1) Or does this head of velvet lack fantastic beam but have long brow, bez and trez tines with an obviously elongated heart piece? (Figure 2) If it does, you should consider growing the bull out in hard antler and competing in one of next year's hard antler classes. With great tine lengths and long beams, you may find that your bull is unbeatable in his hard antler class. If you decide to allow him to grow out, remember to keep feeding him well so that his antler growth doesn't fizzle.

Fig 1. Is the velvet short and stocky, with good beam measurement?

Fig 2. Does it lack fantastic beam, but have long tines?

If you've decided to harvest velvet, you first want to snap some pictures of your bull in velvet that show a clearly legible tag. Remember, beginning with the 1999 calf crop which will qualify for the 2001 antler competitions, all bull calves must have visible, dangling, tamperproof ear tags (preferably two), obtained from NAEBA or your state or province. The bulls must be included in the NAEBA inventory system or any state/provincial government, age-verified inventory system or be registered with NAEBA prior to their regional competitions. Prior to the international competition, all bulls must be NAEBA registered. These requirements will improve the age verification process that is so vital to fair play in sanctioned antler competitions. Finally, to avoid any disappointments after velvet removal, ensure that you have the pictures developed one or two days before velveting to ensure that the bull in velvet can be clearly identified with his tamperproof or dangle ear tags. (1998 or older bulls are grandfathered in on tagging requirements.)

CUTTING VELVET

Obtain a medium-toothed bone or miter saw which will leave a fairly coarse cut. One producer recommends a 14" TASK with 8 teeth per inch, obtainable from Totem building supply stores. Fine-toothed saws give the cut surface a nonporous appearance that appears more calcified. Leave the high-powered, electric cut-off saws locked securely in your garage. They have no place on an elk farm which harvests velvet for competition. Cutting height is absolutely crucial! Each year velvet judges examine velvet entries that have been cut perilously low to the coronet. The lower you cut, the more calcified the base of the antler will appear! Don't get greedy! Finally, just before you cut the velvet, pull some hair out of the bull's mane for DNA analysis, just in case you want to enter the Breeder's Three class.

Velvet your bull using safe and humane velveting procedures, paying particular attention to back cutting each antler to eliminate any chance of stripping a "banana peel" of velvet if a bull jerks his head at the last second. Other than a barely discernible back cut to protect an antler from peeling, there should be only one cutting plane on each antler base. I've personally witnessed up to six different cutting planes on one antler base, suggesting that some trimming was done to give a less calcified appearance. Be prepared for a rather grim calcification score if you start pulling these stunts!

FREEZING COMPETITION ANTLERS

Immediately after harvesting a set of velvet antlers, wash off all blood, dust and foreign material. Take a soft brush and comb out any ruffled velvet. Affix your velvet tags and record the tag numbers for future reference. Place the set of velvet in a freezer with the cut side inclined upwards far enough to prevent any blood from dribbling out of each antler. Place each stick of velvet on 4 - 6 inches of open cell foam to prevent flat spots from freezing into the velvet. Ensure that the velvet does not touch any other antlers, the freezer bottom or sides. Do not wrap! Wrapping the bottom of the antler compresses the velvet hairs, which hurts beam circumference measurements. Wrapping can also seal velvet cuts with blood and ice, giving velvet judges more difficulty in determining accurate calcification scores. Extra effort in handling velvet antlers will maximize presentation scores during the judging process.

Return in 24 hours and weigh each antler if it is frozen solidly. To prevent any sublimation (loss of weight by ice evaporation), place each antler in an extra large garbage bag. Suck all remaining air out of each bag with a vacuum cleaner, before sealing with duct tape. Most ardent competitors double bag each antler. Place a two-gallon pail containing one gallon of food grade antifreeze with two dozen floating ice cubes in each freezer. Large walk-in freezers allow for more air exchange, resulting in increased sublimation or antler shrinkage, so many competitors use smaller 20+ cubic foot chest freezers. Monitor your freezers twice daily. Last year a 43-pound set of velvet antlers didn't make it to a regional competition because a freezer conked out during hot weather!

VELVET TRANSPORT

Elk farmers are an innovative bunch, having transported competition antlers in anything from smaller reefers down to velvet wrapped up with ice in sleeping bags or furniture blankets. The preferred method is to place an operating horizontal freezer in the back of a pickup. For trips longer than an hour or two, most producers plug their freezers into a small, gas-powered generator. Take along some spare gas for those longer trips! If you have a walk-in freezer and choose to transport competition velvet in a chest freezer, plug it in the night before your trip to cool it down to freezing temperatures. Finally, pack each antler individually with lots of foam padding to prevent the antlers from rubbing and chafing each other.

Before you leave for your local competition, ensure that you have all your competition antlers with their appropriate pictures and inventory or registration forms to assist in the age verification process. If you intend to compete in a Breeder's Three class, each entry must be DNA matched to sire before competition time! Last year NAEBA allowed breeders to meet the qualifications between the time of the competition and the publication of the December magazine because of the time frame of the competition changes. With numerous disqualifications due to entries not being Gold registered or DNA matched to sire, there was some grumbling about the process. This year, Breeder's Three entries have to DNA match to the sire before acceptance into any Breeder's Three class, but they do not need to be Gold registered. Last year, several bulls carrying hybrid markers did well in individual age classes, while all Breeder's Three entries had to be Gold registered (tested pure and DNA tested). The apparent discrepancy between individual and Breeder's Three entries will be changed this year. Breeder's Three entries must DNA match to the sire, but can be NAEBA registered (any level) or provincial/state registered, for entrance into any preliminary competition. The entries must be NAEBA registered at any level for entrance into the international competition.

The standard age verification process of a NAEBA registration or inventory record or a state or provincial inventory certificate will suffice for the preliminary competition. However, producers interested in future EBVs for antler production should consider some form of NAEBA registry, as that's the likely database for EBV calculations.

THE REGISTRATION DESK

Being well organized makes the registration process at an antler competition a picnic. Get there early. Bring in only a couple of velvet sets at a time to prevent thawing. Have your photos, registration/ inventory certificates and DNA certificates ready for each bull. Complete an entry form for each velvet or hard antler entry. Print clearly. Statisticians can't spell what they can't read. Pay particular attention to the bull's name and the farm name. Supervise the beam measurement and velvet weighing.

During the registration process, random velvet samples will be collected. These samples will be analyzed for banned substances used as growth promotants. Keep the paired sample provided to you, for future reference. If you're accused of "shopping at a pharmacy" to win competitions, your legal counsel would probably appreciate some exonerating evidence with which he can defend you from the multitude of forthcoming lawsuits. Finally, pay your appropriate entry fees and then go find the barbecue. You've done all you can, and it's now out of your hands!

During registration, hard antler entries have been whisked away to an adjoining room where SCI qualified hard antler judges are immediately put to work, spending an average of 30 minutes measuring each set of antlers. Velvet entries are sorted into age groups in a reefer while a team of statisticians is furiously entering all pertinent data on each entry.

THE PLAYING FIELD

Thorough knowledge of both the CWI scoring system and the velvet market is crucial to keen competitors. Recent velvet harvesting trends have become somewhat excessive in North America, with the majority of poorer antler being overgrown. Many new producers come to preliminary competitions and watch how some of the best antler in the world is grown out and they make the assumption that they can do that at home. Some can, but the majority can't. Nowhere was this more evident than in Alberta's 1999 preliminary competition. Study previous score sheets published in the *North American Elk* and the *Canadian Elk & Deer Farmer*. Astute readers will discern that winners in most velvet classes exhibit fantastic beam and balance, along with better weights within their respective age groups.

THE GLOBAL SUPERMARKET

Our industry has seen a lot of change. The former USSR destabilized. Korean velvet buyers had to run the Russian Mafia gauntlet if they wanted their preferred maral velvet. Some didn't make it. So they looked to Canada and, indeed, most of North America for their second choice of where to buy larger velvet. Older elk producers remember those heady years of the early nineties and $100+ CAN per pound.

Time brings change. Research was being conducted to validate the various pharmacological properties of the world's available velvet stocks. Softening Asian economies brought lackluster global velvet markets in winter 1998 and 1999. New ash content import rules meant that a lot of velvet had brow and bez tines removed before seeing Korean soil. New Zealand responded by taking hundreds of tons of velvet off the market. The Kiwis killed more stags and harvested the majority of their velvet in the "short" grade, ensuring exceptional quality.

In years of stronger world velvet markets, this recent trend to over-calcification wouldn't have been so disconcerting. We've spent all winter listening to the message. North America will lose its preferred nation status if producers continue to overgrow their elk velvet. We live in a global supermarket. Korean buyers will still trip over themselves if offered a chance at well-cut wapiti velvet. Overgrown, calcified velvet will be passed over as Koreans spend their working capital on better cut New Zealand antler. It's risk versus returns, and velvet buyers are good at their game.

Expect more conservative calcification scores at this summer's competitions. Some judges may have been somewhat lenient in previous years as everybody adapted to the scoring system. Now everybody is aware of market trends and nobody wants to see all of North America's 160 tons of velvet discounted $5 or $10 a pound across the continent due to over-calcification. The cost to our industry could total 2 or 3 million!

THE JUDGING PROCESS

Once producers have learned how to humanely harvest velvet antler and are aware of global market forces, additional insight into the judging process will make them better competitors.

Weight

Historically, 50% of the emphasis on the velvet scoring system was on velvet antler weight, with the remainder on quality and style. The new CWI scoring system is more sophisticated. It was designed to more accurately reflect the demands of the marketplace and puts a much greater emphasis on quality and style in the younger age classes. Consider the following examples from the 2-year-old (Figure 3) and mature (Figure 4) class winners at NAEBA's 1999 International Antler Competition:

Fig. 3

Fig. 4

FIGURE 3	FIGURE 4
19.48 lbs. x 2 pts./lb = 38.96 weight pts.	50.98 lbs. x 2 pts./lb = 101.96 weight pts.
38.96 weight pts. = 42.6% is **"Weight"**	101.96 weight pts. = 68.0% is **"Weight"**
91.47 CWI	149.91 CWI

Producers harvesting younger bulls for competition would do better to worry more about quality than quantity.

Beam

Beam circumference in our industry is somewhat understated. Bulls exhibiting better antler beams within an age class frequently carry better weight into the velvet heart piece while calcifying somewhat more slowly. The CWI scoring system calculates beam points by first measuring the minimum beam circumference (in cms) between the bez and trez tines on each antler (done by the statistician at the registration desk), taking an average, and then subtracting 13, 14, 15, 16, 17, or 18 in the case of 2-year-old, 3-year-old, 4-year-old, 5-year-old, 6-year-old and mature class entries.

Beam Points =

$$\frac{\text{Right Beam cms + Left Beam cms}}{2} - 13 = \text{Beam Points (2 year olds)}$$

Since a 20 cm beam on a 2-year-old is huge, a score of 7 is great. In the mature class, a score of 7 means a bull with 25 cm beams, which is awesome. One of the pitfalls in selecting for improved beams, is where the trez tine falls on the beam. Lower trez tine placement occasionally influences beam measurements, as do extra non-typical tines. Producers should compare competition score sheets with the actual antlers or photos to pick out any incongruities. Study Figure 4 again, and you'll see that the extra tine on King's right antler results in a beam measurement of 27.5 cms, while the cleaner, left antler measures 25.2 cms. Remember those Chinese sika deer that Tony Pearse flashed on the screen at the 1996 NAEBA convention in Edmonton. Centuries of selecting for increased beam circumferences and increased velvet weights resulted in Chinese sikas cutting 8 - 10% of their body weight in velvet antler each season!

Tines

Tine length and their placement on the beam are worth a maximum of 5 points. Tine length is judged by drawing a line from the tip of the brow tine to the top of the velvet antler. Tines falling short or exceeding that line are penalized (Figures 5 & 6). Bez tines that grow further up the beam result in a gap on the main beam between the brow and bez tines. In times of softer velvet markets, bottom brow and bez tines occasionally get trimmed off and sold into the Chinese or Taiwan marketplace for significantly less money. This is an undesirable trait and consequently gets penalized in the CWI judging process. Points are also docked for extra "kicker" or non - typical tines.

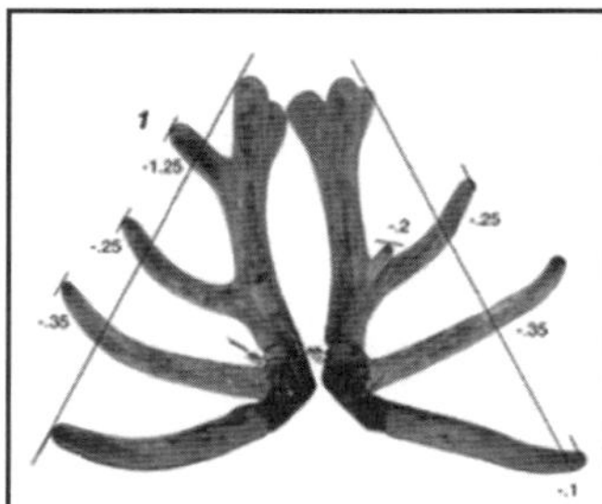

Fig. 5

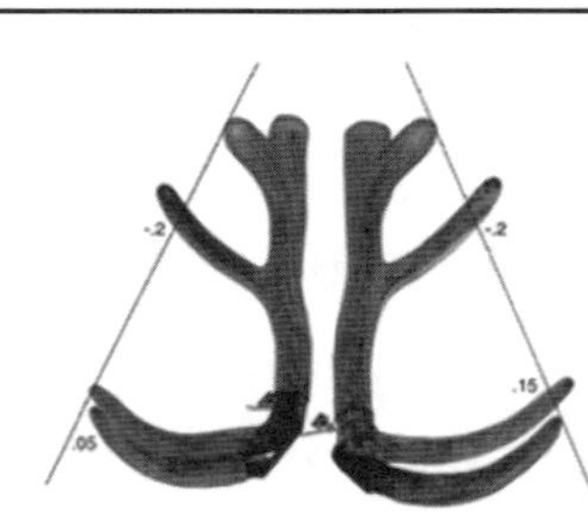

Fig. 6

JUDGING THE "TINES"

	Deductions (-) Figure 5	Figure 6
Nonconforming tines	1.30	4
Irregular nonconforming tines	1.45	2
Total Deductions	**2.75**	**.6**
Score	**2.25**	**4.4**

Mirror Image

The simplest subjective judging criteria, mirror image, is how well the right and left antlers match, irrespective of other quality and style considerations. Beam length, tine placement and lengths, tops and minor irregularities are all considered (Figures 7 & 8). A maximum of 5 points is available for those picture-perfect, matched sets.

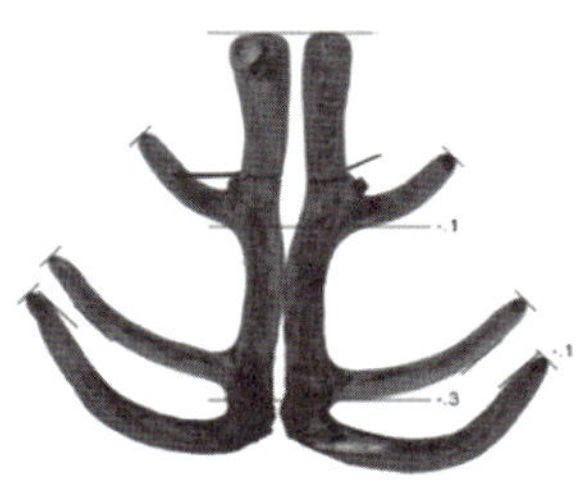

Fig. 7

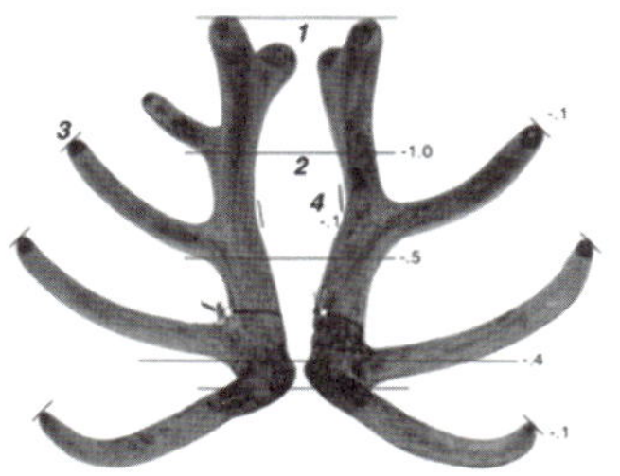

Fig. 8

JUDGING "MIRROR IMAGE"

	Deductions (-) Figure 7	Figure 8
Are the beams even in length	0	0
Are the eyeguards, bez & trez tines evenly placed?	.1	1.9
Are the lengths of the eyeguard, bez & trez tines even?	.3	.2
Are there miscellaneous irregularities?	.1	.1
Total Deductions	**.5**	**2.2**
Score	**4.5**	**2.8**

Balance

Balance is probably the hardest thing for rookie judges to assess. New judges used to set velvet sticks on the edge of their hand to try to determine a midpoint. I find that method solves nothing. Recent trends in judging balance seemed to be the selection of webbed tops and weaker brow and bez tines (Figures 9 & 10). While those two factors influence balance scores, the real determination of balance is whether you are looking at an unbelievable stick of velvet within an age class, that exhibits that wedge shape (Figure 11), or are you really looking at a trophy bull that gets spindly the further up the antler you go?

Fig. 11

Fig. 9

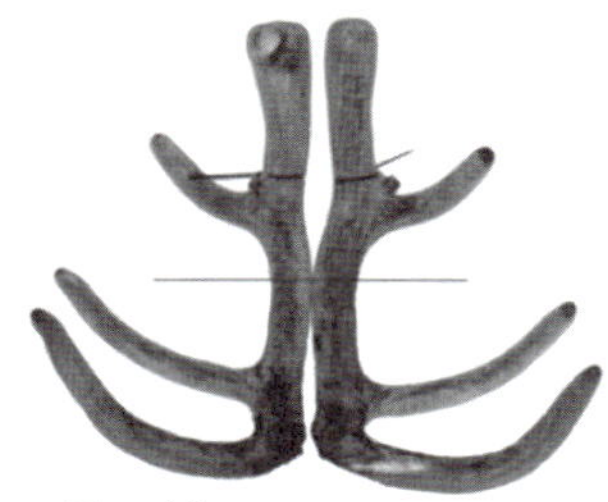

Fig. 10

JUDGING "BALANCE"

Figure 9 - These antlers score a solid 10; the beam continues to increase in width to a very large web on top.

Figure 10 - This set scores a 3.5. Note the poor top weight with heavy bottoms.

Examine the Southern Region's 2-year-old class (Figure 12). The first-place entry has fantastic balance, scoring the maximum 10 points, as the beam gets stronger the further up the stick you go (Average: 20.2 cms as a minimum!). The second-place entry, scoring 5 points for balance, might be considered a potential hard antler candidate, being weaker beamed (a 16.55 cm average), but exhibiting a lot of stronger brow and bez tines. Being well cut, it still made a very respectable velvet entry. The third-place entry has better beams again (17.65 cm average.), along with longer brow and bez tines, but lacks that curved wedge shape, scoring a 4.8 for balance. With a very conservative harvesting strategy, it received the maximum 25 points for calcification and consequently placed very well. Stronger trez tines and elongated heart pieces (tops) would push the balance score down even further, as in Figure 3, which scored a 3. Don't despair - bulls with poor

Fig. 12 Texas two year old class winners

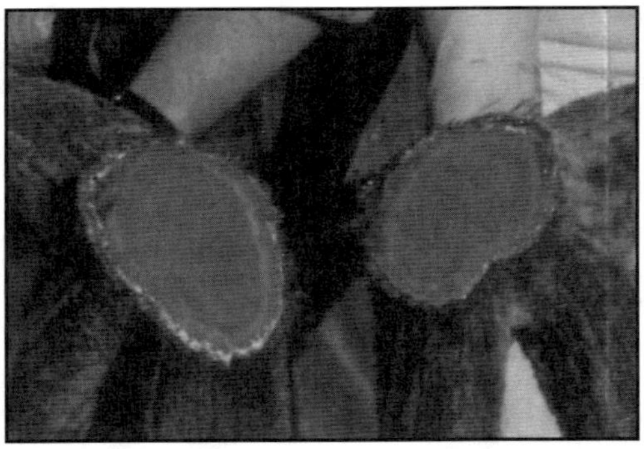

Fig. 13 This set of antlers shows very little calcification

balance usually make phenomenal hard antler candidates, and if cut conservatively, still make sellable velvet!

Calcification

Variation in calcification scores between the preliminary and international competitions has caused much consternation and criticism. As a general trend, judges have been too generous with calcification scores in previous years, and have been afraid to use the full 25-point scale. In 1999, the variation in the ranges and averages among the seven preliminary competitions was evident. With Korean velvet buyers criticizing the current trend to over-calcification, competitors thinking that weight wins, and copycats cutting all their mediocre velvet like phenomenal competition winners-a collision is inevitable! Be forewarned-while the handful of world class velvet entries might look the same as in 1999, judges will be more critical of over-calcified entries in the 2000 competitions.

Since calcification is worth a maximum of 25 points, this category can make or break any entry. As a general rule of thumb, judges look at how pointed the brow, bez and trez tines are in comparison to each other and to other entries within a particular class. If the trez tine is rounded, is it attached to the top or middle of the antlers? Higher trez tines have rounded, sausage ends, giving the appearance of being less calcified. What type of top does it have? Is it webbed or slingshot? Does it exhibit any ribbing on the base of the beam? Once the velvet thaws for 15 minutes, judges confirm the overall calcification appearance by examining the cut base of each antler, looking at the porosity and the extent of calcification rings. Antlers cut too low to the coronet will look more calcified. Using fine-toothed saws or wrapping the velvet base immediately after cutting will result in the pores getting filled in, making accurate assessments tougher.

Remember that calcification creeps in from the outside edge, showing up first around the base of the brow tine. Scores of 24 and 25 mean that there's little calcification during an outward inspection of the velvet or in studying the base of the antlers (Figure 13). Moderately calcified antlers will have a solid, bony, white surface extending across the cut base at the narrowest point between the brow and bez tine and opposite edge of the antler. Antler bases exhibiting an overall white shiny base are very calcified and will be docked accordingly.

As the velvet industry has become more sophisticated, and as our velvet scoring system has matured, judges are now assessing the overall style of the velvet antlers before determining calcification. If they are looking at a phenomenal velvet entry in any particular age class, with fantastic beam and that strongly inverted wedge style, calcification scores will be significantly higher than entries with spindly beams and similar outward appearance! This is easily confirmed when comparing a brace of 2-year-old velvet entries (Figure 12), one of which has 20 cm beams and the other 16 cm beams, but similar brow, bez and trez tine pointing. The bigger beamed 2-year-old will likely be less calcified, as confirmed by examining the cut velvet antler bases.

Presentation

This category is worth a maximum of 5 points and considers factors like damage (nicks, cuts, scrapes, peels, blood, etc.), bulbing (from hanging the antler upside down) and smell (spoiled antlers). Producers need to know that they are dealing with a food grade product, and an important part of any quality control and quality assurance program is handling their product.

Be a Good Winner

The judges have made their decisions, and the score sheets are passed out. If you are called to the podium to accept an award, remember to do a good job of displaying your award winning entry. Be photogenic. Pat Cooper and Richard Patterson could give the rest of us lessons in displaying velvet or hard antler in any winners circle (Figures 14 & 15). If you are provided an opportunity to speak about your bull, describe the pedigree and lineage to interested spectators.

Be a Better Runner Up

Finally, although our industry enjoys a gracious winner, realize that class placements are only the judges' opinions for that day. Judges and, indeed, all of our industry appreciate good sportsmanship at antler competitions.

As the competition gets stiffer in the future, elk producers should be overjoyed to consistently place among the top ten. Remember, any antler that comes to a preliminary competition probably represents the top 1 to 2% of genetics in that region, for a given age class! Most quantitative geneticists would agree that using the top 1 to 2% of males for breeding would result in substantial genetic progress for velvet production being made in each successive generation. Good luck and have fun.

Further Reading:

Pearse, A.J. 2000. *Improvements in Genetic Selection and Recording Systems-Production of Breeding Values*. Minnesota Elk Conference for Veterinarians Proceedings. University of Minn. (See this issue of North American Elk.)

Wang, Z, R-C Yang, L.A. Goonewardene, and C. Huedepohl 1999. *Genetic Analysis of Velvet Antler Yield in Farmed Elk*. Can. J. Anim. Sci. 79: 569-571.

Management of Males in Hard Antler

Many producers choose not to velvet their top herd sires in order to see full antler development, as the ultimate expression of their genetic potential, or for pleasure and promotional work. Market opportunities for such animals for trophy hunt farms and game parks are increasing.

The management of animals in hard antler is not difficult especially if the group is a stable cohort of similar age that came through the velvet growth and hardening season together.

As the rut approaches a choice has to be made. If animals are to be sold and transported in hard antler they should be shifted just prior to antler stripping, one per pen with plenty of head room and loading width. The antler tips should be formed and hardened or damage may result.

As velvet strips, elk need suitable material (trees, shrubs, etc.) or rubbing posts to clean and colour antler. Promotional photography takes place at this stage and then the hard antler is removed prior to the rut.

If antler is to be carried through the rut it is advisable to separate bulls into their breeding groups well in advance and never mix antlered and non-antlered animals at any time. In spite of apparent paddock quietness the farmer must treat antlered bulls with great respect and caution during the breeding season as character can change on a whim and there is potential for accident or worse at any time.

Man O' War at 13 years - full hard antler

The cutting of hard antler requires more caution than in velvet removal simply as the bull is more aggressive and the activities of yarding heighten that aggression, particularly in the company of other animals. Head restraint is crucial and once mineralization is complete the antler takes time and effort to cut. While there is no pain sensation, the buildup of heat in the bone may cause discomfort and mild tranquillizing is appropriate.

Antlered bulls are both impressive and often at the heart of a vision, but must be treated with great respect and caution for their own and other's safety. It is a mistake to introduce a new individual into stable company.

Antlered spikers, or animals that regrow small spikes or partially formed antlers should have these removed as soon as they are stripped, and before any transport. The stress and proximity of others at close quarters invariably creates fighting and damage.

REGROWTH

Older and early cut bulls will often regrow a variable amount of regrowth antler. This can take the normal antler form or be of any shape, often non-branched spikes that grow from the outer edge of the cut surface. The antler grows normally and is vascular and sensitive like normal antler but hardens quickly towards the end of the growth season. There is some resale value if harvested in time, but too often smaller growth is left to clean and harden.

In this form they represent a lethal danger to other stock and handlers and should be removed in August as soon as they harden prior to stripping.

The bulls can still be easily handled at this time and it pays to smooth all cut buttons, remove all regrowth to avoid potential damage and injury to breeding stock. It should be noted, however, that there still may be pain relief required for this operation.

MOST COMMON CO-PRODUCTS

TAIL
Applicable to sika, red deer and elk only because of the presence of a large sebaceous gland on the under surface of the tail.

In Oriental countries, dried deer tail is the most commonly consumed part of the deer next to velvet antler. They are valued based on the size, and relative ease of processing. The finished product is jet black in colour, hairless and has a very distinctive odour. Its value per kilogram makes it one of the most expensive forms of medicine. The dried tail is soaked in a special processed alcoholic liquor for 30 days, the liquid drunk and the tail cooked with or without other herbal or animal medicines and eaten, served with chicken or duck soup. Uses are varied but mainly for sexual dysfunction, uterine disorders, lumbar and leg pain and as a tonic to replenish the Yang system enriching bone marrow and kidney function.

At slaughter the tail should be removed at the base of the spine and held with animal I.D. until the animal has passed all inspection procedures, then frozen hygienically for later processing.

PIZZLE
The pizzle - penis, prepuce & testis is processed to a hard dry form with a distinctive translucent red colour. Appearance and size are critical to product marketing. The pizzle is usually air dried with a portion of pelvic bone attached to authenticate origin as this is particularly important for export.

It is consumed exclusively by men, soaked and sliced in yellow wine liquor or ground and used in tonic form. It is used for impotence, sterility and to improve sexuality.

At slaughter it is necessary to remove the entire penis, testes and scrotum plus the erector muscle and leave attached to a small slice of the pelvic haitch bone. The pizzle should be thoroughly cleaned and frozen for processing.

SINEWS
The sinews are dried to a bright natural colour with or without the inclusion of the rear (dew) claws. Muscle and fat is removed and the sinew straightened. Sinews are used as a tonic soup for vital energy restoration and for muscle and eyesight ailments.

HARD ANTLER
Hard antler is washed, dried and ground and reprocessed in the form of antler glue and used to promote blood flow, lactation deficiency in women and body weakness.

SKINS
The fashion attributes of elk skins have been mentioned earlier and the strength, unusual grain character and ability to shave or split to provide a soft supple leather, combine to increase use in the garment industry both as nappa leather and a buck skin suede. The product does, however, suffer from a variety of faults, mainly mechanical damage due to antler scoring or hoof damage plus the myriad of abrasions and cuts that can arise from poor handling and yard and fence construction. Bacterial damage from poorly preserved skins and tick and fly damage can also occur.

At slaughter, skins can be frozen, salted or treated in the pickled pelt form. In salting, defatting and an adequate salting is essential with the skin left spread to cool prior to treatment.

Quality farm handling is, however, the prerequisite to improving returns and the acceptance of farmed elk skin. The size of red deer and elk skins allow large panel cuts to be made from a single skin and add flexibility to form and design.

These additional products from elk should not be considered as cast offs or by-products, but are a significant and valuable part of elk farming returns in their own right.

Processed Elk Products

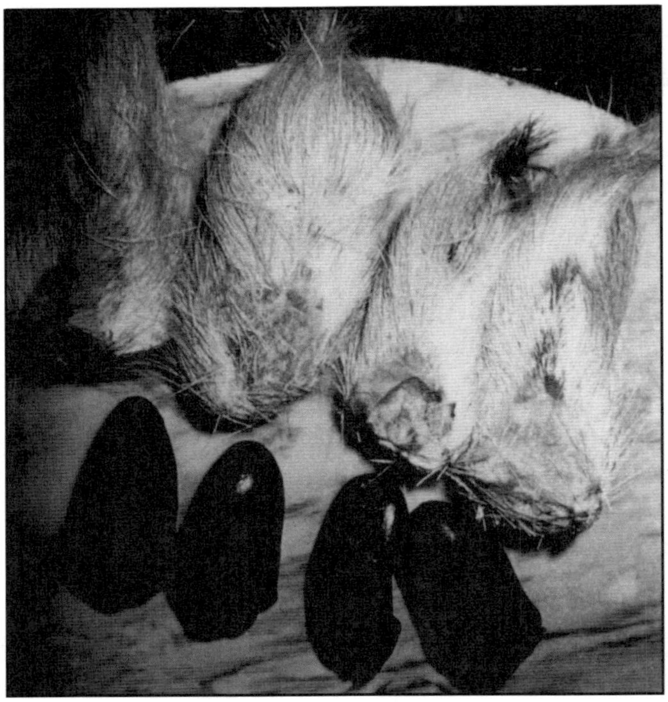

Fig. 1 Tails - in green (frozen fresh) form, and after processing

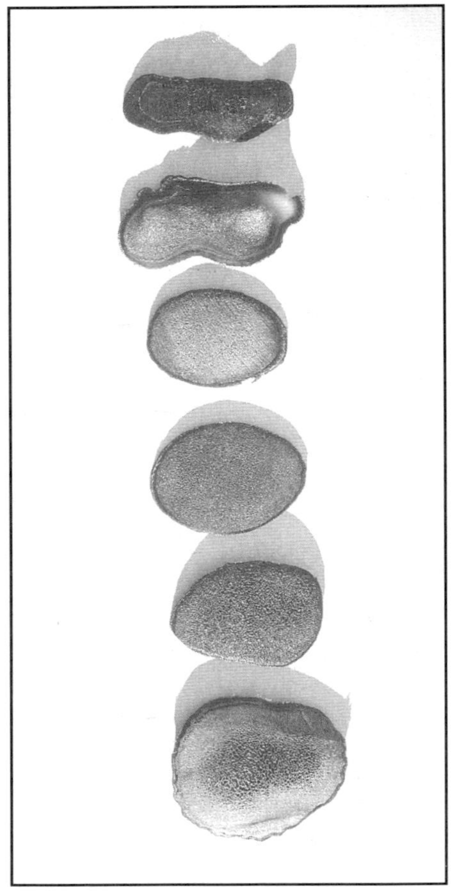

Fig. 3 Antler slices - bottom to tip

Fig. 2 Processed velvet antler - dried and sliced

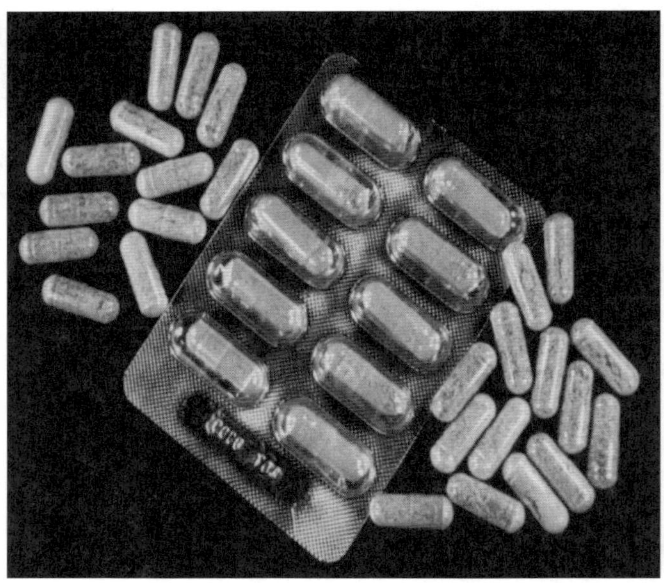

Fig. 4 Processed velvet antler - dried, ground and put into capsules.

Transfer of ownership of livestock, and making a profit in the process, are what livestock sales are all about. To accomplish this, we need to involve all three of the activities mentioned above: promotion, marketing and sales. These are all closely connected, and much discussion could take place regarding what makes them the same, and what makes them different. Ultimately, promotion attracts attention, marketing identifies and secures customers, and sales transfers the product. These three activities can be undertaken in many ways, depending much upon your personality and product base, plus those of your customers.

The elk industry is still a rapidly expanding breeding stock market. In comparison to the cattle industry, where most producers run "commercial" operations, most elk farmers are "purebred breeding stock" producers. This reality presents some special challenges to today's elk farmers. Successfully marketing purebred livestock requires careful genetic selection coupled with top-quality management, but equally important are people skills. We meet people who share our interests, become friends, sell them breeding stock and then compete with them! We need to be able to do all that and still stay both financially successful and friends. To accomplish that, we need to improve our livestock and our marketing skills, to give us that "edge" and keep our customers coming back as they expand and improve their herds.

The "High Road" is the only path to follow if you plan on staying in the breeding stock market for the long haul. Our industry is growing every day, but bad news travels fast. Besides, most of your customers are your friends, and they all deserve to be treated honestly and fairly. Never criticize another person's stock - you can do a good job of promoting and selling your own instead. Build a skill for differentiating your stock from everyone else's without criticism.

The first thing you need to do is to decide *who you are*. Are you an outgoing, gregarious type of person who communicates well with people, especially strangers? Or are you a more quiet, stay - to - yourself kind of person. Think about how you will fit into the group you are planning to sell to, and adjust your target market according to who you work best with. Maybe you aren't a showperson, but you can still sell high quality stock. Maybe the Select Auction is your best venue to get some attention and build your profile. My friend Rob Pek has a great saying: "Where you buy is where you sell", and the more you think about it, the more it rings true. If you pay the *big bucks* at Select Sales, the odds are that the people who frequent those sales will pay those same *big bucks* to buy from you, especially if they become your friends and business associates. If you buy a few head on farm each year, those people will likely be back at your place next year, buying a breeding bull. But don't think that you can grind them down in price every year then ask them for a top-of-the-market price for the bull!

Once you've decided where your personality, skills, pocketbook and background will best fit, then start building a herd that also fits. *Be honest with yourself.* The decisions you make at this stage will influence how successful you are, because what you are doing is *creating an identity*. Once that identity has been created, you have to stick with it, or make changes very slowly. People have to remember you and what you stand for and what you have to offer. That identity is your Market Profile.

Your identity needs a name, one that is memorable, one that reflects you and your program. Black Antler Ranch, Roosevelt Range, Swampy Valley Ranch, Big Bull Elk Farm - think about each of those names and what pictures they conjure up in your mind. Now think what you want people to think of when they consider your farm or herd name.

A logo is next - and it also needs to be carefully designed and presented. It should be simple, and easy to reproduce in colour, in black and white, in silhouette, on clothing. Use just two colours, and stick with those two on every bit on promo material you produce. Your truck, your trailer, your business cards, your hat, your jacket, even your dog if you can find the right one, they should all be in those two colours.

Be highly visible. Wear your hat and shirt or jacket whenever you are out and about. Build a display and take it not just to elk shows, but to other types of livestock shows, country fairs, conventions. Remember that there is plenty of room for expansion in the elk farming business. Display at Holstein shows, Canola Days, Sportsman's shows. Remember the young people in your community in particular. They are the ultimate long term investment, plus you know that every one of their moms and dads want to do their best for them, and help them into an industry with a promising and expanding future. Interviews with the media can also be very productive, both at the local level and beyond. Be careful before you jump in too far with the media. Sign up for a one-day course in how to present yourself in a most positive light.

Your identity is not easy to build, and it takes plenty of care to maintain it. You need to promote for years in advance, especially if you are planning a change to your current program. Keep repeating the message - the experts estimate that you need to reiterate your message many times before it really sinks in.

Now that you have an identity, a profile, and a program, you have to get ready to sell. This sounds too simple, but it is true. When someone decides they want to buy, you don't want to be unprepared. Start with your farm. You start selling as soon as the buyer starts driving down the road beside your farm, then down your lane. Your farm should be presentable, but not necessarily ostentatious. If you are selling to movie stars and software executives, maybe you do want a very expensive looking spread. But if your prospective customers are like you and me, you don't want them to drive into your yard and say "No way, I can't afford this!" Your farm should show excellent management in your pastures, your fences, facilities. That reassures your prospective client, and begins the process of building confidence.

Hand feed your stock! They have to be easy to call, willing to approach a vehicle and a new voice - easy to show. Binoculars are not an effective selling tool. Elk rumps are cute, but the client needs to see more than a bunch of them heading over the ridge!

Group your stock for both good management and for show and sale. You'll find that some groups you put together just don't work well. Elk live in a maternal society, and mothers, daughters, nieces are more comfortable if you keep them together. The same applies to bulls. Some become good buddies, and prosper in each other's company. Others are oil and water. For sale, group similar bunches together. Ages, genders, colours, body types, genetics all can contribute to forming and presenting a nice uniform group. If you have a big heifer you want to sell, put her in with a group of smaller ones you don't want to sell. If you want to sell the smaller ones, don't put the big one in.

Sell the customer what he or she wants. In the process of getting to know them, you'll get a feel for what they are looking for. Then don't spend lots of time going through other groups of livestock. Show them the right ones, and only show them more if they ask. Spend the rest of the time socializing and getting comfortable.

If your customer is not focussing on any of your stock, tell him a story. You need to catch and focus his attention on the elk you want to sell. I think names are a good start. A descriptive name that leads into a story about why that elk has that name. Think of something special about each one or each group. This could be pedigree, relatives, performance statistics like body weight, velvet cuts, hard antler scores, breeding records, temperament as well as more basic stuff like health management, parasite control and vaccinations. Verify your records with DNA genotypes and parentage matches, CWI scores from velvet competitions, printed sale records and registration certificates. Calves produced by Artificial Insemination carry a distinct advantage over most natural bred calves. When you buy that semen, you also buy a piece of the owner's identity and promotion program.

How will your customer make money with these animals? Tell him how! Build confidence in his prospective investment. You might consider offering to take some of the risk by offering marketing or financing assistance, but stay a percent or two above the bank.

Is your visitor not ready to buy? "Tire kickers" can chew up a lot of time. If the visitor needs to learn a lot about the elk farming business, have a few Elk Farming Handbooks around, and offer to sell him one. Tell him that once he's read it, you'll be glad to spend more time with him and you'll give him a hundred dollars back on the first elk he buys from you. Does he need help designing and building his farm? Go ahead and help him, but get a non-refundable deposit on his first group of animals first. While you're helping him, keep building confidence - tell him stories, show him information about how well you've done with the genetics he's buying.

Once you've got a client ready to buy, but not quite prepared to sign the cheque, give him options. "Would you rather have this group, or that one?" "Would you prefer them delivered tomorrow or after they're bred?" Put an offer on the table, and let the customer take time to respond to one of the options - not just "take it or leave it". Don't play too many games like "Oh, I don't know - what do you think they're worth?"

You're both intelligent people, so treat each other that way. Some people like to play the haggling game, and if you do, that's fine. But get to know your client before you jump into that situation. Have a price in mind for every elk, based on what it would cost you to replace it, plus a factor for how badly you want to sell. Guarantees are a popular item in our industry right now, but I think you should give a lot of thought to making such offers. Sometimes the resolution of a guarantee can damage future relationships more than cement them. Maybe a realistic price with a bit of buyer risk accepted would be a better approach, and much more clear than a guarantee a year away.

Be ready to sell - have your herd health management up to date, so no one has to wait for testing. Have your truck and trailer ready to go, or know someone who has one for hire. When the market is right, sell some of your better stock at public auctions to establish the validity of your pricing.

Find new customers - this is an expanding industry! Have a brochure made that describes your program, and circulate it widely. Keep a current inventory of your elk, and have it in an easily readable form that new customers can understand. Educate your local banker, Ag Rep, feed dealers, other livestock groups. Once a year, have an open house at your farm, with farm tours to give people a clear understanding why you're in the business, and that they *should* be. Circulate at trade fairs, conventions, meetings. Don't always stick with the same crowd. Network with those whose breeding programs are similar to yours. You can trade genetics and stick to your goals without a big expenditure, plus you might interest that other fellow in a few extra heifers.

Above all, take care of your customers. They are your most precious resource, and your most effective advertising device. Build a file or database on each one, noting their background, family and connections (remember the spouse and kids' names), business interests, hobbies, likes, dislikes, lifestyle, type of livestock farmed and needed. Read that all over to refresh your memory before you call and visit. Follow up on sales - how are the elk doing, are you breeding them like you planned, where are you selling this winter? Keep them happy, even if it costs. Bad luck? Help them out - replace that bull for free - you'll get it all back in the future. The worst/best advertising is an unhappy/happy customer!

Continually improve your breeding stock, Plan years ahead, Build your Profile, Stay Happy, Be Fun, Go Fishing and Hunting Together, Socialize, Build Bonds and Friendships.

Always Maintain Your Integrity.

An emphasis on quality, with the records to back it up, give you a "story". That story is what you deliver to capture the interest of your customers. These following seven steps, borrowed from Ken Aylesworth, give an excellent guide to turning that interest into sales, and to turning your customers into lifelong friends.

1. GREETING YOUR CUSTOMER

- Be sincere.
- Remember, first impressions are lasting, they could be your last.
- Learn about your customer and his business.

Use open ended questions: Who? What? Where? Why? How?

2. THE QUALIFICATION

Ask pertinent questions and listen to the answers.

- Do you presently have any deer, elk, etc.
- What type are you looking for?
- Please give me specific details.
- How many?
- When do you need them?
- What are the most important things to you (eg. performance, feet & legs, antler, body mass)?
- Are there specific bloodlines you are looking for?

Listen for concerns:

- Resolve those concerns - reduce any stress.
- Use open ended questions to answer a concern. Who? What? Where? When? Why? How?
- Record/Remember the answers.

3. **THE DEMONSTRATION**

Acquaint the customer with your product.

- Show him the animal that fits his needs.
- Let the customer look and feel.
- Do not worry if he changes his mind at this point.
- Always give options.
- Always have promotional material available.
- Performance data
- Pictures of dam/sire
- Be unique

Pricing

- Group
- Individual

Trial Closes

- Would you like an animal like this?
- Would these animals fit your needs?
- How soon would you like them?

Yes!

4. **THE CLOSE**

Always close with an option.

- Do you prefer?
- Can I deliver him on Tuesday or Thursday?
- Always get a commitment/get something signed.

Ask the closing question and shut-up!

- The first person to speak loses.
- Do not give up.
- Do not be defeated.
- Find out what is wrong and find the answer.

5. **PRODUCT DELIVERY**

Resell and ensure satisfaction.

- Repeat the benefits of your animals.
- Reinforce the decision to purchase from you.
- Use this time to prospect.

6. **THE FOLLOW-UP**

Follow-up is the most important step in the seven point process.

- Call, write, stop in
- Always keep in touch

Follow-up ideas:

- Keep a list on each of your customers, include name, education, family, business background, special interests, lifestyle, livestock raised.

Use your information list to:

- Send short notes.
- Develop promo items (eg. hockey tickets, magazines)
- Magazine promo shows commitment and adds validity to your program.
- Correct any problems or concerns your customer may have.
- Develop new leads for new customers.

7. **PROSPECTING**

How to look for new business:

- Friends, relatives, and neighbours.
- Telephone, letters, meet individuals personally.
- Target an area.
- Cold calls.
- Develop an ongoing program.

QUESTIONS TO ASK IN A LIVESTOCK TRANSACTION

The following list, adapted with thanks from Jill Wood of Bryarwood Ranch in Texas, will help to protect the buyer's investment and ensures both parties have an understanding of expectations and responsibilities.

ABOUT THE SALE

- Is there a written sales contract that has been agreed to by both parties?
- Is there a Bill of Sale?
- What does a disclaimer mean to the buyer? What does it mean to the seller? Does it mean buyer beware?
- How long has the seller been in business and what is their reputation? Have you checked their references?
- Who will pay to ship the animal?
- Will there be travel insurance and who will pay for it?
- What happens if the animal does not survive the journey?

- If there is a fertility test, who will pay for the test?
- How is payment going to be handled? Cheque - certified cheque - money order?

ABOUT THE ANIMALS

- Are the animals from a herd that has been sent through the working pens on a regular basis or are they from a ranched or open range situation where they are not used to being worked through a facility?
- Were the animals in good health when they left the seller's and arrived at the buyer's?
- Were the animals counted when they were off-loaded?
- Does the animal have permanent or temporary means of identification? Tattoo - ear tags - USDA tags - H. of A tags?
- What is the Tuberculosis status of the seller's herd? Have the animals been tested? How many times have they been tested? What were the results of the test? Is the animal from an Accredited Tuberculosis-Free Cervidae Herd and was the animal included in the testing?
- What is the Tuberculosis status of the buyer's herd, in the case that the animals need to be returned to the seller?
- Has the animal been Brucellosis tested?
- Has the animal been in contact with animals other than those from the seller's herd?
- Is the animal being brokered from another herd? Is there a brokerage fee and who is going to pay it? What is the Tb status of that herd?
- What is the weaning rate, herd health, breeding, fertility and genetic history of the seller's herd?
- What diseases, if any, have there been in the seller's herd?
- What shots and tests has the animal had? When were those shots and tests run?
- Are there shots that the animals should have before they enter your herd?
- What is the weight of the animal? What is the weight at time of transit? What is the weight at the time of arrival?
- What is the age of the animal?
- What is the average carcass weight of the seller's herd at slaughter for both males and females and at what age was this weight taken?
- What are the genetics of the animal and has there been any particular problem with that bloodline?
- What have the animals been fed? What kind of hay/grass or type of feed are they used to?
- Have they been provided minerals and salt?
- What are the legal requirements for moving deer within or between the areas or jurisdiction involved?

ABOUT BREEDING STOCK - MALES

- Has the male been used as a herd sire in the past?
- Was he used in a single or multisire pasture?
- What percent offspring did he produce?
- Has the male been fertility tested?
- If he is going to be tested, where will he be tested?
- Who will perform the fertility test and who will pay for it?
- Will more than one animal be tested so that comparisons can be made in the semen samples? (remember, elk are seasonal breeders and this can only be done in season)
- How many females can the male be expected to cover?
- At what age should the male be able to breed?
- How many pounds of velvet have been harvested from this male in the past or how much velvet has been produced by the sire of this male?
- What did the antlers of the bull or his sire look like?

ABOUT BREEDING STOCK - FEMALES

- Is she open, guaranteed bred or just exposed to a sire?
- Has the female had an ultrasound or rectal examination or blood serum test to determine pregnancy and if so, at what stage of her pregnancy?

- What is her birthing history? What is her dam's history?
- At what age should she be able to be bred?

As you can see from these lists, there are many things to consider before you either buy or sell elk. It also might be helpful to check with other elk farmers to see what safeguards they have used in buying and selling livestock.

THE INVESTOR OPTION: SOME CONSIDERATIONS

The following excerpt is adapted from an article by Jerry and Kathy Perkins, Delta, CO.

One option for marketing breeding stock is to sell to an investor, then manage the livestock under a contractual arrangement. This arrangement can be mutually beneficial to both parties if properly structured. The farmer can generate additional income on his operation without an additional capital investment in breeding stock. The investor can receive a better-than-average return on investment without the investment in facilities, labour, feed and other cost factors. Freedom from the confining day-to-day operations is also a major consideration for the investor.

There are numerous existing agreements. Agreements vary widely from farm to farm; therefore, to develop a uniform contract that fits every situation would be practically impossible. The specifics in forming the partnership vary greatly from one operation to another. Two typical partnership agreements appear to be favoured: (1) a percentage split of the calf crop; (2) a cost per day per head agreement; (3) a fixed cash lease arrangement.

A well-thought-out agreement with which both parties are happy can pay dividends by saving unnecessary misunderstandings. Agreements can be as simple or complex as the preparers choose to make them. Below are some suggested items to be considered or negotiated before a partnership is formed:

1. Is there trust between the partners? Do the partners have similar goals and expectations? This is of paramount importance! If the answer to these questions is not a resounding "yes", abandon the thought immediately.

2. Has there been a whole herd TB test completed on the farm? Make sure, as an investor, you do not invest money into a herd that could have health problems. Ask for a complete history.

3. What type of fencing/working facility does the farm have? Is animal safety a top priority? Closely inspect the facilities on the farm. A suggestion is to invest the time to look at other operations in the area to see how your animals' prospective home compares.

4. What type of veterinary practices are conducted on the farm? Ask the owner to go over the herd health program.

5. What type of nutrition program does the farm have? Observe the condition of animals on the farm; and again, compare to the surrounding farms. Look at the body condition, hair coat, and growth of the herd.

6. What type of records are kept on the farm? Ask what kind of records are kept, how detailed, if computerized, and also ask if performance records are kept? What type of herd improvement program is used?

7. Outline each parties duties, obligations and responsibilities. Spell out each party's obligations regarding such operating expenses as: advertising, unusual veterinary costs, cost of sales, feeding, record keeping, labour and any other operating expense that is of concern.

8. Who is responsible for sales, and who makes sales decisions?

9. What will be the percentage and method of dividing the calf crop or income between the investor and farmer? If a daily rate is charged, what is the price and when is it paid?

10. Who pays for insurance? Is it required by the farmer or investor in order to keep or purchase the livestock? What type of coverage is desired and what is the cost?

11. If livestock is not insured, who pays for death loss or injuries should they occur?

12. Will investor require outside expertise in purchase of breeding stock to be placed with farmer, or will the farmer purchase them? What will be the cost of the purchased animals and the commission involved? What transportation costs will be involved?

13. How many investor-owned elk will be allowed to be placed on the farm? Is there a limit that the farmer will allow?

14. Is there a time frame for the partnership to exist (in years)? How is contractual agreement terminated by either party? A farmer may expand the present operation with additional fencing,

which is a considerable investment. Will the investor stay long enough for the expansion to be cost effective?

15. Is there a stipulated annual date or deadline for each party to commit for another year of operation or continuation? A farmer will not winter a pregnant cow and the investor decide to remove the animal prior to calving!

16. Who funds expenses involved in facility/ fencing construction?

17. What are the visitation rights of the investor? This can become very frustrating to an elk farming family if friends and family of the investor arrive when not expected.

18. Are elk purchased by an investor required to be free of liens that could interfere with sales of animals or payments to farmer? Most bank security agreements will state "all elk owned or hereafter acquired," meaning the cow and calf are part of the bank's collateral.

How are increases in breeding stock numbers (above the initial purchases) distributed when contract is terminated?

19. When can an animal be culled? Who gets the payment? Who supplies the replacement?

20. How are death losses and strays handled?

21. What is the dispute resolution process?

22. How will the arrangement be terminated, and how are increases in breeding stock numbers (above the initial purchases) distributed at that time?

The win-win situation between the elk farmer and the elk investor can only occur if the process is well-planned and documented. This not only ensures safety and comfort for these two parties, but can be beneficial for families, and other business and investments of each that could be affected by unforeseen circumstances such as disability or death of the parties involved. The rewards can be great for both the farmer and the investor; however, thoughtful consideration, development of a plan, and finally a contractual agreement will lend itself to financial rewards and satisfaction combined with an effort to dismiss and prevent misunderstandings.

CHALLENGES & OPPORTUNITIES FOR THE CANADIAN ELK FARMING INDUSTRY IN 2000 AND BEYOND

by Ian Thorleifson, Viking Livestock, Inc.

Stop for a moment, and think ... it is only twenty years - two decades - since elk farming really began in Canada. In 1976, Rick Alsager was granted the first permit issued in Canada to operate a commercial elk farm. At the time, not half of one percent of all North Americans had ever heard of velvet antler, or ever considered that elk could be successfully and profitably raised as livestock.

There has certainly been a dramatic pace of development since then! The current estimates are 1,100 elk farms and 50,000 elk in Canada, and 1,400 elk farms and 80,000 farmed and ranched elk in the United States. The estimated value of the livestock plus the dedicated facilities and fencing on these farms is currently over 1.2 Billion Canadian dollars. Gross receipts for the Canadian elk farming and velvet antler (EVA) industry in 1999 were estimated at $80 million dollars, putting our industry alongside some of the most significant components of Canada's agricultural industries.

How did we get here? The stimulus to begin commercial development originally came from contact with Asian buyers of velvet antler, and people familiar with harvesting preserves in the United States and Europe. Although a number of "game farms" were already operating profitably in Canada and the United States, New Zealand farmers and entrepreneurs like Sir Timothy Wallis and John Barber really led the way in development and establishment of modern cervid farming.

Management systems for farmed elk in North America were not difficult to develop. Elk were a natural fit for North American climates and existing livestock management systems. Anyone who raised them quickly realized how well suited they were for extensive livestock agriculture. Markets were waiting for EVA and Trophy bulls, and a strong demand for wapiti meat is still not being touched by our industry. In response to these existing market demands, the breeding stock market has been very lucrative all the past twenty years. Prices have certainly fluctuated, but we must be aware of two very key points regarding these price fluctuations:

1. The prices of other commodities have varied just as widely - think of market hogs, with prices paid per pound to farmers varying from twenty cents to over a dollar! Or canola – varying from three to twelve dollars per bushel! And have hogs ever sold for twenty five thousand dollars for a breeding female?

2. Every time the prices drop, our critics say "I told you so…pyramid scheme…ostriches…" you've heard it all. But then the prices bounce back, just like any other agricultural product that is sold to long established markets.

Elk velvet antler (EVA) as a food supplement and nutraceutical remains the strongest driving force for our elk farming industry. It is interesting to notice how prices offered for breeding stock follow a year or two behind the price variations for EVA. (See Table 9, page 207)

$35.00 (CDN) / pound is considered by many producers to be the breakeven level. We have not often needed to sell at a loss. North American production of unprocessed antler is estimated to have surpassed 220,000 pounds in 1998, with the 1999 harvest down to an estimated 155,000 pounds in response to lower prices. Early in the year 1999, there was a substantial amount of EVA sitting in freezers in North America. As of today, April 2000, there is very little in storage anywhere in the world. The rise from $24 to $60 from 1998 to today reflects increasing demand and ability to pay for product over that period of time. The $60. price is, of course, not the final average for the year 2000. In New Zealand, the farmers tend to respond to lowered antler prices by slaughtering all their bottom end males and putting them into the venison trade. In North America, we responded to the latest price drop by developing new markets!

At this point, we should consider the current trends in the North American elk farming industry. I characterize the most important of these as listed below:

- Identification of our elk livestock as privately-owned rather than as "Wildlife"
- Development of Animal Health Management Systems
- Expansion of markets for Managed Hunts in fenced Preserves
- Development of the North American Market for EVA products

These four Trends create our biggest Challenges and Opportunities for the year 2000 and beyond. We will consider and discuss them one after another:

The "WILDLIFE" label has caused more problems for our industry than any other single source of trouble. Our North American history of commercial slaughter of Bison, Passenger Pigeons and other wildlife in the 1800's led to elaborate systems of public ownership of land and wildlife resources and powerful wildlife management lobbies. These lobby groups continually use the "public wildlife" rallying cry, along with far - fetched predictions about diseases, parasites and hybrids to interfere with further development of our elk and deer farming industries. One of the most important questions being asked right now in Western provinces is "do we own our livestock, or just have a Permit to keep them?"

The challenge our industry must rise to is to pressure politicians and other regulators to clarify the difference between Private livestock and Public wildlife.

This essential difference must be clearly addressed in legislation governing both wildlife and agricultural livestock. It must be completely clear that privately owned livestock are excluded from those regulations that address management of wildlife.

Fear of disease and parasites has also held up development of our industry. In fairness to our critics, science did not keep pace with industry growth in the 1970's and 1980's. Testing techniques are still not adequate for TB and CWD, but there is an important difference between these two diseases - better technology for testing for TB has been available since the early 1990's, in the LST and ELISA blood tests. The Canadian Food Inspection Agency has obstinately refused to validate and implement that technology!

CWD is another matter. In spite of the best efforts of brilliant scientists, a test that will identify animals likely to develop this type of disease has still not been developed. Our best test only works after the animal is already dead, and after that animal has probably had an opportunity to infect others in the herd.

The challenge here for our industry is to support development of better testing procedures.

The best approach is probably blood tests for both diseases. We as industry must support the effort to validate and make available the tests I mentioned above by providing CFIA research laboratories with funding and samples as required. Then we need to pressure CFIA to stop skin testing and use the more effective blood tests, combined with examination of the carcasses of elk and deer that die on-farm or are harvested in other ways.

A National Program for CWD eradication in Canada is only months away. Industry will play a large role in developing the details of the program, as will other interest groups like the Canadian Wildlife Federation and Provincial officials. This program will include a mandatory requirement for Post Mortem examination of all adult deads that display certain behaviours or symptoms.

The challenge for our industry is to maximize the benefits of this program. If all deads are examined for TB as well as CWD, could we reduce or eliminate the need to test whole herds for TB? Could we also look for Johne's disease, Brucellosis, and parasites?

Here is another Health Challenge for our industry:

We must compile all the information we have regarding the ***absence*** of P. tenuis, the meningeal worm of white-tailed deer in eastern North America, in farmed Elk, and publish that information in a peer reviewed and scientifically correct journal. Then, maybe we can get rid of the ridiculous restrictions regarding Meningeal Worm and movement of our farmed elk.

Another essential aspect of the National CWD Program, as well as being essential for other health management programs and a Quality Assurance scheme, is an effective Identification and Inventory program for all farmed, ranched, zoo or research herds of cervids in Canada. Be clear what I am saying here - this is NOT a Registry!

ID and Inventory is both a Challenge and an Opportunity for Industry - we can take over management of the ID and Inventory databases currently maintained by CFIA and the Manitoba, Saskatchewan and Alberta governments, make them more effective and efficient and use that new database for:

- keeping wildlife concerns reassured,
- tracing any disease problems
- ensuring consumers of the health and safety of our products
- supplying information to Registry systems
- generating equitable funding for operation of programs for industry development and promotion

We must have good ID and Traceback systems for QA/QC/HACCP requirements. Most people in the industry do not appreciate how very real the threat is that we may lose access to EVA and wapiti markets in the US and Europe without an acceptable Food Safety Program for our entire industry. This is not just a Challenge, it is an Absolute Necessity!

Markets for EVA products are changing rapidly. Until a few years ago, most EVA was exported as fresh frozen product from North America to South Korea or Hong Kong. That pattern has changed - in 1999, of the 70 tonnes estimated produced in Canada and the United States, at least 45% was consumed here in North America. Consumers are increasingly from non-Asian backgrounds, and include athletes, arthritis sufferers, and owners of performance horses and other pets.

The Antler Science and Product Technology Symposium, held in Banff in early April, 2000, was the most significant single event to date in the history of use of EVA. There were large delegations from the United States, New Zealand, Korea, China and Russia - 32 Presentations, 40 Posters. The Proceedings of the ASPT Symposium is the current "Bible" for EVA researchers and marketers as we build on the foundation of scientific knowledge presented there.

The Challenge for industry is to be sure that every possible opportunity to provide informative and educational material regarding the nature and benefits of EVA is maximized, and that researchers are encouraged and supported in their efforts to increase our knowledge and confidence regarding this amazing product.

The Trophy market is also a huge Opportunity for the industry. Texas is a good example - hunting, most of it on fenced ranches, generates several Billion dollars per year in gross income for that state. Cattle operations are quickly taking a back seat to the very lucrative hunting industry. Michigan hunting ranches alone need several thousand adult elk bulls and whitetail bucks every year. Opposition from hunter groups, and in some cases misguided animal welfare promoters, has been strong and vocal. Most of this opposition is based on emotion and such concepts as "Fair Chase" and "Real Man" ideas about hunting. The opposition fades when people realize that the animals are comfortable and well cared-for and humanely and efficiently harvested. They need to be educated, to understand that a properly located and managed preserve can offer a challenging experience most hunters would like to repeat. Politicians and bureaucrats need to appreciate that the economics of harvest preserves are very attractive and strong demand exists for this type of opportunity.

The Challenge for our industry is to educate and inform interest groups and the public, to slowly change perceptions, attitudes and regulations regarding this and all aspects of the elk and deer farming industries.

To sum it all up, the future is very bright for the North American cervid farming industry:

- Opposition groups have been soothed by improved management and production techniques
- Markets that existed long before industry development have still not been satisfied
- Well-organized promotion and marketing campaigns will ensure that market development continues

In a quick two decades, elk and deer farming have become an integral and highly profitable part of North America's agricultural economy

Imagine where we will be in another two decades, and beyond!

BIBLIOGRAPHY AND FURTHER READING

Adam, C.L., and Moir, C.E., *Effect of Winter Nutrition of Young Farmed Red Deer on their Subsequent Growth at Pasture*, Anim. Prod. 40 : 135 - 141, 1985.

Alexander, T.L. and Buxton, D. (eds), *Management and Diseases of Deer.* The Veterinary Deer Society, 250 pgs. 1994 ISBN 0951082612

Bothma, J. du P., *Game Ranch Management*, Van Schaik (Pty) Ltd. Pretoria, South Africa, 672 pp. 1989.

Boyce, M.S. and Hayden-Wing, L.D. eds., *North American Elk: ecology, behaviour and management*, University of Wyoming, Laramie, 1979.

Brelurut, Alain, and Pingard, Aude, *Le Cerf et son Elevage.* Editions du Point Veterinaire, Maisons-Alfort, 143 pgs. 1990 ISBN 2-86326070-7

Burton, B.A., *Code of practice for the care and well-being of farmed fallow deer.* BC Fallow Deer Farmers Assoc., booklet, 1991.

Canadian Venison Council, *Code of Practice,* 1996.

Caton, J.D., *Antelope and Deer of America: A Comprehensive Scientific Treatise upon the Natural History, including the Characteristics, Habitats, Affinities, and Capacity for Domestication of the Antilocapra and Cervidae of North America,* Hurd & Houghton, NY, 426 pp., 1877.

Corah, Dr. Larry, *Body Condition - An Indicator of the Nutritional Status of Beef Cows*, Agri-Practice, Fall, 1989.

Clutton-Brock, J., *A Natural History of Domesticated Mammals*, University of Texas Press, Austin, Tx, 208 pp., 1987.

Edmonson et al, *Body Condition Scoring*, Journal of Dairy Science, Vol. 72, No. 1, 1989.

Fennessy, P.F., Moore, G.H., Corson, I.D., *Energy Requirements of Red Deer*, Proc. N.Z. Soc. Anim. Prod. 41 : 167 - 173, 1981.

Fletcher, T.J., *The early development of deer husbandry*, In: Alexander, T.L. and Buston, D., Management and diseases of deer, 2nd edition, A Veterinary Deer Society Publication, London. 3 - 4, 1994.

Friedel, B.A. and Hudson, R.J., *Productivity of farmed wapiti in Alberta (Cervus elaphus canadensis)*, Canadian Journal of Animal Science, 74: 297 - 303, 1994.

Haigh, Jerry C. and Hudson, Robert J., *Farming Wapiti and Red Deer.* Mosby-Yearbook Inc., 368 pgs. 1993 ISBN 0-80166787-9

Harwell, G., Robinson, R.M., Davis, D.S., Hughes, D.M. and Weishuhn, L., *Texas game ranching: overview of a new industry*, In: Proceedings, First International Conference on Zoological and Avian Medicine, Oahu Hawaii. 435 - 445, 1987.

Hobbs, N.T., *Rearing and Training Elk Calves for use in Food Habits Studies*, J. Wildl. Manage. 43(2) : 568 - 570, 1979.

Hudson, R.J. and Adamczewski, J.Z., *Effect of supplementing summer ranges on lactation and growth of wapiti (Cervus elaphus)*, Canadian Journal of Animal Science, 70: 551 - 560, 1990.

Hudson, R.J. and Burton, B.A., *Wildlife industry*, In: Animal production in Canada, Martin, P.J., Young, B.A. and Hudson, R.J., eds. Fac Extension, University of Alberta, pp 151 - 172, 1992.

Hudson, R.J., Drew, K.R. and Baskin, L.M., *Wildlife production systems: Economic utilization of wild ungulates,* Cambridge University Press, Cambridge and New York, 469 pp. 1989.

Hudson, R.J., Kozak, H.M., Adamczewski, J.Z. and Olsen, C.D., *Reproductive performance of farmed wapiti (Cervus elaphus nelsonii).* Small Ruminant Rev., 4: 19 - 28, 1991

Hudson, R.J. and Watkins, W.G., *Foraging rates of wapiti on green and cured pastures*, Canadian Journal of Zoology, 64: 1705 - 1708, 1986.

Huttu-Hiltunen, V., Kopivupera, N. and Nieminen, M., *Reindeer husbandry in Finland*, Association of Reindeer Herding Cooperatives, Rovaniemi, 22pp., 1990.

Jarman, M.R., *European Deer Economics and the Advent of the Neolithic*, In: Higgs, E.S. (ed.) Papers in Economic Prehistory, Cambridge University Press, Cambridge, pp. 125-147, 1972.

Jiang, Z. and Hudson, R.J., *Estimating Forage Intake and Energy Requirements of Free Ranging Wapiti (Cervus elaphus)*, Can. J. Zool., 70 : 675 - 679, 1991.

Jiang, Z. and Hudson, R.J., *Seasonal maintenance and growth requirements of wapiti*, Can. J. Zool., 74 : 97 - 102, 1994.

Kozak, H.M., Hudson, R.J., and Renecker, L.A., *Effects of supplemental winter feeding on performance and foraging behavior of wapiti*, Rangelands, 16 : 153 - 156, 1994.

Kozak, H.M., Hudson, R.J., French, N., Renecker, L.A., *Effects of winter feeding on lactation, and calf growth of farmed wapiti (Cervus elaphus)*, Rangelands, 17: 116 - 120, 1995.

Knox, D.P., Reid, H.W., and Peters, J.G., *An Outbreak of Selenium Responsive Unthriftiness in Farmed Red Deer (Cervus elaphus)*, Veterinary Record, 120 : 91 - 92, 1987.

Kyle, R., *A Feast in the Wild*, Kudu Publishing, Oxford, 202 pp., 1987.

Lantz, D.E., *Deer Farming in the United States*, USDA Farmer's Bulletin, 330, 1908.

Matthews, L. R., Ingram J., Cook, C., Bremner, K. and Kirton, P., *Induction and assessment of velvet analgesia,* New Zealand, Veterinary Association Deer Branch Course 9: 69-76, 1992.

McCabe, R.E., *Elk and Indians: historical values and perspectives,* In: Elk of North America, Thomas, J.W. and Toweill, D.E. (eds), Stackpole Books, Harrisburg, PA, pp 61-123, 1982.

Ministry of Agriculture and Fisheries (MAFCORP), *Deer welfare: Code of minimum standards*, Aglink FPP 880: 1st Edition, Primedia, Wellington, New Zealand, 1987.

Ministry of Agriculture and Fisheries (MAFCORP), *Code of recommendations and minimum standards for the welfare of deer during the removal of antlers*, Code of Animal Welfare No. 5, Wellington, New Zealand, 1992.

Ministry of Agriculture, Fisheries and Food, *Codes of recommendations for the welfare of livestock: Farmed deer.* Leaflet 940, Publication, Lion House, Willowburn Estate, Alnwick, Northumberland, England, NE66 2PF, 1989.

Ministry of Agriculture, Fisheries and Food, *Guidelines for the transport of farmed deer,* Leaflet 945, 1989.

Mould, E.D. and Robbins, C.T., *Nitrogen metabolism in elk,* Journal of Wildlife Management, 45: 323 - 334, 1981.

Murie, O.J., *The elk of North America,* Stackpole Company, Harrisburg, Pennsylvania, 1951.

New Zealand Game Industry Board, *Velvet training programme*, Farmer Update, Bulletin No. 3. 1993.

Parker, K.L. and Robbins, C.T., *Thermoregulation in mule deer and elk*, Can. J. Zool., 62: 1409 - 1422, 1985.

Prescott-Allen, C. and Prescott-Allen, R., *The First Resource: Wild Species in the North American Economy*, Yale University Press, New Haven and London, 529 pp., 1986.

Reinken, G., *Deer farming: A practical guide to German techniques*, Farming Press, Ipswich, UK, 289 pp., 1990

Renecker, L.A. and Hudson, R.J., *Digestive kinetics of moose (Alces alces), wapiti (Cervus elaphus) and cattle*, Anim. Prod., 50: 51 - 61, 1990.

Sniffen, Dr. Charles, Michigan State University, Ferguson, Dr. James, University of Pennsylvania, New Bolten Centre, *Body Condition Scoring Guide*, Church & Dwight Co., Inc., 1993.

Suttie, J.M., Fennessy, P.F., Veenvliet, B.A., et al., *Energy Nutrition of Young Red Deer (Cervus elaphus) Hinds and a Comparison with Young Stags*, Proc. N.Z. Soc. Anim. Prod. 47 : 111 - 114, 1987.

Thomas, Jack Ward, and Toweill, Dale E. (eds), *Elk of North America, Ecology and Management.* Wildlife Management Institute, 697 pgs. 1982 ISBN 0-81170571-4

Verme, L.J., and Ullrey, D.E., *Feeding and Nutrition of Deer*, in: Digestive Physiology and Nutrition of Ruminants, Vol. 3, Ed. D.C. Church. Pb. Oregon State Univ., USA, 1972.

Von Kerckerinck, J., *Deer farming in North America*, Phanter Press, Rhinebeck, NY., 1987.

Wairimu, S., Hudson, R.J. and Price, M.A., *Catch-up growth of young wapiti stags*, Canadian Journal of Animal Science, 72: 619 - 631, 1992.

Watson, Carol, *Australian Deer Farming Manual.* ABRI, University of New England, Armidale, New South Wales. 1993 ISBN 1863890696

White, R., *Big Game Ranching in the United States*, Wild Sheep & Goat International, Mesilla, NM, 355 pp., 1987.

Whitehead, Kenneth G., *The Whitehead Encyclopedia of Deer.* Voyageur Press, 597 pgs. 1993 ISBN 1853103624

Yerex, David and Spiers, Ian, *Modern Deer Farm Management.* Ampersand Publishing, 175 pgs. 1990 ISBN 044700001-0

Yerex, David, *Wapiti Behind the Wire.* GP Publications, 62 pgs. 1992 ISBN 0-47700055X

Yorks, T.P., *Ranching native and exotic ungulates in the United States,* In: Hudson, R.J., Drew, K.R., & Baskin, L.M. (eds) Wildlife Production Systems, Cambridge University Press, Cambridge, pp. 268-285, 1989.

1999 STATE / PROVINCE REGULATIONS ON ELK

Although this report represents a best effort at accuracy and completeness, be advised to check with the appropriate agencies before finalizing plans. Regulations can be subject to change.

ALABAMA

Information Contact for Requirements to Own Elk:

Dept. of Fish and Game, (334) 242-3469, Gary Moody

No person shall possess, sell, offer for sale, import, bring or cause to be brought or imported into the State of Alabama any deer from any area outside the State of Alabama. However, it is permitted to own elk which are purchased in Alabama.

Information Contact for Requirements to Import Elk into State:

Alabama Dept. of Agriculture and Industries
Animal Industry Division
P.O. Box 3336, Montgomery, AL 36109
(334) 240-7255, Dr. J. Lee Alley

In the event that elk are allowed to be imported into the state, the Department of Agriculture has set the following health requirements: All elk shall have an official Certificate of Veterinary Inspection issued by an accredited veterinarian and must have a permanent individual identification. All elk 6 months of age or older must have a negative brucellosis test within 30 days of entry. All elk regardless of age must have a negative TB test within 30 days of entry.

ALASKA

Information Contact for Requirements to Own Elk:

Dept. of Fish and Game
Division of Wildlife Conservation
P.O. Box 25526, Juneau, AK 99802-5526
(907) 465-4190

An elk farming license must be obtained; the biennial fee is $250. Only elk of North American origin may be imported or possessed for elk farming. Elk may be raised and bred as domestic stock for commercial purposes, including the sale of meat, by a person who lawfully owns the elk and who holds a current valid game mammal farming license. A live elk may not be captured from the wild or released into the wild without an appropriate license or permit from the department. Licensed elk farmers must possess facilities for maintaining elk under positive control. At a minimum, fencing shall be 8 feet high, constructed of 12 1/2 gauge, galvanized woven wire fencing. It is recommended that posts have a 4" minimum diameter, if wood, and 3" minimum diameter, if steel. Posts must be 12 feet tall and set 4 feet deep. Wood posts must be treated to prevent rot. Post spacing shall be a maximum of 25 feet if high-tensile wire is used, 15 feet if conventional wire is used, 60 feet if 10-foot steel "T" posts are spaced every 15 feet between wood or steel line posts. Closer post spacing may be necessary and should be considered to allow for uneven ground topography, unstable soil conditions, or heavy animal pressure. The perimeter fence corridor should have gradual changes in elevation to allow for a tight fit to the terrain. Posts should be stout enough to securely support the wire in the given soil conditions and maintain proper wire tension. Fence right-of-way shall be kept clear, and fence shall be maintained in game-proof condition at all times. Facilities and fencing are subject to annual inspection.

A USDA tag is required. Although a dangle tag is not required, it is recommended.

Pending legislation would delete the requirement for a license from the Dept. of Fish and Game and would consolidate all regulatory control under the Dept. of Agriculture.

Information Contact for Requirements to Import Elk into State:

State Veterinarian
Alaska Dept. of Environmental Conservation
500 South Alaska St., Palmer, AK 99645
(907) 745-3236, Dr. Bert Gore

Elk imported, exported, or possessed for the purpose of elk farming are subject to provisions and regulations adopted by the commissioner of Natural Resources or the commissioner of Environmental Conservation for domestic animals and livestock. They must have a health certificate and import permit, valid for 30 days. All elk must originate from a brucellosis/TB-free herd. In addition, they must test negative for brucellosis and TB within 30 days of importation. They must also test negative to bluetongue and anaplasmosis within 30 days of importation. They must be negative for ectoparasites.

ARIZONA

Information Contact for Requirements to Own Elk:

Dept. of Fish & Game (602) 789-3309

No elk are permitted into the state.

Information Contact for Requirements to Import Elk into State:

Dr. Richard D. Willer, State Veterinarian
1688 W. Adams, Phoenix, AZ 85007
(602) 542-4293

No elk are permitted into the state.

ARKANSAS

Information Contact for Requirements to Own Elk:

Game and Fish Commission
#2 Natural Resources Dr., Little Rock, AR 72205
(501) 223-6359

A Commercial Game Breeder's Permit ($25) is required if animals are raised for sale. A bill of sale must be maintained on all stock purchased and sold, and applicable USDA permits must be obtained. Game breeders must submit a monthly inventory to the commission.

Information Contact for Requirements to Import Elk into State:

Arkansas Livestock & Poultry Commission
P.O. Box 8505
One Natural Resources Dr., Little Rock, AR 72215
(501) 225-5138
Dr. Conley Byrd

All elk and other cervidae over 6 months old must have a negative test for brucellosis within 30 days of shipment. They must also be negative to a cervical test for TB, conducted in accordance with the Veterinary Services Notice of December 31, 1990, within 30 days of shipment. All cervidae must be individually identified. Individual identification and an entry permit number are listed on a Certificate of Veterinary Inspection.

CALIFORNIA

Information Contact for Requirements to Own Elk:

Dept. of Fish & Game (916) 227-2244

No elk are permitted into the state.

Information Contact for Requirements to Import Elk into State:

Dept. of Fish & Game
(916) 227-2244

Dept. of Food and Agriculture
1220 N. St., Room A-107, Sacramento, CA 95814
(916) 654-1447
Dr. Richard Breitmeyer

No elk are permitted into the state unless authorized by the California Department of Fish and Game, Permit and Wildlife License Branch 916) 227-1305.

COLORADO

Information Contact for Requirements to Own Elk:

Dept. of Agriculture
Colorado State Board of Stock Inspection
201 Livestock Exchange Bldg.,
4701 Marion St., Denver, CO 80216-2139
(303) 294-0895

Colorado State Veterinarian
700 Kipling St., #4000, 2nd Floor,
Lakewood, CO 80215
(303) 239-4161
Dr. Jerry Bohlender

Dept. of Natural Resources
Division of Wildlife
6060 Broadway, Denver, CO 80216
(303) 297-1192

Breeders must be licensed under the Colorado State Board of Stock Inspection (Brand Board) as an Alternative Livestock Farm or under the Division of Wildlife as a Commercial Wildlife Park. A very simple rule of thumb may be used to determine which state agency is the licensing authority. If you sell or move live elk from your facility, you must license with the Colorado State Board of Stock Inspection. If you have an exhibitor park or hunting park and no live animals will leave your facility, you may license with the Colorado Division of Wildlife.

Department of Agriculture Requirements:

The minimum requirements of the facility, concerning the perimeter fences and internal animal handling equipment and structure, will be sent by mail when requested. The proposed location will be reviewed by the Division of Wildlife. (It is advisable not to begin construction of your facility until site approval has been gained.) Annual license fee is $300, and the entire process for obtaining a new license will take 30 to 60 days.

Animals are inspected by an authorized Colorado inspector when any change of ownership takes place, when any movement of alternative livestock outside the perimeter fence takes place, and upon arrival from an out-of-state point of origin. A proper facility to inspect the animals must be in place; each animal must be tattooed; and proper proof of title, blood purity papers, and health documents must be available. The minimum requirements of the facility will be sent by mail when requested, but perimeter fences shall meet the minimum standards as established by the Colorado Wildlife Commission. Accurate records must be maintained, including the inventory; animal purity documents as designated by the Wildlife Commission; and herd or individual animal health documents.

Division of Wildlife Requirements:

City or county ordinances which would prohibit the possession of wildlife should be checked; the most restrictive law applies. Contact your local wildlife officer to discuss plans and facility construction. Remain in contact with the local officer during the construction phase. After submitting your application, site plan, and $100, a final facility inspection will be completed within 30 days. After obtaining a license for a Commercial Wildlife Park, elk can be obtained. Out-of-state purchases require an importation permit from the State Veterinarian's office. Records must be maintained per requirements of Division of Wildlife. Each animal must be marked with USDA tags, and inventory information, as well as disease testing and vaccination information, is recorded. Parks shall have the capability to handle and mark elk without relying solely on drugs. Fences are a minimum of 8 feet high, and requirements exist for wire gauges, gates, and posts, as well.

A moratorium on new commercial wildlife park licenses for any species in the family cervidae exists in a delimited area of northeastern Colorado. A mandatory Chronic Wasting Disease surveillance program exists in approximately 34 game management units in northeastern Colorado, although the disease is currently known to exist only in the wild in Colorado and Wyoming and in one location in Canada.

Emergency Rules:

A mandatory CWD surveillance program shall be applicable to all Colorado licensed alternative livestock facilities. For complete rules, contact State Veterinarian. The emergency rules include:

- Brain tissue and any other tissue shall be submitted for examination, as directed by the State Veterinarian, of alternative livestock 15 months of age or older that die for any reason.
- The tissues shall be submitted for testing to a laboratory designated by the State Veterinarian.
- The costs of sampling and testing shall be borne by the Colorado licensed alternative livestock facility.
- A positive diagnosis of CWD shall subject the herd to quarantine by the State Veterinarian. The quarantine shall not be released until a herd plan has been successfully completed to the satisfaction of the StateVeterinarian. The herd plan shall be established by the State Veterinarian in consultation with an epidemiologist designated by the State Veterinarian, the herd owner, and a veterinarian selected by the herd owner.

Information Contact for Requirements to Import Elk into State:

Dept. of Agriculture, (303) 294-0895
Colorado State Veterinarian, (303) 239-4161

All animals that are being shipped to a Colorado destination must be held at a licensed facility. They must have a Certificate of Veterinary Inspection and must have a permit number issued by the Colorado State Veterinarian's Office. (Gametes, eggs, sperm or other genetic material, as well as live animals, must obtain an importation permit.) Proof of ownership shall accompany them, and they must be marked with USDA alphanumeric test tags or with an official Canadian alphanumeric test tag. Any alternative livestock imported into Colorado must originate from a herd determined to be tuberculosis monitored, qualified or accredited under the UM&R, as specified in the following sentence, or standards deemed by the Colorado state veterinarian to be equivalent. Otherwise, all importation of alternative livestock into Colorado from out of state requires compliance with the Tuberculosis Eradication in Cervidae Uniform Methods and Rules, effective May 15, 1994, as approved by the USDA, APHIS, U.S. Federal Building, Room 729, Hyattsville, MD 20782. Certified copies of the UM&R are available for public inspection, and copies may be obtained or examined by contacting the Colorado state veterinarian at Colorado Dept. of Agriculture, 700 Kipling Suite 4000, Lakewood, CO 80215, during regular business hours. The state also has requirements for importing or moving elk of lower TB status into an existing herd with a higher TB status. Call (303) 239-4161 for specific information.

If 12 months of age or over, they must be tested for brucellosis. They must have tested negative for bangs within 30 days, and should have had IVOMEC injected within the last 30 days. Additional disease testing may be required at the discretion of the commissioner. Minimum specific disease testing results and / or health statements must be included on the Certificate of Veterinary Inspection. The certificate must include a signed statement about Johne's Disease.

All elk imported into the state must be tested for hybridization with red deer. Written documentation for each animal, by USDA tag and / or division tag, must be provided to the division showing that each animal has been tested or is the offspring of pure parents, or the animal must be registered with the North American Elk Breeders Association and have a certificate. The Division of Wildlife shall issue annual certification of herd purity based on this documentation. A copy must accompany all elk transported in Colorado. The offspring of any female elk must be tested for red deer hybridization if the calf results from a pregnancy which existed prior to the female elk being imported into the state.

Movement within the state: An appropriate Alternative Livestock License must be in place. They must be marked with USDA official ear tags or tags approved by the Colorado Brand Board. They must comply with the UM&R for Tuberculosis. They must have tested negative for TB within 60 days prior to movement within Colorado, except when originating from herds with a "Colorado TB tested elk herd" or a "Colorado TB accredited elk herd" status issued by the state Dept. of Agriculture. A Certificate of Veterinary Inspection and a copy of certification of purity must accompany them.

CONNECTICUT

Information Contact for Requirements to Own Elk:

Dept. of Agriculture, (860)

An annual single-cervical TB test is required. Breeders must register with the Department of Agriculture and the Department of Protection. Elk can't be housed within 3 miles of a dairy farm (if breeders are starting new operations). One source of identification is required.

Information Contact for Requirements to Import Elk into State:

Dept. of Agriculture
765 Asylum Ave., 3rd Floor
Hartford, CT 06105
(860) 713-2505, Dr. Bruce Sherman

All cervidae require an official interstate health certificate. One copy of the certificate shall be forwarded to the Connecticut state veterinarian. The health certificate shall be originated by an accredited veterinarian who has been additionally trained and approved by a state or federal veterinary medical officer to conduct the single strength cervical tuberculosis test for cervidae and shall be signed by the state veterinarian from the state from which the cervidae is being imported. The certificate shall state that the cervidae is free of contagious and infectious diseases. Cervidae over 6 months of age must be negative to a brucellosis test made within 30 days prior to entry into Connecticut. Cervidae over the age of 6 months must be tested negative to a

0.1 cc single strength cervical tuberculosis test using ppd bovis tuberculin within 60 days prior to entry into Connecticut. Cervidae must originate from a herd that has been proven, through a 0.1 cc single strength cervical tuberculosis test, completely free of tuberculosis for all members of the herd over 6 months of age within the prior 12 months, to which herd no untested animals have been added since that test. Cervidae must be tested for Anaplasmosis and Bluetongue within 30 days prior to importation. Each cervidae must be individually identified by either an official metal ear tag or legible ear tattoo, and must be recorded on the health certificate individually. Prior to importation, an import permit number should be obtained from the Commissioner of Agriculture by the consignee.

DELEWARE

Information Contact for Requirements to Own Elk:

Dept. of Agriculture
2320 S. DuPont Hwy., Dover, DE 19901
(302) 739-4811, Dr. Wesley Towers

The state veterinarian regulates the owning of elk in the state. Any cervidae must be fenced to prohibit escape into the wild animal population.

Information Contact for Requirements to Import Elk into State:

Dept. of Agriculture, (302) 739-4811

Each animal must have clearly identifiable, individual identification such as an official ear tag, tattoo, or registration. The animal or animals must be accompanied by an interstate health certificate signed by an accredited veterinarian, stating that he/she examined the entire herd of origin of the animals within 30 days of their arrival into Delaware. The health certificate shall contain proof that the elk entering the state originated directly from a herd of origin in which all elk over 180 days of age have been tested negative within the last 12 months, using a single cervical tuberculosis test, BTB test, or other USDA approved elk tuberculosis test. No elk originating in herds in which there have been tuberculosis suspects disclosed within the past year will be allowed entry.

FLORIDA

Information Contact for Requirements to Own Elk:

Game and Fresh Water Fish Commission
Division of Law Enforcement
620 S. Meridian St., Tallahassee, FL 323993
(850) 488-6253

A Game Farm License is required. An 8-foot fence is required. Accurate records of sales and purchases must be kept.

Information Contact for Requirements to Import Elk into State:

Dept. of Agriculture and
Division of Animal Industry
Room 325, Mayo Bldg., Tallahassee, FL 32399
(850) 488-8280 Permits,
(850) 488-7747 Leroy Coffman, (850) 488-3021

A health certificate is obtained within 30 days of shipping. Elk should either be from an accredited tuberculosis-free herd, or they should be from a herd which tested negative to tuberculosis within the past 12 months, and shall have individually tested negative to TB within 90 days prior to importation. If neither case is true, they shall have tested negative to 2 tuberculosis tests conducted not less than 90 days apart, with the second test conducted within 90 days prior to importation and with the animals isolated from other members of the herd during the testing period. If over 6 months of age, they shall have tested negative to brucellosis within 90 days prior to importation. A permit number is included on the health certificate.

GEORGIA

Information Contact for Requirements to Own Elk:

Georgia Dept. of Natural Resources
Wildlife Resources Division
Special Permit Unit
2109 U.S. Hwy. 278 S.E., Social Circle, GA 30025
(770) 761-3044

Georgia Dept. of Agriculture
Animal Industry Division
Livestock/Poultry Field Forces
19 M.L.K., Jr., Dr., Room 247, Atlanta, GA 30334-4201
(404) 656-3665

The Dept. of Natural Resources and the Dept. of Agriculture share responsibility for regulating elk in the state in order to ensure the health and safety of wildlife and people and to prevent, control, and eradicate animal diseases. A deer-farming license must be obtained from the Dept. of Agriculture {(404) 656-3665}. The animals must be held for the sole purpose of deer farming and shall not be co-mingled with other wild animals. If the animals are held for other purposes or with other regulated animals, the owner would need to qualify for and obtain a Wild Animal License from the Department of Natural Resources. The operation must be inspected and approved by both departments. Inventory records of the herd are maintained and are subject to review. Facilities must be in accordance with rules established by the Commissioner of Agriculture. An 8-foot fence is required, with the bottom 6 feet made of woven mesh and constructed of a design approved by both departments.

Information Contact for Requirements to Import Elk into State:

Georgia Dept. of Agriculture
Animal Industry Division
Animal Health Section
19 M.L.K., Jr., Dr., Room 105, Atlanta, GA 30334
(404) 656-3667, (404) 656-3671, Dr. Lee Brooks

Health and transportation requirements must meet the health requirements established by rule of the Georgia Dept. of Agriculture. Those animals specifically used for deer farming must meet the requirements of the Uniform Methods and Rules of the Code of Federal Regulations for Tuberculosis and Brucellosis in Cervidae.

HAWAII

Information Contact for Requirements to Own Elk:

Hawaii Dept. of Agriculture
Plant Quarantine Branch
Land Vertebrate Specialist
(808) 586-0844
(808) 586-0859, Dr. Fred Helm

Import permit required. Call for further details.

Information Contact for Requirements to Import Elk into State:

Hawaii Dept. of Agriculture • Animal Industry Division
99-941 Halawa Valley St., Aiea, HI 96701
(808) 483-7119 Dr. Larry Rawson
(808) 483-7118 Dr. Jason Moniz

An import permit is required from the Dept. of Agriculture, Plant Quarantine Branch, (808) 586-0844, to enter the state. A copy of the permit is attached to the health certificate. The certificate, issued within 7 days before shipment, gives a description of each animal, including breed, sex, age, identifying marks, and other identifications. Certification statements shall indicate:

- The described animals are free from external parasites and symptoms of transmissible diseases and have not had recent exposure to these diseases.
- The described animals have originated from a herd not under quarantine for or exposed to tuberculosis, brucellosis, or chronic wasting disease.
- The animals have been found negative to a single cervical intradermal tuberculosis test by a designated accredited veterinarian within 30 days before shipment.
- They have been tested by an accredited veterinarian and found negative to an official test for brucellosis performed at a USDA approved laboratory within 30 days before shipment.

IDAHO

Information Contact for Requirements to Own Elk:

Dept. of Agriculture
Division of Animal Industries
2270 Old Penitentiary Rd., Boise, ID 83720
(208) 332-8540, (208) 332-8560

A domestic cervidae license must be purchased from the Division of Animal Industries. The facility of the applicant will be inspected and approved by the division prior to issuing a license. The initial application is $100, with an annual $50 renewal fee. A per head fee of $5 is assessed to cover the cost of administering the program. A perimeter fence must be constructed of high-tensile, non-slip woven wire or other fencing material approved by the administrator. For elk the fence must be a minimum of 8 feet in height for its entire length. Posts must be spaced no more than 24 feet apart, with stays, supports or braces as needed, and be placed in the ground a minimum of 3 feet. Each enclosure must have adequate gates that prohibit escape. All elk must be individually and uniquely identified by 2 of the 3 following methods: (1) Official USDA ear tag or other ear tag approved by the administrator. (2) Ear tattoo using an alpha numeric tattoo sequence that has been recorded with the Division of Animal Industries. (3) Microchip approved by the Division of Animal Industries, in cooperation with the Idaho Brand Department, with an identifying number / frequency that has been recorded with the division. All progeny of domestic cervidae must be permanently identified by December 31 of the year of birth or upon leaving the elk farm, whichever is earlier. An annual report of all animals held is due no later than December 31 of each year.

Tuberculosis Eradication in Cervidae, Uniform Methods and Rules and Brucellosis Eradication, Uniform Methods and Rules will be used as the standards for tuberculosis and brucellosis eradication in elk. The administrator may require, when sufficient risk exists, that domestic cervidae be tested for brucellosis, tuberculosis, meningeal worm, or muscle worm, and / or for other diseases or parasites determined to pose a risk to other domestic cervidae, livestock or wildlife. The administrator shall determine appropriate testing procedures and methods. Any animals identified as having red deer genetic influence must be destroyed, removed from the state, or neutered.

Information Contact for Requirements to Import Elk into State:

Dept. of Agriculture (208) 332-8540, (208) 332-8560

Domestic cervidae may enter the state provided they meet the following requirements and are accompanied by a Certificate of Veterinary Inspection attesting that they have been inspected within 30 days of date of shipment and that they are free from evidence of infectious, contagious or communicable diseases or known exposure thereto during the preceding 60 days. If they are 6 months of age or older, they shall have tested negative for brucellosis by at least 2 types of official brucellosis tests, one of which must be the rivanol, the PCFIA or the CITE test, within 30 days prior to entry. If animals originate from an accredited herd, they may be imported without further TB testing provided that they are accompanied by a certificate stating that they originated from an accredited herd. If animals originate from a qualified herd, they may be imported if accompanied by a certificate stating that they originated from a qualified herd and have been classified negative to an official TB test that was conducted within 90 days prior to the movement date. If animals originate from a monitored herd, they may be imported if accompanied by a certificate stating that they originated from a monitored herd and have been classified negative to an official tuberculosis test that was conducted within 90 days prior to the date of movement. If animals do not originate from an accredited, qualified, or monitored herd, they may be imported if accompanied by a certificate stating that they have been classified negative to 2 official TB tests that were conducted no less than 90 days apart, that the second test was conducted within 90 days prior to the date of movement, and that the animals were isolated from all other members of the herd during the testing period. Test eligible age is 6 months or older, or less than 6 months of age if not accompanied by a negative tested dam.

Elk must be tested negative for red deer genetic factor by a lab approved by the division of animal Industries and be from a region not known to be endemic with meningeal worm as reported by the Southeastern Cooperative Wildlife Disease

Study. They must be individually identified by an official USDA identification tag or microchip on a Certificate of Veterinary Inspection issued by the veterinarian who conducted the tests. An import permit is issued by the Idaho Division of Animal Industries.

All live domestic cervidae 6 months of age or older moving from one premise to another premise within the state of Idaho shall be accompanied by a copy of an official negative test for TB conducted within the last 90 days or written permission from the administrator. Animals originating from an accredited, qualified or monitored herd, as described in UM&R, shall be exempted from test requirements, if they are accompanied by a certificate signed by an accredited veterinarian or the administrator stating that they have originated directly from such herd.

ILLINOIS

Information Contact for Requirements to Own Elk:

Dept. of Natural Resources
524 S. 2nd St., Springfield, IL 62701
(217) 782-6384

No requirements exist for elk, but to obtain an import permit, applicants must provide proof that their elk will not become a nuisance or cause damage to any existing wild or domestic species.

Information Contact for Requirements to Import Elk into State:

Illinois Dept. of Agriculture
Bureau of Animal Health
P.O. Box 19281, Springfield, IL 62794
(217) 782-4944, Dr. Richard Hull

To import elk into Illinois, a permit from the Director of Agriculture is required. All cervidae shall be accompanied by a health certificate issued within 30 days of entry showing the following:

- Obtain an entry permit number by calling (217) 782-4944. You will be asked the number of animals, name and address of consignor and the name and complete mailing address of consignee.
- All cervidae must be individually identified by ear tag or tattoo.
- All cervidae 12 months of age and over shall be negative to two single cervical tests for bovine TB no less than 90 days apart, with the second test conducted within 90 days prior to entry. They must have been isolated from all other members of the herd during the testing period unless they originated from an accredited, qualified or monitored herd.

Cervidae from an accredited herd may be moved into Illinois without further TB testing provided they are accompanied by a certificate stating that they originated from an accredited herd. Cervids originating from a qualified or monitored herd may enter with a negative test within 90 days prior to entry and a certificate stating that the animals originate from a qualified or monitored herd.

- Elk 6 months of age and older need a negative brucellosis card or PCFIA test within 60 days prior to entry.

INDIANA

Information Contact for Requirements to Own Elk:

Indiana Office of USDA (317) 290-3300

No fencing or licensing regulations exist.

Information Contact for Requirements to Import Elk into State:

Board of Animal Health

Office of the State Veterinarian
805 Beachway Dr., Suite 50, Indianapolis, IN 46224
(317) 227-0300, Dr. Bret Marsh
(317) 227-0316, Permits
(317) 227-0319, Cattle/Ruminant Division

Summary of Tuberculosis Requirements:

- Must have an official health certificate within 30 days
- Must have official identification
- Must call (317) 227-0316 for a permit number
- Give information as to herd status of accredited herd, qualified or monitored herd.
- Qualified and monitored herds must have a negative tuberculosis test within 90 days of entry.
- If there is no herd status, cervidae must have two negative tuberculosis tests, the last one being within 90 days, and must meet the status of the herd requirements they are entering.

Indiana follows the brucellosis UMR in cervidae dated 9/30/98.

- No test required on animals from a certified brucellosis-free herd
- Monitored brucellosis cervid herd recognized
- Animals from herds without status must test negative within 30 days of interstate movement.

IOWA

Information Contact for Requirements to Own Elk:

Dept. of Agriculture
Division of Animal Industry
Henry A. Wallace Bldg., Des Moines, IA 50319
(515) 281-5321
(515) 281-5305, Dr. John J. Schiltz

There are no requirements to own elk in this state. Elk are now classified as livestock in Iowa, and the Dept. of Agriculture foresees no changes in the regulations. Elk in licensed hunting preserves may be subject to DNR regulations.

Information Contact for Requirements to Import Elk into State:

Dept. of Agriculture
(515) 281-5321, (515) 281-5305
(515) 281-5547, Import/Export Clerk

A Certificate of Veterinary Inspection and individual identification are required. All elk 6 months of age and older must be tested negative for brucellosis within 90 days of importation, or must originate from a brucellosis certified free herd. All elk 6 months of age and older must be tested negative for TB, utilizing the single cervical test within 90 days of importation, or must originate from a TB accredited herd or a

TB qualified herd which has been tested within 90 days of importation. Test dates must be included on the Certificate of Veterinary Inspection. Herd status and testing protocols are according to the USDA TB and Brucellosis Eradication in Cervidae Uniform Methods and Rules.

KANSAS

Information Contact for Requirements to Own Elk:

Dept. of Agriculture
Kansas Animal Health Dept.
708 S. Jackson, Topeka, KS 66603-3714
(785) 296-2326, George Teagarden, Livestock Commissioner

A domesticated deer permit ($25-$100) is required. Application for permit is made in writing on a form provided by the commissioner and must be submitted at least 30 days before purchasing any elk. Elk are identified through implantation of microchips, ear tags, ear tattoo, ear notches or any other permanent identification required by K.S.A. 47-2101 (g). Records of the herd shall include:

- The name and address of the person from whom each animal was acquired.
- The date each animal was acquired, or if born on the premises, the year of birth.
- A description of each animal, including the following characteristics the age, the size, the color markings, the sex, the breed, vaccination and testing information, and identification.
- The name and address of the person to whom any elk is sold, given, or bartered, or to whom the elk is otherwise delivered, and the method of disposition.
- The date of death, the cause of death, and the method of disposition if the animal dies, is euthanized, or is slaughtered.

Each owner shall confine elk in a manner designed to prevent escape. Perimeter fencing shall provide an escape-proof barrier; be structurally sound; be in good repair; and be of sufficient height to prevent escape, but not less than eight feet.

Each owner shall provide handling facilities which allow each animal to be physically handled without undue harm to the elk or handler. Each access lane and catch pen shall be designed and constructed to safely contain elk for inspection, identification, testing, quarantine, or other action required by the commissioner.

Elk shall be provided with free access to clean water; adequate feed; appropriate shelter, natural or otherwise; and protection from predators.

Each owner or handler shall conform to all federal and state rules and regulations for contagious and communicable disease. The Uniform Methods and Rules for Tuberculosis Eradication in Cervidae is adopted by reference.

The livestock commissioner shall inspect any premises where a domesticated deer herd has been issued a permit upon receipt of a written, signed complaint that such premises is not being operated, managed or maintained in accordance with rules and regulations. The owner of any herd infected with brucellosis shall take one of the following steps: quarantine and depopulate the herd, or agree to and implement a plan approved by the livestock commissioner to eradicate the disease. The owner of any herd infected with tuberculosis shall take one of the following steps: depopulate the herd or agree to and implement a plan approved by the commissioner to eradicate the disease.

Information Contact for Requirements to Import Elk into State:

Kansas Animal Health Dept., (785) 296-2326

Each cervidae entering Kansas shall be identified by one of the methods stated above and shall be accompanied by a permit obtained from the animal health department and by an official health certificate or certificate of veterinary inspection on a form approved by the livestock commissioner. Cervidae originating from free-roaming herds located within Wind River Reservation or Yellowstone National Park or the 20-mile zone surrounding the park shall be prohibited. Each cervidae transported within the state of Kansas shall be accompanied by a transportation permit on a form provided by the animal health department. Applicants for transportation permits shall possess a current cervidae permit and provide written proof of ownership for each animal being transported.

Each animal six months of age or older which is entering Kansas from a state that does not have a high incidence of brucellosis, as determined by the livestock commissioner, or which is changing ownership within Kansas, and that is not intended for immediate slaughter shall meet one of the following requirements:

- Originates from and is a part of a certified brucellosis-free herd.
- Tests negative for brucellosis, using an official test within 30 days before entry or change in ownership.
- Has been tested for brucellosis with a test confirmed at a state-federal cooperative laboratory.

Each animal six months of age or older which is entering Kansas from a non-quarantined herd from any state having a high incidence of brucellosis, as determined by the livestock commissioner, shall test negative for brucellosis, using an official test, within 30 days before shipment into Kansas.

Unless originating from a certified brucellosis-free herd, each cervidae six months of age or older which is imported into Kansas from a state having a high incidence of brucellosis, except those animals consigned directly to slaughter or to a licensed feed lot, shall be placed under quarantine upon arrival at the Kansas destination and shall be held in isolation from all other susceptible species. Retesting, at the owner's expense, will occur no less than 45 days and no more than 180 days after arrival.

Each animal six months of age or older that is entering the state or changing ownership within the state and that is not intended for immediate slaughter shall meet one of the following requirements:

- Originates from and be a part of an accredited tuberculosis-free herd.
- Tests negative for tuberculosis, using an official test, within

60 days before entry into the state or change of ownership. Any imported cervidae may be quarantined for a re-test for tuberculosis by order of the commissioner.

Elk from a free-roaming herd must have two negative TB tests, 90 days apart, up to 150 days prior to entry.

KENTUCKY

Information Contact for Requirements to Own Elk:

Dept. of Wildlife
Arnold L. Mitchell
Bldg., #1 Game Farm Rd., Frankfort, KY 40601
(502) 564-4406, John Phillips
(502) 564-3956, (606) 824-7507

Information Contact for Requirements to Import Elk into State:

Dr. D. L. Notter
State Veterinarian
100 Fair Oaks Ln., Suite 252, Frankfort, KY 40601
(502) 564-3956

A permit and Certificate of Veterinary Inspection, obtained within 30 days of entry, shall accompany elk entering the state. The certificate will attest that they have been tested negative for brucellosis within 30 days, using the PCFIA or CF tests. They shall have been tested negative for tuberculosis within 30 days, using the single cervical test.

LOUISIANA

Information Contact for Requirements to Own Elk:

Dept. of Agriculture & Forestry
Office of Animal Health Services
Livestock Sanitary Board
P.O. Box 1951, Baton Rouge, LA 70821
(504) 925-3980, Dr. Maxwell Lea, Jr.

A license from the Department of Agriculture and Forestry is required ($50). A written application is submitted, and the premises are inspected by personnel of the department. The premises must be located in a rural area, and the area must include at least 5,000 square feet for the first animal and 2,500 square feet for each subsequent animal if the total area of the premise is less than 5 acres. The fences must be at least 8 feet high and must consist of wire that is a minimum gauge of 12 1/2. The fencing material must be chain link, woven wire, solid panel, or welded panel. The premises must have sufficient drainage to prevent extended periods of standing water. Records of all sales, trades, purchases, or transfers must be maintained for at least 36 months. All elk must be identified by means of an implanted microchip, placed under the skin at the base of the left ear or by a permanent ear tattoo and ear tag. Contact the department for regulations pertaining to premises where elk will be hunted.

Information Contact for Requirements to Import Elk into State:

Dept. of Agriculture
Office of Animal Health Services, (504) 925-3980

All elk shall be accompanied by a Certificate of Veterinary Inspection, including an entry permit number. They shall be permanently and individually identified by means of a microchip implanted at the base of the left ear or by a permanent ear tattoo and ear tag. They shall be tested for brucellosis in accordance with the Brucellosis Eradication in Cervidae Uniform Methods and Rules as published by USDA, APHIS. All imported elk 6 months of age and older shall be tested negative for brucellosis within 30 days of entry unless they originate from a herd which has been officially declared as a certified brucellosis free herd by the state of origin. All elk shall be tested for tuberculosis in accordance with the Tuberculosis Eradication in Cervidae Uniform Methods and Rules as published by USDA, APHIS. Contact the department for details, or see Idaho's regulations for tuberculosis testing, which are also in accordance with the UM&R.

MAINE

Information Contact for Requirements to Own Elk:

Dept. of Agriculture
Division of Animal Health and Industry
28 State House Station, Augusta, ME 04333-0028
(207) 287-3701, Dr. Chip Ridky or Dr. Donald Hoenig

The state has licensing requirements for owning elk, and potential breeders should also check with nearby towns for their requirements pertaining to fencing, facilities, etc.

Information Contact for Requirements to Import Elk into State:

Dept. of Agriculture, (207) 287-3701

A health certificate (Certificate of Veterinary Inspection) must be obtained within 30 days of entry. A negative brucellosis test and a negative bluetongue test are required within 30 days. A negative TB test is required within 60 days. An import permit is needed.

MARYLAND

Information Contact for Requirements to Own Elk:

Dept. of Natural Resources, (410) 974-3195

Elk are not allowed in the state of Maryland.

Information Contact for Requirements to Import Elk into State:

Dept. of Natural Resources, (410) 974-3195
Dr. A. B. Park, Asst. Secretary
Food Safety and Consumer Services
50 Harry S. Truman Pkwy., Annapolis, MD 21401
(301) 841-5782

Elk are not allowed to be imported into the state of Maryland.

MASSACHUSETTS

Information Contact for Requirements to Own Elk:

Division of Fisheries and Wildlife (617) 727-3151

No elk are permitted into the state. Fallow deer or red deer may be commercially propagated for sale, with a valid license.

Information Contact for Requirements to Import Elk into State:

Dr. Anne Mann, Chief
Bureau of Animal Health
Leverett Saltonstall Bldg., 100 Cambridge St., Boston, MA 02202
(617) 727-3018
(617) 727-1614

Consignor shall also contact Fisheries and Wildlife. Animal shall be individually identified by either ear tag or tattoo, which shall be listed on state of origin health certificate. A prior entrance permit issued by the Massachusetts Bureau of Animal Health is required. Testing requirements: 1) brucellosis–negative test within 30 days, 2) tuberculosis–negative test within 60 days (single cervical site, according to USDA-APHIS recommendation), 3) bluetongue–negative test within 30 days if from endemic area (accept test being virus isolation negative).

MICHIGAN

Information Contact for Requirements to Own Elk:

Dept. of Natural Resources
Box 30028, Lansing, MI 48909-7528
(517) 373-1263
(517) 373-9329, Jim Janson

A Permit to Hold Wildlife in Captivity is required. Holding facilities must be constructed before applying for the permit. The permit must be in possession before acquiring the animals.

An escape proof, 8-foot fence and 1,500 square feet for the first animal, plus 1,000 square feet for each additional animal, are required. Adequate shelter overhead shelter or protected area of at least 8 x 12 for each pen is required. A monthly inventory report must be completed.

At press time, legislation was pending. Check with the state veterinarian and the Dept. of Natural Resources for changes in regulations.

Information Contact for Requirements to Import Elk into State:

Michigan Dept. of Agriculture
Animal Industry Division
P.O. Box 30017, Lansing, MI 48909
(517) 373-8205, Dr. Mike VanderKlok

The import requirements are currently being revised. Call for changes. The present requirements are:

1. All cervidae imported into Michigan must be accompanied by an official interstate health certificate and be officially identified with a USDA alpha-numeric ear tag.

2. All cervidae 6 months of age or older must receive a negative brucellosis test within 30 days prior to movement. This information must be present on the health certificate.

3. All cervidae 6 months of age or older must meet the requirements of one of the following:

- Originate directly from a tuberculosis accredited herd.
- Originate directly from a tuberculosis qualified herd and receive a negative tuberculosis test within 90 days prior to movement.
- Originate directly from a tuberculosis monitored herd and receive a negative tuberculosis test within 90 days prior to movement.
- Receive two negative tuberculosis tests no less than 90 days apart with the second test being within 90 days prior to movement. The cervidae must be isolated from all other members of the herd during the test period, and both tests must take place prior to movement. Verification of both tests and the isolation must be present on the health certificate.

Cervidae less than 12 months of age born and originating from qualified or monitored herds may be imported without further tuberculosis testing provided they have not had contact with cervids of a lesser status herd. This information must be present on the health certificate. Cervidae imported from accredited, qualified or monitored herds must have the accredited, qualified or monitored herd number and current qualifying test date present on the health certificate.

MINNESOTA

Information Contact for Requirements to Own Elk:

Dept. of Natural Resources
500 Lafayette Rd., St. Paul, MN 55155
(612) 296-4771 Division of Enforcement
(612) 296-4506, License Bureau
(612) 296-1325 Section of Wildlife, Dave Shad

Dept. of Agriculture
Board of Animal Health
119 Agriculture Bldg.
90 West Plato Blvd., St. Paul, MN 55107
(612) 297-2942, Dr. T.J. Hagerty
(612) 297-5510, Paul Hugunin

Breeders choose whether to be under the jurisdiction of the Dept. of Natural Resources or the Dept. of Agriculture.

Requirements of Dept. of Natural Resources:

A Game Farm License is required to breed, propagate, sell or commercially dispose of elk. The license fee is $16.50. Applicant must have:

- Privately owned or leased land that is enclosed to sufficiently confine the animals.
- Adequate knowledge to properly care for elk.
- A clean and healthy facility.
- Adequate food, water, shelter, and sanitation for the animals.
- Facilities that will protect elk from injury and prevent overcrowding.

Every sale of an animal or animal parts or products must be accompanied with a completed game farm sales receipt. Complete and current records are required, and an annual report is filed.

Requirements of Dept. of Agriculture:

Fencing to prevent escape is required after 8/1/95, fence must be at least 96 inches high. Permanent ID is required, which can be U.S. Dept. of Agriculture metal ear tag, electronic implants, or other approved by the Board of Animal Health in consultation with the commissioner of Natural Resources. Newborn elk or imports must have ID by March 1 of each year. No red deer in native elk area without written approval of Natural Resources.

Information Contact for Requirements to Import Elk into State:

Board of Animal Health (612) 296-2942, ext. 16 or 27

An import permit is required, accompanied by a certificate issued by an accredited veterinarian. A negative brucellosis test is required within 30 days of entry; a negative TB test is required within 90 days of entry.

MISSISSIPPI

Information Contact for Requirements to Own Elk:

Dept. of Wildlife (601) 364-2214

No regulations concerning elk exist at this time.

Information Contact for Requirements to Import Elk into State:

Mississippi Board of Animal Health
P.O. Box 3889, Jackson, MS 39207
(601) 359-1170, Dr. James A. Watson

A health certificate is required. A permit or waybill shall accompany shipment; individual identification is required. Negative brucellosis and TB tests are required within 30 days; animal must be from non-quarantined herd in a modified accredited area or tuberculosis-free area.

MISSOURI

Information Contact for Requirements to Own Elk:

Missouri Dept. of Agriculture
Jefferson City, MO 65102
(573) 751-3377

No requirements exist for domesticated elk except those to own a certified herd. Call for details. Because elk are not indigenous to Missouri, the Conservation Department regulates only elk that are placed on game farms to be hunted. Chronic Wasting Disease must be reported immediately to the State Veterinarian. All cervids from infected or source herds will be quarantined.

Information Contact for Requirements to Import Elk into State:

Missouri Dept. of Agriculture
Animal Health Division
P.O. Box 630, Jefferson City, MO 65102,
(573) 751-3377
Permits: (573) 751-4359, Dr. John W. Hunt, Jr.

An entry permit is required. Permanent ID (ear tag, tattoo, micro-chip, or brand) is required. Two negative brucellosis tests within 30 days prior to shipment are required (one the CF or PCFIA test and the other a negative standard plate or BAPA test) for animals 6 months of age or older. A negative TB test (single cervical method) is required within 30 days prior to shipment. If the animals are going to a sale in Missouri, check with the sale on additional test requirements.

MONTANA

Information Contact for Requirements to Own Elk:

Montana Fish, Wildlife & Parks
1420 East 6th Ave., P.O. Box 200701
Helena, MT 59620-0701
(406) 444-2452
(406) 444-4039 Karen Zackheim
(Game Farm Program Coordinator)

Dept. of Livestock
Montana Dept. of Livestock
Animal Health Division
P.O. Box 202001, Helena, MT 59620
(406) 444-2043 Dr. Arnold Gertonson
(406) 444-9525 Luella Schultz

Montana has split regulatory jurisdiction over game farms between two state agencies. The Department of Fish, Wildlife & Parks has primary jurisdiction over game farm licensing, reports and recordkeeping, exterior fencing, classification of exotic species, inspection of perimeter fencing, and enforcement of the preceding functions. A game farm license ($200-400) is necessary to operate a game farm. Before fencing, all existing game animals must to the extent possible be driven from the land by the applicant. Perimeter fences must be 8 feet tall, and the FWP must approve the post spacing and fencing materials. (Call FWP for additional requirements for wire, braces, and supports.) Applicants must demonstrate that they or their principal managers have the necessary skills to properly care for game farm animals. Game farms must file bi-annual reports to FWP, and the game farm record book must be complete and accurate. (Game farm rules may change. Contact NAEBA; the Department of Fish, Wildlife & Parks; or the Department of Livestock.)

The Department of Livestock has regulatory jurisdiction over game farms relating to the marking, inspection, transportation, importation and quarantine of animals; hold orders; interior facilities; health; and enforcement of the preceding functions. Elk must be identified with a DoL approved tag and tattooed with a whole herd mark assigned to each animal owner by DoL. Newborns must be tagged by January 1 following their year of birth, and imported animals must be marked within 30 days of import.

Legislation is pending at press time. Check with the Dept. of Livestock and Montana Fish, Wildlife and Parks for changes.

Information Contact for Requirements to Import Elk into State:

Dept. of Livestock (406) 444-9525 (406) 444-2043

Prior to selling, transporting, or disposing of one or more elk, the licensee shall contact the Department of Livestock to request an inspection by a Department of Livestock agent. All elk transported within the state must be properly identified with tags and tattoo, and the animals must meet all disease testing requirements. The animal shipment must be accompanied by a copy of the bill of sale and the certificate of inspection issued by the department agent. Elk may be transported through Montana if animals proceed directly through the state with no intent to unload and if accompanied by an official health certificate from the state of origin. Elk ova and semen must meet all health, sale and testing requirements of the Department of Livestock.

A game farm licensee licensed to own elk may import elk into Montana after obtaining an import permit from the Department of Livestock. The animals must be accompanied by an official health certificate issued by an accredited veterinarian of the state of origin, and the animals must be individually identified with an official USDA tag (alpha numeric National Uniform Ear tag System) or the Canadian equivalent. Animals must meet all Montana disease testing requirements. As of May 1997, test requirements include tests for TB and brucellosis. The tuberculosis test requirements are for 2 negative single cervical tests 90 days apart or a whole herd test within 90 days of shipment. Brucellosis test requirements include 1 brucellosis test within 30 days of shipment and another 30-120 days after entry into the state. Each brucellosis test requires that 2 laboratory diagnostic procedures be completed on the sample: CF or PCFIA and one other (rivanol, BAPA or the plate test). Two disease status statements are required on the health certificate and must be certified by the accredited veterinarian of the state of origin: 1) The animal has not been exposed to Johne's Disease and Johne's Disease has not been identified in the herd of origin, and 2) no CNS symptoms or deaths have been observed in the last 5 years in the herd of origin. The entire shipment must receive anthelmentic treatment 20-30 days prior to shipment.

No red deer-elk hybrids may be imported into Montana. All elk must be tested and a copy of the laboratory test results must be mailed by the veterinarian of the state of origin to the State Veterinarian of Montana.

Quarantine of animals upon arrival in Montana is required until the animals are tagged and marked as required by Montana laws and until the animals have met all testing requirements. Animals must go directly to the destination identified on the import permit and may not be moved from that site until released from quarantine by the DoL.

NEBRASKA

Information Contact for Requirements to Own Elk:

Nebraska Dept. of Agriculture
Bureau of Animal Industry
P.O. Box 94787, Lincoln, NE 68509-5064
(402) 471-2351, Dr. Larry Williams

Domesticated elk are those legally obtained and raised in a confined area for breeding stock, exhibition, or companionship, or for the carcass, skin, or other part of the animal, but not for hunting or sport. The Department of Agriculture maintains a registry; owners register by Dec. 31 of each year and within 30 days after change in ownership or location of a domesticated herd. It is unlawful to buy, sell, or barter an elk or a part of an elk, unless the animal or part is appropriately marked for proof of ownership (including the herd number) with one of the following:

- Ear tag which conforms to the alpha numeric National Uniform Ear tagging System.
- Owner applied plastic ear tags greater than or equal to size of aforementioned ear tag.
- Neck collars, ankle bands, or other ID which cannot be removed without destroying the device.
- Freeze brand, when registered by recognized brand registry.
- A tattoo in addition to other ID.
- Electronic microchips, in addition to other ID, when properly recorded and implanted.

Parts of the animal that will be bought, sold, or bartered should be tagged or labeled for proof of ownership with designated information.

The Nebraska Game and Parks Commission prohibits red deer and hybrids of red deer. They cannot be imported into or owned in Nebraska.

Information Contact for Requirements to Import Elk into State:

Bureau of Animal Industry (402) 471-2351

A Certificate of Veterinary Examination is required. A permit for entry must be obtained from the Bureau of Animal Industry. An individual ID with approved ear tag is required. Negative tests for brucellosis and TB within 90 days are required. The elk must qualify as one of the following:

- Originate as a group which has been separated from untested animals and subjected to 2 single cervical TB tests 90-365 days apart with the last test within 90 days.
- Originate from a herd which has been whole-herd tested within 1 year.
- Quarantined at destination to be retested 90-365 days after entry.

NEVADA

Information Contact for Requirements to Own Elk:

Division of Agriculture
Dept. of Business and Industry
350 Capitol Hill Ave., Reno, NV 89502
(702) 688-1180, Dr. David Thain

A permit for importation, possession and propagation of alternative livestock is necessary. An application for a permit must be obtained from the above address. The application fee is $200. The facilities must pass an inspection by the division. The facilities must include an enclosure with a restraining device for the safe and humane restraint of individual animals. The top of the fence must be not less than 8 feet above the ground, with the bottom 7 feet constructed of smooth woven high-tensile mesh wire, which is no smaller than 12 1/2 gauge, with vertical spacing that does not exceed 6 inches. The fence posts, if constructed of wood, should have a diameter of not less than 4 inches or an equivalent strength if of steel. They should extend from not less than 3 feet below the ground to the full height of the fence. Each gate must be of the same height as the fence and must be secured by a latching device and padlock. No fee is charged for an initial inspection of the facilities, but a fee of $100 is charged for each additional inspection. Each animal must be identified by an identification device within 60 days of its birth. A complete record must be maintained and made available for inspection by a representative of the board.

Information Contact for Requirements to Import Elk into State:

Dept. of Business and Industry (702) 688-1180

To import, a permit which describes the animals' destination is necessary. Each animal is identified by an identification device and entry number. A health certificate which identifies the animal by its identification device and entry number is necessary. Each animal is certified by a veterinarian as having originated from a herd that for the 12 months before the entry of the animal has been free from tuberculosis and brucellosis and is not known to have been exposed to infection by Mycobacterium johnes or Parelaphostrongylus tenuis. Each animal 6 months of age or older is certified as having reacted negatively within 30 days before its entry to 2 official tests for brucellosis, including a complement fixation test, and an official single cervical test for tuberculosis. If the animal is Rocky Mountain elk, the animal is certified not to be a hybrid with any other cervid species or subspecies by a laboratory for genetic testing which is recognized by the chief of the Bureau of Animal Industry of the division.

NEW HAMPSHIRE

Information Contact for Requirements to Own Elk:

State of New Hampshire
Fish & Game Dept.
Law Enforcement
2 Hazen Dr., Concord, NH 03301
(603) 271-1438

Elk are designated as controlled. No live wildlife designated as controlled, the eggs or progeny thereof, shall be possessed without a permit to possess wildlife. Application for a permit to possess shall be submitted to the department with a current health certificate no less than 30 days in advance of the date possession commences. The permit fee is $20. If wildlife are from outside New Hampshire, a permit to import shall be held or applied for at the time application is made for a permit to possess. The executive director or his agents shall inspect elk, facilities, permits and / or records. The application is signed by a conservation officer after a satisfactory inspection. Elk require an 8-foot-high enclosure and a minimum of 2,000 square feet per animal. They require a permanently affixed means of ID. Requirements for adequate feeding, watering, handling and health exist. A veterinarian must inspect the herd annually for disease and general health and condition. Records must be maintained. To propagate elk, a permit to propagate wildlife must be obtained. The application is signed by a conservation officer after an inspection of the facility and approval by the executive director or his agent. The premises are posted, and an annual report is filed.

Information Contact for Requirements to Import Elk into State:

Fish & Game Dept.
Law Enforcement
(603) 271-1139

New Hampshire
Dept. of Agriculture, Markets & Food
25 Capitol St., P.O. Box 2042, Concord, NH 03302
(603) 271-2404, Dr. Clifford McGinnis

A permit shall be obtained prior to shipment, and the permit number shall appear on the certificate of veterinary inspection. A certificate of veterinary inspection shall be required for the importation of cervidae into the state of New Hampshire.

All cervidae entering the state shall have permanent identification of each animal with a metal ear tag, legible tattoo or microchip, providing a reader is available. Cervidae shall also:

- be from a classified-free herd;
- be over 6 months of age, have been tested negative for brucellosis within 30 days of shipment and be from a herd tested free of brucellosis within the last year; or
- if under 6 months of age, be identified as progeny of a brucellosis tested dam.

All cervidae entering the state of New Hampshire shall:

- be from a tuberculosis accredited-free;
- be over 6 months of age, have been tested negative for tuberculosis within 60 days of shipment and be from a herd tested free of tuberculosis within the last year; or
- if under 6 months of age, be identified as progeny of a tuberculosis tested dam.

No cervidae shall enter the state from any premise where CWD was diagnosed or which has received cervidae from a premise where CWD was diagnosed.

NEW JERSEY

Information Contact for Requirements to Own Elk:

Division of Fish, Game and Wildlife
P.O. Box 430, Trenton, NJ 08625-0430
(908) 735-5450

A permit is required. Elk must be properly fed; facilities, including fencing, must be adequate. Animals must be kept free of infectious diseases and parasites which may be dangerous to the animal, livestock or people of the state. (The division may request certification from a veterinarian or other qualified person.) All imported elk must be brucellosis and TB tested or from a certified TB-free herd.

Information Contact for Requirements to Import Elk into State:

Dept. of Agriculture
Division of Animal Health
P.O. Box 330, Trenton, NJ 08625-0400
(609) 292-3965, Dr. Ernest Zirkle

A permit from the Division of Fish, Game and Wildlife is required. The animal must be individually identified, and consignor and consignee are listed. A health certificate is required, and the animal must be free of infectious diseases and parasites. (The state veterinarian states that he would like the animals to be TB and brucellosis tested, but it is not a requirement.)

NEW MEXICO

Information Contact for Requirements to Own Elk:

New Mexico Department of Game & Fish
Law Enforcement Division
Special Use Permits Program
P.O. Box 25112, Santa Fe, NM 87504

A Class "A" Park License is required. Application is in writing, and the fee is $501. The park can not exceed 3,200 acres and must be enclosed by a game-proof fence which conforms to State Game Commission specifications. The overall height of the fence shall be at least 7 1/2 feet from ground level to the top wires. For detailed specifications of materials, construction, right-of-way, etc., contact the State Game Commission or NAEBA. A report, at a time specified by the director, will provide information regarding the kind, number, age and sex of the game.

Information Contact for Requirements to Import Elk into State:

New Mexico Livestock Board
7013 Central N.E., Albuquerque, NM 87108
(505) 841-4000, Dr. Steven England

A prior permit is required from the New Mexico Dept. of Game and Fish. Elk must have a negative test for brucellosis and tuberculosis within 30 days of entry. The permit and health certificate must accompany shipment. The importation of elk which may transmit disease to species of wild animals or pose a threat to human health is prohibited.

NEW YORK

Information Contact for Requirements to Own Elk:

Dept. of Agriculture
Division of Animal Industry
1 Winners Circle, Albany, NY 12235
(518) 457-3971, Dr. John Huntley

Elk are administered under the New York Dept. of Agriculture, not the Dept. of Conservation, per both departments. The Dept. of Agriculture states that the only regulations concerning elk are the disease requirements below.

Information Contact for Requirements to Import Elk into State:

Division of Animal Industry (518) 457-3502

An Approved Health Certificate is required. The certificate is issued by an accredited veterinarian and is approved by the chief livestock health official of the state or country of origin. The individual identification on the certificate shall include all ear tags and tattoo carried by the animal and its species, breed, age, sex, and registration number, if any. The certificate includes the full name and address of the consignor and consignee; date of issue; dates and results of tests; date of entry; and statement that animal has been inspected by an accredited veterinarian and is not showing signs of infectious, contagious or communicable disease and that results of the tests are as indicated. No elk affected with any infectious, contagious, or communicable disease shall be brought into New York State except with written permission of the New York State Department of Agriculture and Markets.

Elk 6 months of age and older must have tested negative for brucellosis within 30 days prior to entry. They must have tested negative for TB (cervical test or other U.S. Dept. of Ag. approved test) within 60 days prior to entry. If elk originates in a state or country where bluetongue or anaplasmosis is endemic, it must have tested negative for anaplasmosis within 30 days of entry and must be accompanied by a certificate issued by a veterinarian attesting to the absence of known evidence of bluetongue or EHD in the herd of origin for the previous 12 months. Elk less than 6 months of age must originate from a herd in which all eligible elk have had negative results to a cervical tuberculosis test in the preceding 12 months; all eligible elk added to the herd have had negative results to cervical tuberculosis test prior to addition; and in which there has been no evidence of tuberculosis in the preceding 12 months.

NORTH CAROLINA

Information Contact for Requirements to Own Elk:

Wildlife Resources Commission
P.O. Box 29613, Raleigh, NC 27626
(919) 661-4872

The North Carolina Wildlife Commission requires a captivity license ($5.00) for all cervidae. All cervidae require a permit prior to importation. The applicant must first construct or acquire a facility for keeping elk that complies with the standards set by the commission. The enclosure shall be on a well-drained site containing trees or brush for shade. The minimum size of the enclosure shall be not less than one-half acre for the first 3 animals and an additional one-fourth acre for each additional animal held. The enclosure shall be surrounded by a sturdy fence at least 10 feet high, dog-proof to a height of at least 6 feet. No exposed barbed wire or protruding nails shall be permitted within the enclosure. A roofed building large enough to provide shelter for each elk in both a standing or a lying position must be provided. The building shall be closed on 3 sides and shall have a wooden floor. It shall be constructed at least 10 feet from the fence. An ample supply of clear water and salt must be provided. Food shall be placed in the enclosure not less than 3 times weekly. Straw and leaf litter, replenished every week, shall be used as a floor covering in the shelter. An effective program for the control of insects, ectoparasites, disease, and odor shall be established and maintained. The animal must be protected against fright. Domestic livestock and dogs shall be excluded from the enclosure.

Information Contact for Requirements to Import Elk into State:

Dept. of Agriculture
Veterinary Division
P.O. Box 26026, Raleigh, NC 27611
(919) 733-7601, Dr. John K. Atwell

- All animals must be accompanied by an official health certificate issued within 30 days showing individual identification.
- All animals 6 months of age or older must have tested negative for brucellosis within 30 days prior to importation.

• All animals 6 months of age or older must have tested negative for tuberculosis within 60 days prior to importation, and all cervidae must meet the requirements of the Uniform Methods and Rules: Tuberculosis in Cervidae.

• Submit a statement that the herd of origin has had no brucellosis or tuberculosis diagnosed within the past 12 months.

NORTH DAKOTA

Information Contact for Requirements to Own Elk:

Dept. of Agriculture
Board of Animal Health
600 E. Boulevard Ave., Dept. 602, Bismarck, ND 58505-0020
(701) 328-2657, Dr. Susan Keller

Tags or ID numbers are obtained from an agent of the board. A non-traditional livestock license is required. Fences must be 7 feet high. Posts must be of sufficient strength and must extend to upper limits of height, spaced no more than 24 feet apart. The gates in the perimeter fence must be locked. Adequate holding and handling facilities must be in place. Adequate water, food, and shelter must be provided. Reports are recorded and filed annually.

Information Contact for Requirements to Import Elk into State:

Board of Animal Health (701) 328-2657

An importation permit and a North Dakota non-traditional livestock possession license are required. Genetic testing for purity is required for zones 1 and 2. A certificate of veterinary inspection is required, which attests the following: The animal is negative for TB within 30-90 days; the entire herd of origin is negative for TB (single cervical test) within the last 12 months; or the animal must be from an accredited herd. The animal is negative for brucellosis after 2 tests within 30 days (1 test must be the complement fixation test). Several import statements are required:

• This herd has never been under quarantine for a reportable disease.

• I hereby certify that these animals did not originate from a state where tuberculosis has been identified. (owner's signature)

• I hereby certify that the animals in the whole-herd test are all of the animals which I own. (owner's signature)

• To the best of my knowledge, animals listed herein are not infected with paratuberculosis (Johne's Disease) and have not been exposed to animals with paratuberculosis. To the best of my knowledge, the premises of origin have not been the site of a significant disease outbreak in the previous 24 months that was not contained and extirpated using recommended disease control.

• This elk and the herd from which it originates have no history of emaciation, depression, excessive salivation or thirst, or neurologic disease. Or in the event of these symptoms, appropriate diagnostic measures were taken to rule out a transmissible spongiform encephalopathy. These animals have not been exposed to an elk or deer diagnosed positive for a TSE.

Additionally, a copy of the medical history of the herd of origin for the past two years shall be faxed to the North Dakota Board of Animal Health prior to issuance of the permit for entry. (701) 328-4567

OHIO

Information Contact for Requirements to Own Elk:

Dept. of Agriculture
Division of Animal Industry
8995 E. Main St., Reynoldsburg, OH 43068
(614) 728-6220, Dr. David Glauer

Elk are considered agricultural animals administered by the Department of Agriculture; there are no state regulations other than health requirements. (See below.) Shooting preserves will continue to be under the Division of Wildlife.

Information Contact for Requirements to Import Elk into State:

Division of Animal Industry (614) 728-6220

No non-domestic animal shall be imported into the state of Ohio unless it is in full compliance with all other state and federal agency rules and regulations and it is free of evidence of any contagious or infectious diseases or parasites harmful to humans or animals. Animals should be accompanied by an entry permit and Certificate of Veterinary Inspection. Animals are accompanied by documentation to prove they had legal residence in the state or country of origin (i.e., license numbers of appropriate state and federal permits or tenure status on certificate of veterinary inspection.) Elk may move on a brucellosis-free herd status that is acceptable to the chief of the Division of Animal Industry or on negative test results from an approved state/federal brucellosis laboratory conducted within 30 days prior to entry. They may move on a tuberculosis-free herd status that is acceptable to the chief or have a negative test which was conducted within 60 days of entry or in accordance with the stipulation on the entry permit.

OKLAHOMA

Information Contact for Requirements to Own Elk:

Dept. of Wildlife Conservation
1801 N. Lincoln, P.O. Box 53465,
Oklahoma City, OK 73152
(405) 521-3719

A Commercial Wildlife Breeder's License is required ($48). Breeders are required to keep permanent records; records are inspected by the conservation department. Adequate fence and shelter are required. Contact county game warden for pen and fencing requirements.

Information Contact for Requirements to Import Elk into State:

Dept. of Agriculture
Animal Industry Division
2800 N. Lincoln Blvd., Oklahoma City, OK 73105
(405) 524-6404
(405) 521-3891, Dr. Burke Healey

The holder of a current Commercial Wildlife Breeder's License may sell and transport certain wildlife alive for propagating purposes. Import permits issued by the Dept. of Agriculture are required for any cervidae brought into the state. Phone number to contact: (405) 524-6404. A Certificate of Veterinary Inspection is issued within 30 days of entry. Except for nursing offspring accompanying a negative tested dam, all animals must be tested negative for tuberculosis within the 30 days prior to entry. Animals 6 months of age and older must be tested negative for brucellosis within 30 days of entry. The following statement must be attested to: Animals originate from a premise where tuberculosis or brucellosis has not been diagnosed during the previous 12 months. Animals must be individually identified by official metal ear tag or other visible, permanent identification.

OREGON

Information Contact for Requirements to Own Elk:

Dept. of Fish & Wildlife (503) 872-5260

No new license will be issued for elk. Only those which have been "grandfathered in" will be licensed.

Information Contact for Requirements to Import Elk into State:

Oregon Dept. of Agriculture
Animal Health & Identification Division
635 Capitol St. N.E., Salem, OR 97301-2532
(503) 986-4680

An import permit is required. A health certificate is issued no more than 30 days prior to entry. Negative tests required:

- Brucellosis no more than 30 days prior to entry
- Tuberculosis no more than 60 days prior to entry

They are examined and certified as free of external parasites by a veterinarian or qualified entomologist at the point of origin, or they are certified as having been subjected to treatment with a residual insecticide approved by federal authorities. Elk originating from any state east of the Mississippi River, in Oklahoma, Texas, or any Canadian province must be certified free of meningeal worm. The elk must test negative for hybridization with red deer.

PENNSYLVANIA

Information Contact for Requirements to Own Elk:

Pennsylvania Game Commission
2001 Elmerton Ave., Harrisburg, PA 17110
(717) 783-8164

A Wildlife Propagation Permit is required. Fences must be 10 feet high in order to confine elk and to exclude surrounding wildlife. Elk must be marked for identification. Facilities must be adequate, providing for the comfort and health of the animal. Inspection is required.

Information Contact for Requirements to Import Elk into State:

Dept. of Agriculture
Bureau of Animal Industry
2301 N. Cameron St., Harrisburg, PA 17110
(717) 783-6677, Dr. John I. Enck Jr.

A health certificate must be obtained. A negative TB test is required within 90 days of entry in animals 12 months of age and older. Negative brucellosis and bluetongue tests are required within 60 days. Elk must be identified with an ear tag or tattoo.

RHODE ISLAND

Information Contact for Requirements to Own Elk:

Department of Agriculture and
Resource Marketing
Animal Health Section
235 Promenade St., Room 370, Third Floor, Providence, RI 02908
(401) 222-2781

Information Contact for Requirements to Import Elk into State:

Animal Health Section (401) 222-2781

Persons wishing to import native species of cervidae must first obtain a permit from the Rhode Island Fish and Wildlife and comply with Title 20 of Rhode Island General Laws requirements prior to importation.

A Certificate of Veterinary Inspection is required. One of these conditions must be met:

- They must be tested negative to two official tuberculosis tests, which were conducted no less than 90 days apart–the second tuberculosis test being conducted within 90 days prior to the date of importation and the cervid isolated from all other animals during the testing period.
- They must originate from a federally accredited tuberculosis-free herd in which all captive cervids in the herd that are eligible for testing have been tested negative to at least three consecutive official tuberculosis tests conducted at 10-14 months intervals. Any cervids added to this herd must (1) move directly from another accredited herd; or (2) move directly from a federally qualified cervid herd which has undergone a negative official tuberculosis within the previous 12 months and test negative to an official tuberculosis test within 90 days prior to movement to the premises of the accredited herd; or (3) test negative to two official tuberculosis tests conducted at least 90 days apart, with the second test conducted within 90 days prior to movement to the premises of the accredited herd.
- They must originate from a federally qualified cervid herd in which all eligible cervids have tested negative to an official tuberculosis test within the previous 12 months, with the cervid to be imported testing negative to an official tuberculosis test conducted within 90 days prior to date of importation.

All imported cervidae are subject to quarantine and retest upon importation. Cervidae over 6 months of age must individually test negative to a USDA officially recognized brucellosis test within 30 days immediately prior to entry. They must test negative to a USDA officially recognized test for anaplasmosis and bluetongue within 30 days prior to importation.

SOUTH CAROLINA

Information Contact for Requirements to Own Elk:

Dept. of Natural Resources
1000 Assembly St. - Suite 261
P.O. Box 167, Columbia, SC 29202
(803) 734-3898 or (803) 734-8738

There are no restrictions on owning elk at this time, but import permits are presently being denied. The department is reviewing the situation.

Information Contact for Requirements to Import Elk into State:

Clemson University Livestock-Poultry Health Division
P.O. Box 102406, Columbia, SC 29224-2406
(803) 788-2260, Dr. Jones Bryan

It is unlawful to bring cervidae into the state without a permit from the Dept. of Natural Resources and the State Veterinarian's office. A permit may be granted only after the investigations and inspections of the wildlife have been made as the department considers necessary and the department approves the possession, transportation, or importation into the state.

SOUTH DAKOTA

Information Contact for Requirements to Own Elk:

Animal Industry Board
411 S. Fort St., Pierre, SD 57501-4503
(605) 773-3321, Dr. S.D. Holland

A permit is required ($10 per animal up to $100 maximum). An inspection is required for approval. Adequate fence and facilities are required. Animals must be marked for identification purposes. Possession of red deer and hybrids is restricted geographically to the area of South Dakota east of the Missouri River. Surveillance for CWD shall be maintained for all cervid herds. All herds of origin, adjacent herds, and herds having contact with affected animals or exposed animals shall be quarantined. A herd having contact with affected or exposed animals may be removed from quarantine after five years of compliance with the provisions or as directed by the State Veterinarian in consultation with the designated epidemiologist. Herd plans in affected herds shall be developed by the herd owner and the designated epidemiologist.

Information Contact for Requirements to Import Elk into State:

Animal Industry Board (605) 773-3321

Before intrastate movement or importation, other than intrastate movement to a state or federally inspected slaughter establishment, all cervidae shall be accompanied by an official certificate of veterinary inspection and a telephone entry permit number requested by the licensed, accredited veterinarian signing the certificate and issued by the State Veterinarian. One of the following statements shall appear on the certificate:

1. All cervidae identified on this certificate originate from a herd in which all cervidae have been kept for at least one year or into which they were born. There has been no exposure to or additions from any other source in the past year. There have been no diagnosis, signs or epidemiological evidence of CWD in this herd for the past year. Records and causes of death for the past year in this herd of origin are available to the animal health official of the state or origin.

2. All cervidae identified on this certificate originate from a herd which has been determined to have the stated monitored status by the animal health official of the state of South Dakota.

3. All cervidae identified on this certificate originate from a herd which has been determined to have certified CWD cervid herd status by the animal health official of the state of South Dakota.

In addition, all animals of the cervidae family must:

- Be tested negative for brucellosis by 2 official brucellosis tests within the 30 days prior to entry.
- Be tested negative to bluetongue and anaplasmosis within the 30 days prior to entry.
- Be negative to a 0.1cc SC (single cervical) test conducted using PPD bovis tuberculin within the 30 days prior to entry.
- Be from a herd which has had a negative complete herd tuberculosis test of all eligible animals, using the 0.1cc SC, PPD bovis tuberculin test, within the past 90-360 days, to which there have been no additions since the test.
- Be individually identified, by official identification tags, on a Certificate of Veterinary Inspection issued by the veterinarian who conducted the test.
- Enter on an import permit issued by the board.

TENNESSEE

Information Contact for Requirements to Own Elk:

Dept. of Agriculture
P.O. Box 40627, Nashville, TN 37204
(615) 837-5120, Dr. Robert L. Hartin

No requirements exist for elk at this time.

Information Contact for Requirements to Import Elk into State:

Dept. of Agriculture (615) 837-5130

One source of identification and a health certificate are required. The animal must have been tested negative for brucellosis and TB within 30 days of entry, and it must originate from a herd which has been tested for TB within the past 12 months, or it must have 2 negative TB tests at least 90 days apart.

TEXAS

Information Contact for Requirements to Own Elk:

Parks & Wildlife Dept.
4200 Smith School Rd., Austin, TX 78744
(512) 389-4775, (512) 389-4800

The legislature has declared elk to be exotic livestock, which are not regulated, throughout the state.

Information Contact for Requirements to Import Elk into State:

Texas Animal Health Commission
2105 Kramer Lane, P.O. Box 12966, Austin, TX 78711
(512) 719-0777, Dr. Terry Beals

An entry permit and a Certificate of Veterinary Inspection are required. The accredited veterinarian must state on the certificate that the animals have been inspected and are free of external parasites and evidence of contagious and communicable disease. The certificate must state that they have been tested negative to two tuberculosis tests, which were conducted no less than 90 days apart; that the second test was conducted within 90 days prior to the date of movement; and that the animals were isolated from all other members of the herd during the testing period; or that the animals meet one of the following conditions:

- Originate from an accredited free herd
- Originate from a qualified herd and have a classified negative 90 day tuberculosis test (If the qualifying herd test was administered within 90 days of movement, the animal(s) do not require an additional test.)
- Originate from a monitored herd and have a classified negative 90 day tuberculosis test
- Be less than 12 months of age, originate and were born in accredited, qualified, or monitored herds and have not been exposed to cervids from a lower status

UTAH

Information Contact for Requirements to Own Elk:

Dept. of Agriculture
Animal Industry Division
P.O. Box 14650, 350 N. Redwood Rd.,
Salt Lake City, UT 84114
(801) 538-7160, Dr. Michael Marshall
(801) 538-7166, Terry Menlove

The Dept. of Agriculture is responsible for enforcing laws and rules relating to:

- The importation, possession, or transportation of domesticated elk into the state or within the state.
- The inspection of domesticated elk facilities.
- Preventing the outbreak and controlling the spread of disease-causing pathogens among domesticated elk in domesticated elk facilities.
- Preventing the spread of disease-causing pathogens from domesticated elk to wildlife, other animals, or humans.
- If necessary, quarantining any domesticated elk pursuant to Livestock Inspection and Quarantine.

A license is required. The department inspects and approves the facility. Following the issuance of a license, the licensee shall have each domesticated elk inspected immediately before releasing into the facility. To renew a license, the licensee must submit to the department an inspection certificate that shows the elk and facility have been inspected and certified by the department for health, proof of ownership and genetic purity; the facility has been properly maintained during the preceding year; and proper records have been kept.

All perimeter fences and gates shall be a minimum of 8 feet tall and constructed of high-tensile steel. Gates shall be locked, with consecutive or self-closing gates when animals are present. There are additional specifications for the posts and wire; call for details. Internal handling facilities shall be capable of humanely restraining an individual animal and facilitating the application or reading of any animal identification; the taking of blood or tissue samples; and any other required or necessary testing procedure. The facility shall be properly constructed to protect inspection personnel while they are handling the elk. The owner shall provide ample signage around the facility indicating that it is a domesticated elk facility, so that the public is aware that the animals are not wild elk. Elk must be marked within 30 days of a change in ownership or by January 31 for weaned calves.

Each domesticated elk facility licensee shall have the elk inspected by the department prior to any transportation, sale, removal of antlers, or slaughter.

A recent Senate bill removes the prohibition of domestic elk harvesting and provides rule-making authority to the Utah Dept. of Agriculture and Foods to regulate the possession, transportation, and documentation of domesticated elk carcasses.

Information Contact for Requirements to Import Elk into State:

Animal Industry Division (801) 538-7160, (801) 538-7166

To prevent introduction of the meningeal worm, no domesticated elk that originates east of the 100 degree meridian will be allowed into the state.

For the importation, transportation, or change of ownership of any domesticated elk, the department shall require:

- Proof of genetic testing to ensure the purity of the domesticated elk herds and prevent the introduction of red deer or hybrid non-native species into domesticated elk herds in Utah by showing evidence of the purity of live animals, gametes, eggs, sperm, or other genetic material.
- Proof that the domesticated elk originates from a legal source.
- Identification of each elk by tattoo or microchip. (Call for details.)

To import domesticated elk, one must obtain:

- A domesticated elk facility license from the department.
- An entry permit from the state veterinarian's office.

The entry permit shall include:

- A health certificate with an indication of the current health status, including proof that the elk are tuberculosis and brucellosis free.
- Proof of genetic purity.
- Name and address of the consignor and consignee; origin of shipment; destination; total number of animals in the shipment.

Transportation within the state requires a brand inspection certificate. (Call for details.)

VERMONT

Information Contact for Requirements to Own Elk:

Dept. of Agriculture, Food & Markets
116 State St., Drawer 20, Montpelier, VT 05620
(802) 828-2421, Dr. Samuel Hutchins III

Cervidae must be contained within an escape-proof enclosure at all times except when they are used as work animals or for exhibition under conditions appropriate for the species involved. The perimeter fences shall be conventional or hi-tensile of which at least the bottom 6 feet must be mesh. Minimum wire gauges are 12 1/2 gauge for a conventional fence and 14 1/2 gauge for woven hi-tensile. Call the NAEBA office or Dr. Hutchins' office for more complete fencing regulations. All owners must construct or maintain suitable facilities for restraining and testing cervidae. Rules on the Transportation of Animals, 12 VSA Chapter 8, Subchapter 7, apply to the transportation of cervidae. Cervidae with antlers not in velvet must be "dehorned" or shall be separated from antlerless deer or shall be so confined as to prevent injury to each other from the antlers.

Information Contact for Requirements to Import Elk into State:

Dept. of Agriculture, Food & Markets (802) 828-2421

The required documentation for importation is an import permit and a Certificate of Veterinary Inspection dated within 30 days prior to the date of importation. Official individual identification of each cervid imported is required. It is the importer's responsibility to construct and maintain suitable facilities at the premises of destination for restraining and testing cervidae.

The health certificate shall certify that all cervidae have tested negative to a USDA officially recognized test for brucellosis dated within 30 days prior to the date of entry.

Cervidae that originate from accredited free herds may enter Vermont without further tuberculosis testing provided that they are accompanied by a Certificate of Veterinary Inspection stating that they originated from an accredited herd. Cervidae that are not known to be affected with or exposed to tuberculosis and that originate from qualified herds may enter Vermont if the animals are accompanied by a Certificate of Veterinary Inspection stating that they originate from a qualified cervid herd and have been classified negative to a USDA approved tuberculosis test that was conducted within 90 days prior to the date of movement. If the qualifying test was administered within 90 days of movement, the animals to be moved do not require an additional test. Cervidae from monitored herds that are not known to be affected with or exposed to tuberculosis may enter if accompanied by a health certificate stating that the animal originated from a monitored herd and has been classified negative to a USDA approved tuberculosis test that was conducted within 90 days prior to entry.

Cervidae not known to be affected with or exposed to tuberculosis, which originate from all other herds, may enter if they are accompanied by a Certificate of Veterinary Inspection stating that they have been classified negative to 2 official tuberculosis tests that were conducted no less than 90 days apart, that the second test was conducted within 90 days prior to the date of movement, and that the animals were isolated from all other members of the herd during the testing period. Cervidae that are less than 12 months of age and which originate from and were born in qualified or monitored herds may enter Vermont without further tuberculosis testing, provided they are accompanied by a certificate stating that and provided that they have not been exposed to cervidae from a lower status herd.

The certificate must certify that cervidae over 120 days of age have tested negative to a USDA officially recognized test for epizootic hemorrhagic disease when the herd of origin is in an area determined by the USDA to be endemic.

All cervidae 30 days of age or older originating in one of a list of states (call for list) shall have been tested negative for anaplasmosis with a USDA officially recognized test conducted within 30 days prior to entry.

All cervidae 30 days of age or older originating in one of a list of states (call for list) shall have been tested negative for bluetongue with a USDA officially recognized test conducted within 30 days prior to entry. Cervidae testing serological positive will be allowed entry if they are negative on a virus isolation test conducted in an approved laboratory.

Any cervidae originating from a state with a case positive premises within the previous 30 days or from a state adjacent to a state with a case positive premises within the previous 30 days must have import statements concerning vesicular stomatitis on the certificate.

Cervidae must be free of nematodes of the subfamily Elaphostrongylinae at the time of importation. Testing and isolation are not required for native born cervids that are imported directly from states and provinces where the nematode is not known to exist. All cervids from areas where the nematode is known to exist must first be isolated and tested prior to importation into Vermont according to standards pre-approved by the commissioner.

All elk must be born in, and originate from, herds in which CWD has not been diagnosed in elk, and in which no individuals within the herd have been exposed to CWD unless the importer applies in writing to the commissioner for an exception to this rule prior to importation. A voluntary program for the surveillance, control and eradication of CWD is in place. Call for details.

VIRGINIA

Information Contact for Requirements to Own Elk:

Dept. of Game and Inland Fisheries (804) 367-6878

Elk are not allowed in the state, except in zoos.

Information Contact for Requirements to Import Elk into State:

Dept. of Agriculture and Consumer Services
Division of Animal Industry Services
P.O. Box 1163, Richmond, Virginia 23209
(804) 786-2483
(804) 786-2481, Dr. William Sims, Jr.

Elk are not allowed to be imported into the state. To transport elk across the state, a permit is issued by USDA or Virginia Department of Game and Inland Fisheries. An official health certificate is required, and the animal is tested for brucellosis, tuberculosis, and bluetongue no more than 30 days prior to entry.

WASHINGTON

Information Contact for Requirements to Own Elk:

Dept. of Fish and Wildlife (360) 902-2380

Only zoos and research labs are allowed to own elk, and they are not allowed to propagate elk.

Information Contact for Requirements to Import Elk into State:

Division of Food Safety and Animal Health
P.O. Box 42577, Olympia, WA 98504
Permits/Information, (360) 902-1878
(360) 902-1881, Dr. Robert Mead

Elk are not allowed to be imported into the state. To pass through the state, permission must first be obtained from the Dept. of Fish and Wildlife. At that point, an official interstate health certificate is required. A brucellosis test in accordance with state/federal brucellosis protocol must be conducted within 30 days prior to entry. A tuberculosis test must be conducted in accordance with federal tuberculosis protocols within 30 days prior to entry. Elk must be from herds not known to be affected with or exposed to TB and must comply with the following:

- Be negative to a single cervical tuberculin test within 30 days and originate from a herd with a negative whole-herd test within 12 months. All additions to the herd must also have tested negative.
- Originate from a state with an approved state program and meet the requirements of a herd status plan and interstate testing requirements.
- All elk must be examined prior to entry for meningeal worm and must have a negative fecal exam for dorsal-spined larvae (if originating from specified western states).

WEST VIRGINIA

Information Contact for Requirements to Own Elk:

Dept. of Natural Resources (304) 558-2771

Nothing in the state regulations prohibits the entry of elk into the state. However, the present officials are reluctant to permit elk.

Information Contact for Requirements to Import Elk into State:

Dept. of Agriculture
Animal Health Division
State Capitol
1900 Kanawna Boulevard, East, Charleston,
WV 25305-0172
(304) 558-2214, Dr. Lewis Thomas

The Animal Health Division allows import, but the Dept. of Natural Resources' reluctance to admit elk has been a problem. For elk to pass through the state, a valid official health certificate issued by an accredited veterinarian and a special permit from the commissioner are required. The certificate indicates that the animal is from a herd which has been whole-herd tested for tuberculosis within the last 12 months and that the animal has been tested negative for tuberculosis within 60 days and has been tested negative for brucellosis within 30 days.

WISCONSIN

Information Contact for Requirements to Own Elk:

Wisconsin Dept. of Agriculture
Trade and Consumer Protection
Division of Animal Health
2811 Agriculture Dr., P.O. Box 8911,
Madison, WI 53704
(608) 224-4872, Dr. Clarence Siroky

A registration fee may be charged. Farm-raised deer are livestock which means that all the rules and regulations pertaining to livestock apply to elk, including rules pertaining to markets, dealers and truckers. Modeled after the federal UM&R on Tuberculosis in Cervidae, Wisconsin has a cervidae tuberculosis certification program. Captive cervidae cannot be commingled with bovine animals on the same premises or in the same building, enclosure or vehicle.

Local municipalities regulate fencing requirements, which are outlined in state statutes. The fence shall be at least 7 feet 10 inches high of woven, high tensile wire. The top 6 inches may consist of a single strand of smooth, high-tensile wire. If the fence is made with wood posts, the posts must be at least 12 feet long. The line posts must be at least 3.5 inches in diameter, and the corner and gate posts must be at least 5.5 inches in diameter. The wires are fastened securely to the posts using 9-gauge staples of at least 1.5 inches size. If the fence is made with steel or iron posts, the posts are at least 11 feet long; or if the posts are of the type known as T post, the posts are at least 10 feet long. Call for additional requirements concerning specifications for woven wire or high-tensile fence.

A complete record and inventory must be kept and made available to DATCP for inspection upon request.

Information Contact for Requirements to Import Elk into State:

Division of Animal Health (608) 224-4872

Any cervid moved within the state must be accompanied by a valid interstate health certificate or a certificate of veterinary inspection. The animal must have tested negative on the single cervical tuberculin test not more than 90 days prior to the intrastate movement. Requirements for importing elk into the state follow UM&R specifications. No cervid may be imported into Wisconsin if it originates from a herd that has had a member respond to any tuberculin test. The animal must be accompanied by a valid interstate health certificate or a Certificate of Veterinary Inspection. The certificate must reflect all of the following:

- A negative brucellosis test report, conducted not more than 30 days prior to the import date.
- A negative tuberculosis test report, based on two pre-import tests conducted not less than 90 days apart with the second test conducted not more than 90 days prior to the import date. The single cervical tuberculin test must be used unless the Dept. of Agriculture, Trade and Consumer

Protection authorizes a different tuberculin test. On or before the first test date, the animal must be isolated from all other animals by at least 10 feet and it cannot share common feeders, water utensils or handling facilities during the isolation period. These requirements are exempted if the animal is imported directly from an accredited tuberculosis-free herd, qualified tuberculosis-free herd, or monitored tuberculosis-free herd and meets other requirements. (Contact the division for details.)

• A confirmation / permit number. This number is assigned by the Division of Animal Health. Call (608) 224-4878 at least 5 days prior to import to notify the DATCP of the proposed import and to obtain the permit number.

WYOMING

Information Contact for Requirements to Own Elk:

Wyoming Game and Fish Commission (307) 777-4597

Domesticated elk are not permitted in the state, with the single exception of NX Bar Game Farm.

Information Contact for Requirements to Import Elk into State:

Wyoming State Veterinary Laboratory
Box 3312, University Station
Laramie, WY 82071
Beth Williams, (307) 766-5629

Elk cannot be imported into the state except by NX Bar Game Farm. To transport elk through the state, a permit is required from the Wyoming Game and Fish Commission. A certificate of veterinary inspection is required. The certificate must indicate that they are not affected with and have not been exposed to any infectious, contagious or communicable disease or the following parasites: meningeal worm, tissue worm, and large American liver fluke; and that they have not originated from a quarantined area. An interstate transportation permit shall only allow transportation of live elk through the state of Wyoming for a period not exceed 24 consecutive hours from the time of issuance. (Consult with Wyoming Game and Fish Commission for exact testing procedures / protocol.)

MEXICO

Information Contact for Requirements to Import Elk into Country:

United States Dept. of Agriculture
Missouri Office
P.O. Box 104418, Jefferson City, MO 65110
Pam Dawson, Export Documents Examiner
(573) 636-3116

Import Requirements for Captive-Raised Cervidae Exported from the U.S.:

The animals must be accompanied by a U.S. Origin Health Certificate issued by a veterinarian authorized by the USDA and endorsed by a Veterinary Services veterinarian. The certificate shall contain the name and address of both the consignor and the consignee and individual identification, age and sex of the animals to be exported. Additional information shall include:

Certification Statements

• The animals were examined and found to be clinically free of infectious diseases.

• The herd participates in the tuberculosis program for cervidae.

• The animals were inspected prior to exportation and were found to be free of ectoparasites.

• The animals were treated against ectoparasites within three days prior to exportation.

• The animals originate from a tuberculosis accredited free herd or state (or see "Testing Requirement").

Testing Requirement

• Tuberculosis: Single cervical test using ppd mammalian tuberculin (Or Certification #5).

Other Information

• All animals may be inspected by a Mexican veterinarian at a facility in the U.S. at one of the following border points: Brownsville, Texas; Laredo, Texas; Del Rio, Texas; Eagle Pass, Texas; El Paso, Texas; Presidio, Texas; Columbus, New Mexico; San Luis, Arizona; Nogales, Arizona; Calexico, California.

• The U.S. origin health certificate must be typewritten. Handwritten certificates will be rejected by the Mexican Animal Health authorities.

Import Requirements for Wild-Caught Cervidae exported from the U.S.:

The animals must be accompanied by a U.S. Origin Health Certificate issued by a veterinarian authorized by the USDA and endorsed by a Veterinary Services veterinarian. The certificate shall contain the name and address of both the consignor and the consignee and individual identification of the animals to be exported. Additional information shall include:

Certification Statements

• The animals were treated with an approved acaricide at the time of capture or at embarkation.

• The animals originate from zones where during the 6 months prior to exportation no notifiable diseases, including tuberculosis, have been reported to occur.

• The cervidae for export are wild-caught and not captive-raised.

• The cervidae were examined and found to be clinically healthy and free of ectoparasites.

• The wild-caught cervidae meet federal requirements for interstate movement.

Other Information

(See "Other Information" above.)

Information Contact for Requirements to Import Elk into Canada:

Brian Jamison
Canadian Food Inspection Agency
59 Camelot Drive • Nepean, ON KIA 0Y9
(613) 225-2342 • Fax: (613) 228-6630
e-mail:bjamieson@em.agr.ca

Instructions and Recommendations to Customs

Importer/Broker Instructions

Conditions of Import:

The original of this permit and any other necessary export documentation pertaining to the shipment of animal(s) or thing(s) must be provided for inspection at the first port of entry or to a Canadian Food Inspection Agency Import Service Center.

The conditions in this permit can only be changed or amended by a CFIA inspector. Any change to the permit by an unauthorized person will render the permit invalid.

Accompanying export documentation must be issued in either English or French.

Historical documentation for the animals being presented for importation must be verified by the certifying veterinarian. From the time of birth or importation into the country of origin until export, continuous records of animal identification and herds of residence of the animal(s) being exported have been verified by the certifying veterinarian.

The animal(s) and donor animal(s) must originate from a herd certified free from brucella abortus.

The herd of origin has been free from clinical, diagnostic and epidemiological evidence of brucellosis during the thirty-six (36) months immediately prior to movement of the animal(s) off the premises or collection.

The animal(s) must have originated from a herd certified free from the transmissible spongiform encephalopathy known as chronic wasting disease.

Chronic wasting disease must never have been diagnosed in the herd of origin nor in any herd where the animals being imported have resided at any time. The herd of origin must not contain any cervid animals derived from the wild of Colorado or Wyoming. The animal(s) or donor animal(s) of the thing(s) presented for importation must not be the progeny of a sire or dam suspected or known to be affected with CWD nor have resided at any time in a herd in which the disease has been diagnosed. The animals presented for importation did not originate from nor have ever resided in the states of Colorado or Wyoming. The herd of origin is located not less than forty (40) kilometers from any wild population known to be infected with CWD.

The animal(s) or donor animal(s) must originate from a premises certified free from tuberculosis. The premises of origin has been free from any clinical, microbiological, pathological, or epidemiological evidence of mycobacterium bovis for the thirty-six (36) months immediately prior to movement of the animal(s) off the premises or collection.

The animal(s) or donor animal(s) must originate from a herd certified free from tuberculosis. The herd of origin must be recognized as a tuberculosis accredited free herd by the central veterinary service of the country of origin. The herd of origin and any herd that animals were sourced from must have been free from any clinical or epidemiological evidence of tuberculosis for the twenty-four (24) months prior to exportation or movement off the premises. Excluding natural additions or animals from herds certified with a similar status by the central veterinary service of the country of origin, animals added to the herd must have been included in a recertification herd test. Animals less than one year of age that are presented for importation must be natural additions to the certified herd. The herd of origin is located not less than forty (40) kilometers from any wild animal population known to be infected with tuberculosis.

There must not be any mycobacterium bovis susceptible ruminants on the premises of origin that are not part of a herd recognized as tuberculosis accredited free by the central veterinary service. The exporting herd must not contain animals sourced from any herd in which tuberculosis has been diagnosed.

The animal(s) or donor animal(s) must be certified free from anaplasmosis. Negative results must have been obtained using the complement fixation test on samples taken from the animals within thirty (30) days of exportation. During post-entry quarantine in Canada, the animals must be retested for anaplasmosis using the complement fixation test negative at 1/5 or an alternate test acceptable to CIFA, with negative results.

The animal(s) must be certified free from bluetongue and epizootic hemorrhagic disease of deer. Negative results of bluetongue and EHD must be obtained using AGID on samples taken from the animals within thirty (30) days of exportation or collection. During the post-entry quarantine in Canada, the animals must be retested for bluetongue and EHD using the AGID test or an alternate test acceptable to CFIA with negative results.

The animal(s) must be certified free of brucella abortus. Animals being presented for importation must not have been vaccinated for brucellosis. Negative results for brucella abortus must have been obtained using the buffered plate antigen test (BPAT) or complement fixation (CF) on samples taken from the animal(s) within thirty (30) days of export or collection. During post-entry quarantine in Canada, the animal(s) must be retested for brucellosis using the BPAT or an alternate test acceptable to CFIA with negative results. If one or more of the animals presented for importation test positive on the BPAT, the affected animals must be isolated from the group and retested using the complement fixation (CF) test. If all BPAT test positive animals test negative to the CF test, the animals with a negative BPAT are eligible for export.

The animal(s) must be certified free from tuberculosis. Negative results must have been obtained using an intradermal test for tuberculosis, deemed equivalent to the Canadian mid-cervical test by the central veterinary service of the country of origin, on each animal presented for importation within thirty days of export to Canada. During the post-entry quarantine in Canada, the test must be repeated with negative results on each animal at a minimum interval of sixty (60) days from the first test. No animals shall have undergone testing for tuberculosis during the ninety (90) days preceding the date of testing for export to Canada. Responders will be isolated from non-responders and subjected to a comparative cervical test (CCT) for tuberculosis not less than sixty (60) days following the date of the original tuberculin injection. If all responders test negative on the CCT, the animals that tested negative on the initial intradermal test for export within thirty (30) days without further testing for tuberculosis.

The animal(s) being presented for importation must be certified as captive animals. The animals must have been born and raised in captivity and, so far as can be determined, must not have had exposure to wild animals.

The animal(s) being presented for importation must be identified by an acceptable electronic or non-electronic means. If utilizing non-electonic identification, the animals to be imported must be uniquely identified by a tattoo, an ear tag, wing or leg bands. The accompanying export certification must correlate the unique identification number to a description detailing the animals' breed, color, sex, age and any identifying marks. Alternatively, the cage(s) containing the animals must be securely fastened with official government shipping container seals or other identifying devices in such a manner as to ensure that animals cannot be added or removed. The numbers of the bands or seals or other identifying devices must be recorded on the accompanying export certification. Immature, unweaned animals traveling at the foot of their dam may be identified by external means correlating to that of the dam, and this identification must be recorded on the health certificate. The imported animals must bear a legible, permanent tattoo that shows the letters USA or a tattoo described under the definition of official calfhood vaccinate, and an official ear tag of the United States Department of Agriculture.

The animals must also be identified with a dangle tag of sufficient size and readability to enable the identification of the animals without restraint or offloading at the port of entry. The numbers of the dangle tag and the USDA/State tag for each animal shall be recorded on the export health certificate.

The animal(s), donor animal(s) or thing(s) being presented for importation must not come into contact with any animals, products or equipment of a lesser zoosanitary health status during the entire required periods of residency, isolation, storage and transportation until arrival at the premises of destination in Canada.

The animal(s), germplasm or thing(s) described on this permit must be shipped by the most direct and appropriate route from the point of export to the address of destination in Canada. Trans-shipment through another country requires written authorization from the Canadian Food Inspection Agency.

Documentation for importation must be presented to a federal inspector at the first point of entry. The shipment of animal(s) or thing(s) must be presented to an inspector designated under the Health of Animals Act at the first point of entry or at destination.

The containers, cages or vehicles used to transport the imported animal(s) or thing(s) must be sealed at the port of entry by an inspector designated under the Health of Animals Act in a manner to preclude opening. The shipment may then proceed under license to the facility of treatment or quarantine. The seals must be removed under the supervision of an inspector designated under the Health of Animals Act.
The animal(s) being presented for importation must be quarantined for a minimum of twenty-eight (28) days or a longer period of time as necessary to complete the tests required to meet the import conditions. The animals must be quarantined at the premises previously approved for such purpose by a veterinary inspector designated under the Health of Animals Act and specified on this permit.

The animal(s) must be imported into Canada through a minimum level quarantine facility approved by the Canadian Food Inspection Agency for that purpose. The import quarantine facility must have been previously approved for use as a minimum security quarantine facility by a veterinary inspector designated under the Health of Animals Act. The facility evaluation shall include: location, fencing, physical structure, lighting, water supply, waste disposal, vector and pest control, movement of people, security, cleaning and disinfection protocols. A report that the facility has been approved must have been issued by the Canadian Food Inspection Agency. No animal may be moved from its respective quarantine premises until duly discharged by an inspector designated under the Health of Animals Act. On completion of quarantine with negative results on all tests, the animals will be released to the importer and/or owner.

The importer is responsible for all costs incurred or associated with any testing or treatment of the animal(s) or thing(s) that may be required under the import permit or under the authority of the Health of Animals Act or the Health of Animals Regulations. The importer shall pay all fees for services required in respect of the importation under the National Animal Health Program Costs Recovery Fees Regulations in place at the time of importation.

Consideration of an application necessary for issuance of a permit to import the described animal or thing is subject to Class 3 fees due to the development of conditions within the last two years based on formal risk assessment.

The issuance of this permit does not relieve the owner or the importer of the obligation to comply with any other relevant federal, provincial or municipal legislation or requirement.

Failure to comply with the conditions contained in this permit or with the provisions of the Health of Animals Act and Regulations may result in the cancellation of this permit and will result in the forfeiture to the Crown of the animal(s) or thing(s) imported or in the removal of the animal(s) or thing(s) from Canada, all without compensation to, and at the expense of the importer. No person shall import any animal(s) or thing(s) into Canada from any country unless the animal(s) or thing(s) meet the conditions that are shown on the export certificate issued by the exporting country. The importer(s) are responsible for the animal(s) or thing(s) imported, their health, fitness, soundness, freedom from disease, active or latent and genetic or other defects. Where the Canadian Food Inspection Agency carries out tests and takes precautions when animal(s) or thing(s) are imported into Canada to reduce the risk of the introduction and spread of disease in Canada, such tests and precautions do not constitute and are not a warranty, guarantee, assurance, undertaking or anything similar that the animals imported are healthy, fit, sound, free from disease, active or latent or genetic or other defects and such tests and precautions do not relieve or lessen the importer's responsibility as set out above. The importer, his heirs, executors, successors and assigns release and discharge Her Majesty the Queen in right of Canada and the CFIA of and from all claims and demands, damages, actions or causes of action arising or to arise by reason of the importation of the animals and agrees to indemnify and save harmless Her Majesty the Queen in right of Canada and the CFIA from and against all actions, damages, claims and demands which may be brought in respect of or arising out of the importation of such animal(s) or thing(s), their health, fitness, soundness, freedom or otherwise from disease, active or latent, genetic or other defects.

Inspection Instructions - Animal Health Division:

At the port of entry, the animals are referred for veterinary inspection by CFIA.

For larger groups, inspection shall be limited to group inspection for absence of indications of clinical disease and humane transportation concerns. In small groups (5 animals or less), the identify of individual animals shall be verified.

The vehicle must be sealed and licensed to the previously pre-approved quarantine site.

The importer shall remove the seal at destination and retain for presentation to CFIA inspector on his/her subsequent inspection.

Information Contact for Requirements to Import Elk into the United States:

USDA-APHIS-VS-NCIE
Import Animals
4700 River Rd., Unit 39
Riverdale, MD 20737
(301) 734-7633
(301) 734-8010

Summary of Federal Regulations for Cervids to Enter United States from Canada. (This is an abridged version of the regulations; you are advised to consult the complete regulations before planning movement of animals.)

1.General Requirements

1.1 An import permit is required from the USDA if the animals were not born in Canada or were legally imported into Canada and were unconditionally released and eligible to move freely within Canada for at least 60 days.

1.2 An official health certificate is required. The official health certificate must be issued by a veterinarian designated by the Canadian Food Inspection Agency (CFIA), and must be endorsed by a veterinarian employed by the CFIA attesting to the certifications and tests required in this protocol.

1.3 Cervids described on the health certificate are all captive farmed. Animals born in the wild are ineligible for entry under this protocol. Properly identified cervids under 6 months of age, travelling at the side of their dam, are exempted from the test requirements.

1.4 All cervids must be identified with an official unique individual ear tag or tattoo, and must have a large, readable, dangle ear tag which allows numbers to be checked without offloading the animals. Tag numbers and the corresponding official tag or tattoo numbers must be recorded on the health certificate.

1.5 The health certificate must contain the certification statements as listed in the following section. The dates and the results of the tuberculosis and brucellosis testing must be recorded on the health certificate.

2. Certification Statements

2.1 Canada is free of foot-and-mouth disease, rinderpest, contagious pleuropneumonia, and surra.

2.2 Continuous records of animal identification and herds of residence of the animal being exported from the time of birth until exported were verified.

2.3 Herd of origin status:

a) Chronic wasting disease has never been diagnosed in the herd of origin.

b) The herd of origin is a tuberculosis herd of negative status according to the provisions of the CFIA captive ungulate program, and there has been no direct or indirect contact between the herd of origin and any known tuberculosis infected source.

c) The herd of origin is located not less than 40 km from any known Myco bacterium bovis (tuberculosis), Brucella abortus (brucellosis), or chronic wasting disease infected wild population.

d) The herd of origin does not contain animals from any herd where tuberculosis or brucellosis has ever been diagnosed.

e) Neither tuberculosis not brucellosis has ever been diagnosed on the premises of origin during the five years preceding the start of testing for export.

f) On the premises of origin there are no tuberculosis susceptible ruminants, that have not attained the status of a herd officially recognized free of tuberculosis or a herd of negative status for tuberculosis under the Canadian Animal Health Program for Farmed Cervids.

2.4 The animal intended for export is not the direct offspring of an animal that has been diagnosed with chronic wasting disease, nor has it resided at any time in a herd in which the disease has been diagnosed.

2.5 If less than one year of age, the animals intended for export were natural additions to the herd of origin.

2.6 Animals intended for export which were added to the herd of origin, except natural additions or animals from other herds of negative status for tuberculosis or U.S. Accredited Free herds, have been included in a re-certification test.

2.7 During the 60 days prior to export to the United States, the cervids selected for export and the herd of origin remained free from symptoms of infectious contagious disease, and free of any known exposure to such disease.

3.Testing Requirements

3.1 Tuberculosis-Within 60 days prior to export, each cervid must be tested with negative results for bovine tuberculosis using the Canadian Mid-Cervical Test (MCT). Any cervid classified as a responder to the MCT is ineligible for entry. However, if the status of all responders can be established by comparative testing, or by post-mortem examination and tissue culture, negative contact animals may be considered for entry.

3.2 Brucellosis-Within 30 days prior to export, cervids must test negative to a buffered plate antigen test for Brucella abortus. Should an animal test positive, it is ineligible for entry. It must be removed from the group and test negative to a complement fixation test at a dilution of 1:5 in order for the test negative contact animals to be considered for entry.

4.Port of Entry Inspection

Cervids for U.S. entry must be presented to the port by appointment. The port veterinarian shall conduct a visual health examination of the cervids and verify individual identification and the correctness of the health certificate.

ALBERTA

Information Contact for Requirements to Own Elk:

Alberta Agriculture, Food and Rural Development
Animal Industry Division
#204, 7000 - 113 St., Edmonton, AB T5K 2M4
(780) 427-4567, John Girvan, Head of Inspection Services, Diversified Livestock

Information Contact for Requirements to Import Elk into Province:

Al Cook
Commercial Wildlife Branch
Alberta Environmental Protection
4th Floor, Bramalea Bldg., 9920 - 108 St.
Edmonton, Alberta T5K 2M4
(403) 427-3344

Alberta will establish import protocols for movement of elk and deer into the province during the year 2000. Contact the above for details.

BRITISH COLUMBIA

Information Contact for Requirements to Own Elk:

Provincial Ministry of Agriculture
Game Department
250) 356-1677

No elk are permitted into the province for purposes of commercial farming.

Information Contact for Requirements to Import Elk into Province:

Provincial Ministry of Agriculture
Game Department
(250) 356-1677

No elk are permitted into the province.

MANITOBA

Information Contact for Requirements to Own Elk:

Animal Industry Branch
Manitoba Agriculture
204 - 545 University Crescent, Winnipeg, MB R3T 5S6
(204) 945-7557, Elk Specialist

A Game Production Farm License is required; the annual fee is $100. The operator must construct the game production facility in accordance with the design specifications approved by the director and permit the facility to be inspected before the license is issued. All animals must be proven to be free of any disease prescribed as reportable or otherwise proscribed under the Health of Animals Act. Game production animals must be registered. An application for registration is submitted to the director, and the director is provided with satisfactory evidence that the animal meets all of the identification, genetic, purity, health and quarantine requirements. On or before December 31, an application for registration is submitted to the director for each living progeny born that year. All elk are identified by means of a tag on each of the animal's ears; tags are provided by the director. The elk must also be tested for red deer genes and a DNA profile determined. Parentage by DNA must also be verified.

The containment area shall be not less than 10 acres in total area of land that is suitable habitat for elk. No more than 2 registered elk shall stock each acre. Fencing must be in accordance to regulations which specify a minimum height of 8 feet and which list specifications for wires, posts, and gates. (Call for details.) A windbreak must be provided in each separate pasture area. An adequate and clean supply of water shall be provided. The handling facility shall consist of at least 2 handling pens leading to a squeeze chute, and it shall have a curved alleyway of at least 50 feet in length leading to the handling pens. (Call for further details about required specifications.) The operator must provide a quarantine facility. All animals shall receive humane care in accordance with the Recommended Code of Practice for the Care and Handling of Farmed Deer. No person other than a veterinarian shall remove calcified hard antlers or velvet antlers unless he or she holds a game production animal antler removal license. All antlers must be tagged, and records relating the antlers to identified bulls must be kept and submitted by October 1 in the year of removal.

Information Contact for Requirements to Import Elk into Province:

For Zoo Purposes:

Wildlife Branch
Dept. of Natural Resources
Box 24, 200 Saulteaux Crescent, Winnipeg, MB R3J 3W3
(204) 945-7775

For Game Production Animals:

Animal Industry Branch
Agricultural Services Complex
204 - 545 University Crescent, Winnepeg, MB R3T 5S6
(204) 945-7557

Information from the Animal Industry Branch:

A person who transports a game production animal shall have in his or her personal possession or in the transportation unit at all times the proper Transportation Authorization Permit issued by Agriculture Canada, a Manitoba Livestock Manifest, and a Manitoba Game Production Animal Registration Certificate.

Animals must be genetically tested because of the prohibition on red deer or hybrids. It is forbidden to bring the semen, ovum, or embryo of elk into Manitoba without first ensuring by a genetic test that the donor of the semen or ovum is not an animal of any of the subspecies of red deer or a hybrid of such an animal, or that neither of the parents of the embryo is an animal of any of the subspecies of red deer or a hybrid of such an animal. The director must be provided with the results of the genetic test and written permission must be received. DNA testing on imported elk is also necessary. DNA testing must also verify parentage of the imported elk.

A person who brings an elk into Manitoba as a game production animal shall comply with the requirements of the Ungulate Movement Regulations. A health certificate is required which indicates the following: The animal has been tested in accordance with, and proven to be free of any disease prescribed as reportable or otherwise proscribed under the Health of Animals Act. The animal originates from premises that are free of any such disease. The animal is free of any symptoms of Chronic Wasting Disease.

NEW BRUNSWICK

Information Contact for Requirements to Own Elk:

New Brunswick Agriculture
P.O. Box 6000, Fredericton, NB E3B 5H1
(506) 453-7978

Dept. of Natural Resources and Energy
(506) 453-2448, Gerald Redmond

A permit under the Captive Wildlife Policy is administered by the Department of Natural Resources and Energy. The policy also has specifications on health testing for importation purposes.

All infectious diseases in captive wildlife must be reported immediately by the owner or attending veterinarian to the chief provincial veterinarian of the Dept. of Agriculture and the executive director of Fish and Wildlife Branch. An annual health inspection by a veterinarian is required. The report must be forwarded to the executive director of the Fish and Wildlife Branch.

A perimeter fence must be erected around all big game operations and must consist of a fence with a minimum height of 1,900 millimeters above the ground, plus additional fencing to bring the total enclosed height above ground to a minimum of 2,133 millimeters. At no location may the bottom of the fence be further than 75 millimeters from the ground. The prefabricated fence must be either a soft, galvanized steel of 4 millimeters diameter or 2.5 millimeters diameter galvanized, high-tensile steel. Fence posts, if wooden, must be of cedar or treated wood and a minimum of 90 millimeters diameter, 3 meters in length and a minimum of 2,133 millimeters above ground, spaced a maximum of 5 meters apart. Strainer and corner posts must be a minimum of 3 meters in length, and 175 millimeters in diameter.

All commercial elk or deer operations must have handling facilities including a corral or paddock equipped with a drive chute and squeeze chute to facilitate the handling of the animals for herd health programs and annual veterinarian inspection. These facilities must be enclosed within the perimeter fence. A perimeter fence and handling facilities must be erected before an operating permit is issued.

NOVA SCOTIA

Information Contact for Requirements to Own Elk:

Department of Agriculture and Marketing
Production Technology Branch
P.O. Box 550
Truro, NS B2N 5E3
Roy MacKenzie, (902) 893-6517

A elk farming license, which is renewed annually, is required. Applicant must own or lease land within the province of at least 5 contiguous hectares. A complete, permanent record must be kept. Unique identification of each deer farm animal is required. Perimeter fences are required to be not less than 2.43 meters high. Adequate shelter and handling facilities are necessary. Applications must be approved before a deer farming license is issued.

Information Contact for Requirements to Import Elk into Province:

Department of Agriculture and Marketing
Production Technology Branch
(902) 893-6517

Obtain a transportation authorization permit from the Health of Animal Directorate office (Food Production and Inspection Branch AAFC). An import permit issued by the provincial Minister of Agriculture and Marketing is also required.

An applicant must meet the guidelines for importing elk farm animals to Nova Scotia before the animals arrive on the farm. Some of the requirements include:

- Origin of the elk to be imported
- Number, their age and identification
- Importation date and copy of transportation authorization permit issued by AAFC
- Proposed quarantine of imported elk and status report from veterinarian for elk herd of origin
- Meet all required TB and brucellosis and fecal testing

Animals will be quarantined for importation to Nova Scotia a minimum of 90 days. Release is contingent on all requirements for the quarantine period being completed. Full details can be given upon request.

ONTARIO

Information Contact for Requirements to Own Elk:

Dr. Elliot Salsberg
Animal Health Program-Ontario
Canadian Food Inspection Agency
(519) 837-9400

No provincial regulations exist. The Code of Practice is recommended.

Information Contact for Requirements to Import Elk into Province:

Animal Health Program, (519) 837-9400

PRINCE EDWARD ISLAND

Information Contact for Requirements to Own Elk:

P.E.I. Department of Agriculture and Forestry
P.O. Box 2000, Charlottetown, PEI C1A 7N8
(902) 368-5087, Dr. Robert Morrison

A license is required. Every person wishing to obtain a license should submit an application to the minister on a form provided by the minister. No person shall obtain game animals unless those animals are obtained from a person who has been issued a license and the animals are obtained from a source approved by the minister. No person who holds game animals shall allow those animals to roam free, escape from captivity or be released to the wild. Every game farm operator shall ensure that enclosures contain shelter sufficient to protect the game animals from the elements and that the perimeter fence for the enclosure is at least 2.1 meters high and is secured so as to prevent public access to the enclosure and the escape of game animals. Every game farm operator shall ensure that adequate handling facilities are available for the examination and testing of the game animals. Every game farm operator shall ensure that all game animals born on the game farm are uniquely identified by November 30 in the year of birth and all game animals on the game farm carry visual identification applied immediately upon the animals' arrival at the game farm or in the case of animals born on that farm, by November 30 in the year of birth. Every person who is required to uniquely identify a game animal shall apply to the minister, in the manner and form required by the minister, for a unique identification tag. Every game farm operator shall maintain a permanent written record, accounting for all births, deaths, purchases and transfers of game animals on the game farm. Every game farm operator shall at least annually and on such other occasions as a veterinarian may require, make all game animals held on the game farm available for inspection and testing.

Information Contact for Requirements to Import Elk into Province:

Department of Agriculture and Forestry (902) 368-5087

In the event that elk are allowed to be imported into the state, the Department of Agriculture has set the following requirements: All elk shall have an official Certificate of Veterinary Inspection issued by an accredited veterinarian and must have a permanent individual identification. All elk 6 months of age or older must have a negative brucellosis test within 30 days of entry. All elk regardless of age must have a negative TB test within 30 days of entry.

QUEBEC

Information Contact for Requirements to Own Elk:

Department of Agriculture (418) 643-2673

Information Contact for Requirements to Import Elk into Province:

Department of Agriculture (418) 643-2673

SASKATCHEWAN

Information Contact for Requirements to Own Elk:

Dept. of Agriculture
Saskatchewan Livestock Land Revenue
(306) 787-4682

A game farming license is required.

The minister shall have the power to take any measures deemed advisable, including quarantine or immediate slaughter, for the suppression or limitation of an outbreak of disease. Game farm operators must make all animals available for inspection at all reasonable times by a game inspector. Complete records are kept and made available to a game inspector at all reasonable times for the purpose of inspection. Individual animals must be uniquely identified prior to entering the province or by November 30 in the year of their birth.

Game farm operators shall ensure that enclosures contain natural bluffs or shelters sufficient to protect the animals from the wind. The perimeter fence must be at least 2.1 meters high and securely locked so as to prevent access or escape. The stocking density may not exceed 10 adult animals per acre, and operators must provide adequate feed and water. Operators must ensure that antler removal and other game animal handling procedures are conducted in a humane manner.

Information Contact for Requirements to Import Elk into Province:

Dept. of Agriculture
Provincial Veterinary Lab
(306) 787-4682, Fax (306) 787-6439

An import license and veterinarian's report are necessary. A veterinarian must examine the animal and determine it to be free of brucellosis, tuberculosis, anaplasmosis, and every other named disease. The animal must be held in quarantine for a period of not less than 14 days to ensure that no named disease is present. If a game animal shows signs of disease during the quarantine period, it must be examined by a veterinarian. When a game animal is imported from another jurisdiction within Canada, the veterinarian conducting the examination of the animal need not test for bluetongue or anaplasmosis. No person shall transport livestock within Saskatchewan prior to its inspection unless he holds in his immediate possession at all times a livestock manifest. No person shall transport livestock within Saskatchewan after inspection unless he holds in his immediate possession at all times a livestock permit.

Regulations for the methods and vehicle used for transportation of elk may be obtained by calling the Dept. of Agriculture. All individual animals must be uniquely identified prior to entering the province.

YUKON

Information Contact for Requirements to Own Elk:

Renewable Resources
Box 2703, Whitehorse, Yukon Y1A 2C6
(867) 667-5838

A game farm license is required ($50). A license may not be issued for a game farm on land belonging to the Crown that is unoccupied, on land where a grazing lease is administered by the minister of Community and Transportation Services or the minister of Indian and Northern Affairs Canada, or within an area where the use of the land for a game farm is prohibited by a regulation made under the Area Development Act or a zoning by-law made under the Municipal Act. The minister shall not issue a license authorizing the operation of a game farm on land which is in a licensing area other than the Whitehorse Periphery Development Area before the consultation process with the affected Yukon First Nation has been completed. No person may hunt a game animal or permit the hunting of a game animal.

All animals born on the farm must be uniquely identified by November 30 following the birth, and all animals arriving at the farm must immediately carry visual identification. Operators shall ensure that the animal enclosures contain natural bluffs or shelters sufficient to protect the animals, that the perimeter fence is at least 2.1 meters high, and that all activities on the farm be contained within the perimeter fence. Fences must be kept securely locked so as to prevent public access to the enclosure and the escape of animals. No more than 5 animals per hectare may be kept. Operators shall provide animals with feed and water adequate to meet normal seasonal maintenance and production requirements. All animal handling procedures must be conducted in a humane manner. Velvet antlers must be removed from the animal in such a way as to minimize discomfort and stress to the animal:

- The animal must be sedated or physically restrained in such a manner as to prevent injury.
- The animal must be injected with sufficient local anesthetic to anesthetize the pedicle area.
- The antlers must not be cut closer than 1.5 centimeters from the pedicle, using a clean and sharp saw.
- Hemorrhaging from the severed blood vessels must be controlled.

Written records are kept; a copy of the record is provided to the minister before the March 31 immediately following the taxation year or with an application for renewal.

Information Contact for Requirements to Import Elk into Province:

Renewable Resources
Box 2703, Whitehorse, Yukon Y1A 2C6
(867) 667-5838

To import or export, a permit is required. To import into the province, a veterinarian must examine the animal using methods and tests acceptable to the minister and certify that the animal is free of named diseases. The veterinarian's report is provided to the inspector who issues the import permit. Animals must be uniquely identified prior to entering.

No person shall transport a game animal in a vehicle on a highway outside the Whitehorse Periphery Development Area without a permit issued by the minister.

Alberta - Yukon Elk Importation Protocol

In order to facilitate the importation of elk from the Yukon into Alberta, Alberta and the Yukon agree to the following operating procedure. For a complete copy of the protocol, as well as an interpretation of the protocol intent regarding disease testing, contact the NAEBA office or Yukon Renewable Resources office.

Alberta will allow the importation of elk from the Yukon only if:

1. The Yukon adopts and enforces during the currency of this protocol, standards in relation to the importation of cervids into the Yukon that are equivalent to or more stringent than those applying generally to the importation of elk into Alberta.

2. The Director of the Agriculture Branch, Yukon Renewable Resources, certifies in an Elk Health Certificate, that:

   - The elk come from game farms where elk have been held separate and apart from all other species.
   - The movement of the elk is permitted by Canadian Food Inspection Agency (CFIA).
   - The herd from which the elk originate has a current CFIA rating of "negative" for tuberculosis and brucellosis and, in the case of a calf being exported, each cow from which a calf originates has tested negative for tuberculosis and brucellosis.
   - All animals (exclusive of calves) have individually tested negative for both TB and brucellosis.
   - The Yukon elk herd remains free of detectable "red deer" genes, as established by documentation from one of the following laboratories: SRC Bovine Blood Laboratory, NZ Deer Blood Typing Laboratory, Wyoming State Blood Laboratory. Elk which originated from Elk Island National Park are exempt from testing requirements.
   - The elk are individually identified by tamperproof metal ear tag.
   - The elk were born in the Yukon or Alberta.

3. All elk to be imported are identified in and accompanied by Schedule A.

4. The importer meets all requirements of Alberta law for the importation.

Chapter 1 **Models for Business Planning**

Analysis of projected costs and returns is a very dynamic process. Prices for inputs change, and the returns on breeding stock of any kind are totally dependent on what the buyer will agree to pay and what the seller is prepared to accept. This situation applies to all types of agriculture, indeed to most businesses. Necessarily, then you must use these charts and worksheets as guides and models. You must establish your own business plan, and input your own appropriate numbers to the calculations.

Every elk farm is a bit different. The following will provide you with the models to use in making your farming proposal attractive to both lenders, investors and yourself. These models and your successful business plan should ensure that you build an elk farming enterprise that will repay your work and investment and go on working for your children as they also become keen and successful elk farmers.

Best of luck, work hard enough, enjoy elk farming.

ELK FARMING COMPANY - BUDGET REPORT Calendar year 2000							US Dollars						
Inventory: 60 velvet / breeding bulls	Jan-100	Feb-100	Mar-100	Apr-100	May-100	Jun-100	Jul-100	Aug-100	Sep-100	Oct-100	Nov-100	Dec-100	Total
70 cows 60 calves/yrlgs	Budget	Budget	Budget	Budget	Budget	Budget	Budget	Budget	Budget	Budget	Budget	Budget	2000
INCOME CATEGORIES:													
LIVESTOCK SALES		36,000						11,000	10,000		40,000	40,000	137,000
MISCELLANEOUS			280		60	280		100		80			800
VELVET & BYPRODUCTS	140	140	210	280	210	140	17,000	11,000	200	250	150	140	29,860
WAPITI MEAT		210	70	140	140	850	280		400				2,090
OTHER INCOME						150	100		1,500				1,750
TOTAL INCOME	**140**	**36,350**	**560**	**420**	**410**	**1420**	**17,280**	**22,100**	**12,100**	**330**	**40,150**	**40,140**	**169,750**
EXPENSE CATEGORIES:													
- ADMINISTRATION-													
LOAN SERVICING		7,000					7,000				7,000	7,000	28,000
MEMBERSHIP FEES	350	250			250					250		75	1,175
OFFICE SUPPLIES	50		100	100			50	50		75		200	625
POSTAGE	50	75	150		75		70		75	100		50	645
PROFESSIONAL FEES			500							300			800
SERVICE CHARGES	50	50	100	50	150		75	75	75		50		675
SUBSCRIPTIONS		150			75			100		50			375
TELEPHONE	250	300	200	200	350	100	100	200	200	300	250	200	2,650
- DIRECT EXPENSES-													
CAPITAL PURCHASES					3,500								3,500
DELIVERY:													
LIVESTOCK		700								300		400	1,400
OTHER		150	250	500	700		50		100	100		50	1,900
EQUIPMENT MAINTENANCE		200	200	100	100	1,800			1,500		750	800	5,450
FEED:													
GRAIN & SUPPLEMENTS	2,000						4,000		2,000		2,000		10,000
HAY & HAYLAGE	1,800	1,000	1,500	1,000	500	2,000			800	800		1,500	10,900
FENCING MATERIALS			2,500					500					3,000
FERTILIZER AND SEED			1,200		3,000		1,000						5,200
LIVESTOCK ACQUISITION		5,000						15,000				4,000	24,000
PROPERTY INSURANCE			2,000										2,000
PROPERTY MAINTENANCE	300	50	300	250	500	1,500	100			400	200	300	3,900
PROPERTY TAX				1,200									1,200
SALARIES & BENEFITS	1,500	1,000	1,500	1,500	1,500	2,000	1,000	1000	2,000	1,000	1,500	1,500	17,000
TOOLS		50	100		200				100			50	500
FARM VEHICLES													
FUEL	100	75	75	75	75	100	75	75	200	125	150	150	1,275
LEASE/PURCHASE	500	500	500	500	500	500	500	500	500	500	500	500	6,000
LICENCE, REG., INS.			975				110						1,085
REPAIRS	200	150		200		400	200		700		100		1,950
UTILITIES:													
ELECTRICITY	150	150	150	150	150	150	150	150	150	150	150	150	1,800
HEATING FUEL	150		150								150		450
VETERINARY EXPENSES:													
DRUGS	70	70	70	800		70		1,200			1,200		3,480
VETERINARIAN	50			100		100	100						350
MARKETING:													
ADVERTISING		500	300	500	100	500		500	300	550		550	3,800
MEALS & ENTERTAINMENT	200		100			150		100		100	100	300	1,050
PROMOTIONAL MATERIALS	600												600
TRADE SHOWS	650		150								400		1,200
PROCESSING			200			150	175			150		150	825
TRAVEL & MEETINGS:													
AIRFARE AND DRIVING	500		100				400		100		150		1,250
MEALS & HOTELS	350		200				200		100		200		1,050
MISCELLANEOUS		100			500					300			
TOTAL EXPENSES:	**9,870**	**17,520**	**13,570**	**7,225**	**12,225**	**9,520**	**15,355**	**19,450**	**8,900**	**5,550**	**14,850**	**17,925**	**151,060**
BOTTOM LINE:	**-9,730**	**18,830**	**-13,010**	**-6,805**	**-11,815**	**-8,100**	**1,925**	**2,650**	**3,200**	**-5,220**	**25,300**	**22,215**	**18,690**
Livestock Inventories (Averages for the year):		190 head	Actual feed costs average per elk per year.			$110.00	Total farm operation costs (ave.) per elk per year						$647.00

Elk Farming Financials - year 2000 - based on a survey of Saskatchewan producers

Average Values

Total # of elk on farm	70	
# breeding stock	50	
# velvet bulls	30	
Total acres fenced	150	
acres per adult elk	2.1	
Equipment and Facilities		
Fence - Total & (per mile)	$55,000	$11,000
Handling Facilities	$16,000	
Hydraulic squeeze	$9,000	
Truck	$18,000	
Stock Trailer	$8,00	
Tractor w/front end loader	$30,000	
Freezer	$1,000	
Water bowls	$1,500	
Trenching	$1,800	
Gates	$3,500	
Dugouts	$3,000	
Miscellaneous	$10,000	
Total per farm	**$156,800**	
Labour - per year		550 hours
$/hour		$15
Total Labour (own or hired)		**$8,250**

Average market value	$6,800
breeding cows	$9,000
breeding bulls	$20,000
velvet bulls	$3,000
male calves	$1,500
female calves	$4,500
Annual costs per head	
Feed costs per head	
Hay	$70
Grain	$40
Supplement	$40
Veterinarian	$30
Custom work	$55
Repairs	$35
Fuel and oil	$25
Utilities	$20
Insurance	$30
Marketing & transportation	$40
Labour	$117
Miscellaneous	$15
Variable costs per head	**$517**
add financing costs:	
Depreciation (over 20 years)	$112
Opportunity Cost (on livestock value @ 7%)	$476
Operating Interest	$40
Total costs per head per year:	**$1,145**

ELK FARMING COMPANY - BUDGET REPORT Calendar year 2000

Inventory: 60 velvet / breeding bulls 70 cows 60 calves/yrlgs	Jan-00 Budget	Feb-00 Budget	Mar-00 Budget	Apr-00 Budget	May-00 Budget	Jun-00 Budget	Jul-00 Budget	Aug-00 Budget	Sep-00 Budget	Oct-00 Budget	Nov-00 Budget	Dec-00 Budget	Total 2000
INCOME CATEGORIES:													
LIVESTOCK SALES		48,000						15,000	16,000		55,000	60,000	194,000
MISCELLANEOUS			400		80	400		100		120			1,100
VELVET & BYPRODUCTS	200	200	300	400	300	200	24,000	18,000	300	300	200	200	44,600
WAPITI MEAT		300	100	200	200	1,200	400		600				3,000
OTHER INCOME						150	100		1,500				1,750
TOTAL INCOME	**200**	**48,500**	**800**	**600**	**580**	**2,000**	**24,400**	**33,100**	**18,900**	**420**	**55,200**	**60,200**	**242,700**
EXPENSE CATEGORIES:													
- ADMINISTRATION-													
LOAN SERVICING		10,000					10,000				10,000	10,000	40,000
MEMBERSHIP FEES	350	250			250					250		75	1,175
OFFICE SUPPLIES	100		125	125			125	50		125		100	750
POSTAGE	50	75	150		75		70		75	100		50	645
PROFESSIONAL FEES			1,200							500			1,700
SERVICE CHARGES	50	50	100	50	150		75	75	75		50		675
SUBSCRIPTIONS		150			75			100		50			375
TELEPHONE	350	500	500	400	450	400	400	300	300	350	250	350	4,550
- DIRECT EXPENSES-													0
CAPITAL PURCHASES					5,000								5,000
DELIVERY:													0
LIVESTOCK		1,100								500		500	2,100
OTHER	350	250	350	400	1,000	350	100		250	200		100	3,350
EQUIPMENT MAINTENANCE		200	200	100	100	1,800			2,000		750	800	5,950
FEED:													0
GRAIN & SUPPLEMENTS	2,600						5,800		2,600		2,600		13,600
HAY & HAYLAGE	2,500	1,500	1,500	2,000	500	2,400			1,200	800		2,000	14,400
FENCING MATERIALS			2,500					500					4,000
FERTILIZER AND SEED			1,200		3,000		1,000						5,200
LIVESTOCK ACQUISITION		8,000						20,000				8,000	36,000
PROPERTY INSURANCE			2,200										2,200
PROPERTY MAINTENANCE	300	50	800	250	1,200	1,500	100			400	200	300	5,100
PROPERTY TAX				1,200									1,200
SALARIES & BENEFITS	2,000	1,500	1,500	2,000	2,000	3,000	2,000	1,000	2,000	1,500	2,000	2,000	22,500
TOOLS		50	100		200				100			50	500
FARM VEHICLES													0
FUEL	200	150	150	150	150	200	150	150	350	250	250	300	2,450
LEASE/PURCHASE	800	800	800	800	800	800	800	800	800	800	800	800	9,600
LICENCE, REG., INS.			975				110						1,085
REPAIRS	250	300		250		500	350		900		200		2,750
UTILITIES:													0
ELECTRICITY	300	350	300	250	200	100	150	200	150	150	100	150	2,400
HEATING FUEL	200		200	100							150		650
VETERINARY EXPENSES:													0
DRUGS	100	100	100	1,200		100		1,900			1,900		5,400
VETERINARIAN	50			100		100	100						350
MARKETING:													
ADVERTISING		580	300	585	300	585		585	500	600		600	4,640
MEALS & ENTERTAINMENT	100		300			250		100		100	100	500	1,450
PROMOTIONAL MATERIALS	600												600
TRADE SHOWS	900		200								500		1,600
PROCESSING			200		250	150			175	150		150	1,075
TRAVEL & MEETINGS:													
AIRFARE AND DRIVING	750		100				500		100		150		1,600
MEALS & HOTELS	500		350				400		100		200		1,550
MISCELLANEOUS		100								300			
TOTAL EXPENSES:	**13,400**	**27,060**	**16,400**	**9,960**	**16,200**	**12,235**	**22,230**	**25,760**	**11,675**	**7,125**	**20,200**	**26,825**	**208,170**
BOTTOM LINE:	**-13,200**	**21,440**	**-15,600**	**-9,360**	**-15,620**	**-10,235**	**2,170**	**7,340**	**7,225**	**-6,705**	**35,000**	**33,375**	**34,530**
Livestock Inventories (Averages for the year):	190 head	Actual feed costs average per elk per year.			$148.00	Total farm operation costs (ave.) per elk per year							$889.00

ELK PRODUCTION - CALCULATIONS FROM SASKATCHEWAN AGRICULTURE

All prices in Canadian Dollars

Operating Requirements

Cash costs include wages and benefits, feed, veterinary fees and supplies, repairs and maintenance, fuel and oil, utilities, insurance, marketing and transportation, property taxes, miscellaneous business costs and interest on cash costs.

Wages and Benefits

The amount of labour is approximately 66% of the requirement for a beef cow-calf enterprise as elk are handled less often and are not assisted during calving. Estimated wages (including the owner's wage) are $50 per head for mature breeding stock, replacements and bulls. For hired labour, benefits should be estimated at an additional 14% of the cost of wages and would include Canada Pension plan and unemployment insurance contributions.

Feed

Breeding stock require approximately 2% of their body weight on a dry matter basis per day to maintain their weight.

Hay requirements:

Depending on the type and condition of pasture available, hay is usually only provided during the winter months.

Based on full winter feed, elk requirements are as follows:

Cows: 10 lbs. per day x 180 days = 1800 lbs. (0.82 tonne)

Bulls: 12 lbs. per day x 180 days = 2160 lbs. (0.98 tonne)

Feed costs for replacements are estimated to be 75% of those for adults.

If forage is available and snow is not too deep, elk will utilize pasture to some extent during the winter, however, common practice is to feed the herd for the entire winter period.

Grain requirements:

Grain is usually used to supplement the hay and to flush elk cows during breeding time to ensure a high conception rate. Oats are the most common choice, with some producers also using barley with caution.

Cows: 4 lbs. per day x 180 days = 720 lbs. (0.33 tonne)

Bulls: 6 lbs. per day x 150 days = 900 lbs. (0.41 tonne)

Supplement requirements:

Often, elk will not consume mineral free choice. Therefore, a pelleted supplement that contains protein, minerals and salt is recommended to be fed year round.

Cows and Bulls: 1.1 lbs. per day x 365 days = 402 lbs. (0.18 tonne)

Yearlings: 0.55 lbs per day x 365 days = 201 lbs. (0.09 tonne)

Livestock Purchases

Velvet bulls or feeder animals could be purchased to create, or expand, a velvet or feeder enterprise.

Breeding Stock Purchases

Additions and/or replacements to the breeding herd are often retained from the herd's annual calf crop. New bulls will be needed as the heifers become breeding age in order to avoid inbreeding that could have a detrimental effect on the long-term genetics of the herd.

Veterinary Fees and Supplies

Veterinary services include deworming medication, blood tests, minor treatments with antibiotics, vaccinations, fecal sample testing, velvetting supplies.

Repairs and Maintenance

Repairs and maintenance is required for vehicles, fences, handling facilities, etc.

Fuel and Oil

Fuel and oil is required for the truck and tractor.

Utilities

Telephone is used to make marketing arrangements and source stock. Hydro costs will be minimal.

Insurance

Insurance on elk cows costs approximately $0.75/$100 value. This translates to approximately $135 per head. Insurance on facilities and equipment will bring the total insurance cost to $140 per head. Normally insurance will only be purchased on the breeding animals.

Marketing and Transportation

Marketing and transportation costs will depend on the marketing system used. A large part of these costs may be the cost of advertising the stock for sale, as most of the sales will be made FOB the farm. The cost is estimated to be $10 per head marketed.

Death Loss

When properly managed, elk may be productive for many years. However, an amount for death loss of 3% is included in the expenses. This amount is calculated as 3% of the value of the total herd and is used to replace animals that die.

Property Taxes

Property taxes are paid on a per acre basis.

Miscellaneous Costs

Miscellaneous costs are items such as license fees, advertising, association memberships, etc.

Operating Interest

Interest on cash costs should be calculated on the average amount of costs for the year.

Table 5: Summary of the Capital and Operating Inputs of an Elk Enterprise

Operating Inputs	**Cost**
Labour	$50/head
Feed	
Hay (180 days/year - forage valued at $60 per tonne)	0.82 tonne/cow ($49/ cow); 0.98 tonne/bull ($59/bull); 0.68 tonne/yearling ($41/yearling)
Grain (oats) (180 days/year - oats valued at $100 per tonne)	0.33 tonne/cow ($33/cow); 0.41 tonne/bull ($41/bull); 0.28 tonne/yearling ($28/yearling)
Supplement (365 days/year - supplement valued at $300 per tonne)	0.18 tonne/cow or bull ($54 per cow or bull); 0.09 tonne/yrlg ($27/yearling)
Total Feed Cost	
Per Cow	$136 per head
Per Bull	$154 per head
Yearling	$96 per head
Livestock/ breeding stock purchases	market price (see Table 7)
Veterinary Fees and Supplies	$30/cow; $25/yearling or bull
Repairs and Maintenance	vary with the size of the operation
Gas, Fuel and Oil	$10/head
Utilities	$15/bull; $10/head for cows and yearlings
Insurance	$0.75/$100 value of breeding cow herd
Marketing and Transportation	$10/head
Death Loss	5% for calves; 3% for all others
Game Farm License	$100/year
Property Taxes	$1.00 - 5.00/acre
Miscellaneous	vary with the size of operation

Table 6: Capital Investment and Operating Costs for an Established 40 Cow/40 Velvet Bull Elk Enterprise

Breeding and Velvet Livestock	**No. of Head**	**Cost per Head**	**Total Cost**
Elk Cows	40	$11,000	$ 440,000
Breeding Bulls	2	$20,000	$ 40,000
Velvet Bulls	38	$4,000	$ 152,000
Total	80		$ 632,000
Land	**Acres**	**Cost per Acre**	**Total Cost**
Pasture	160	$300	$ 48,000
Pasture Development	160	$30	$ 4,800
Total	160		$ 52,000
Equipment, Fence and Facilities			**Total Cost**
Fence	3.6 miles	$12,000 / mile	$ 43,200
Handling System and Hydraulic Squeeze			$ 17,500
Truck (1/2 Ton), Stock Trailer, Tractor and Loader (used); 3 Freezers			$ 46,100
Miscellaneous			$ 2,000
Total			$ 108,800
Total Capital Investment			**$ 793,600**

Table 7: Summary of the Cash Costs for an Established 40Cow / 40 Velvet Bull Elk Enterprise

Operating Inputs	Cow-Calf Cost	Replacement Cost	Velvet Bull Cost	Total Cost
Wages & Benefits	$ 2,275	$ 910	$ 2,276	$ 5,461
Grain	$ 1,277	$ 438	$ 1,636	$ 3,351
Hay	$ 1,956	$ 648	$ 2,356	$ 4,960
Supplement	$ 2,160	$ 422	$ 2,160	$ 4,742
Livestock Purchases	$ 0	$ 0	$ 0	$ 0
Breeding Stock Purchases	$ 0	$ 0	$ 0	$ 0
Veterinary Fees & Supplies	$ 1,199	$ 399	$ 998	$ 2,596
Repairs & Maintenance	$ 599	$ 239	$ 599	$ 1,437
Fuel & Oil	$ 399	$ 160	$ 399	$ 958
Utilities	$ 399	$ 160	$ 597	$ 1,156
Insurance	$ 5,622	$ 80	$ 201	$ 5,903
Marketing & Transportation	$ 247	$ 0	$ 70	$ 317
Death Loss	$ 21,654	$ 3,045	$ 7,170	$ 31,869
Property Taxes	$ 160	$ 0	$ 0	$ 160
Miscellaneous	$ 350	$ 150	$ 200	$ 700
Operating Interest	$ 1,532	$ 266	$ 740	$ 2,538
Total Cash Cost	$ 39,829	$ 6,917	$ 19,402	$ 66,148

Cash Costs of raising and maintaining:

Cash cost of raising a calf to weaning = $39,829 + 36 calves = $1,106/calf.

Cash cost of raising a replacement from weaning until a long yearling = $6,917 + 16 animals = $432/ animal.

Cash cost of maintaining a velvet bull = $19,402 + 40 bulls =$485/bull.

Income

At the present time, the main source of revenue is the sale of breeding stock. Other revenue comes from the sale of velvet, venison, and hunt bulls.

Table 8: Estimated Elk Price List (Saskatchewan), May 2000

Operating Inputs	Price Range CDN Dollars
Bull Calves	$1,000 - $4,000
Heifer Calves	$4,000 - $10,000
Yearling Exposed Heifers	$7,000 - $16,000
Exposed Cows	$6,000 - 20,000
Yearling Velvet Bulls	$1,500 - $5,000
Two Year+ Old Velvet Bulls	$2,500 - $5,000
Hunt Bulls	$3,000 - $20,000
Mature Breeding Bulls	$12,000 ++

Projected Income Statement

The projected income statement is based on the following assumptions:

1. The breeding herd consists of 33 cows and 7 exposed replacement heifers, for a total of 40 animals to calve each year. Calving rate is 95% for cows and 85% for heifers with a 95% weaning rate for both. Equal number of heifer and bull calves are born.

2. The velvet herd consists of 40 bulls, 2 of which are the breeding bulls.

3. Sixty percent of the heifer calves are sold after weaning at approximately 7 months of age for $6,000 each. The remaining heifer calves are kept and either sold as exposed heifers or used to replace herd cows. Exposed heifers have a market value of $10,000 each. In order to maintain cow numbers at 40 it is required that 5 exposed cows be sold to make room for the retained heifers. The exposed cows are sold for $8,000 each. Cull cows for slaughter will also be sold.

4. Fifty percent of the bull calves are retained to upgrade the velvet herd. The remaining bull calves are sold as potential velvet bulls for $2,000 each. Six velvet bulls will be sold at $4,000 each in order to allow young bulls to be retained to upgrade the herd's velvet production. Cull bulls for slaughter will also be sold.

Table 9: Projected Annual Income for an Established 40 Cow / 40 velvet Bull Elk Enterprise

Description	No. of Animals	Value per Head	Total Value
Heifer Calves	11	$ 6,000	$ 66,000
Bull Calves	9	$ 2,000	$ 18,000
Replacement Heifers	0	$ 10,000	$ 0
Herd Cows	5	$ 8,000	$ 40,000
Slaughter Cows	2	$ 1,000	$ 2,000
Slaughter Bulls	1	$ 1,500	$ 1,500
Velvet Bulls	6	$ 4,000	$ 36,000
Cull Herd Bulls	0	$ 10,000	$ 0
Velvet (green)	500 lbs.	$ 50	$ 25,000
Totals	**34 head**		**$188,500**

Income and Cost Summary

The following table provides a summary of the income and cost for a 40 cow/40 velvet bull elk enterprise. The depreciation figure is an average of the depreciation calculated for similar enterprises.

Table 10: Estimated Income & Cost Sumary of an Established 40 Cow / 40 Velvet Bull Elk Enterprise

	Total Amount
Total Cash Income	$188,500
- Total Cash Costs	$ 66,148
= Net Cash Income	$122,352
+ Inventory Change	$6,895
- Depreciation (Facilities & Equipment)	$2,975
= Return to Investment & Management	**$126,272**

Cash Flow Projections

Most producers start an elk enterprise by purchasing exposed heifers and a breeding bull. This approach reduces the amount of capital needed and provides an opportunity for the producer to become familiar with raising elk while the enterprise is small. However, cash flow requirements must be estimated and carefully managed. The accumulated cash flow deficit peaks in Year 6, before starting to decrease in Year 7. The level of management, the rate of expansion and the purchase of breeding stock will affect the rate of growth of the herd and the cash flow requirements.

Cash flow management will be important for an elk enterprise after it is fully developed. While expenses will occur throughout the year, the majority of the cash inflow will occur during a short period each year when stock and velvet are sold.

Financial Model for an Elk Operation:

The financial model following presents an example of the expected financial results of an elk enterprise that is initially established with 12 bred heifers and 1 mature breeding bull.

The purpose of this financial model is to provide information to potential producers, who may be interested in establishing an elk enterprise. This model identifies the resources to be utilized and includes a financial analysis that demonstrates an approach to determine the cash flow requirements and to determine the profitability of establishing an elk enterprise.

The reader is cautioned that some of the assumptions used in the preparation of these financial statements, although considered reasonable, will change as the industry continues to develop, especially in the area of future product prices. Changes in product prices will have a major impact on profitability. The prices used are representative of market conditions at the time of publication, and are not meant to predict future prices. See Graphs 2,3 and 4 for the effects of changes in product prices on accumulated cash flow, equality, and annual income. Most of the assumptions are based on presently known information and are provided by existing producers.

The following assumptions were used when creating these financial statements:

1. It is assumed that the decision to develop an elk enterprise is made in Year 1 of these fiscal projections. The initial capital investment is $295,100. The amount of equity is $200,000 with the balance being borrowed and repaid over a 10 year period at 8% interest. Initial handling facilities and fences will be developed and constructed during Year 1, with expansion occurring during future years as required. Twelve bred heifers and a breeding bull will be purchased in November of Year 1 and will produce a calf crop in year 2 of the budget. The entire calf crop will be retained in order to facilitate herd expansion. The retained heifer calves will be bred at 16 months of age and produce their first calves at 24 months of age. This is consistent with the goals of this enterprise that is to expand to a herd size of 40 cows and 40 velvet bulls. Once the projected herd size of 40 cows is reached, females will be sold in order to maintain consistent herd numbers. An allowance is provided in the estimate of expenses to replace animals that die. This amount is equal to 3% of the total value of the herd.

2. The investment cost of equipment purchased for the elk enterprise is based on the assumption that used equipment would be purchased when available. While this assumption reduces the investment costs, the amount allowed for maintenance is higher.

3. One hundred and sixty acres of pastureland was purchased for this enterprise and 80 acres were improved in Year 1 and 80 acres improved in Year 4. It is assumed that this land will provide the summer grazing requirement for this established elk enterprise of 33 cows, 7 replacement heifers and offspring and 2 herd bulls, as well as the 38 velvet bulls. Initially only a portion of the land will be required to support the elk herd and the remainder may be used in another enterprise. The land will be brought into production as warranted by herd expansion. No attempt is made in this analysis to include any revenue from an alternate use of the extra land.

4. In addition to the initial capital investment, operating loans reach a total of $210,024 in the sixth year of operating. An interest charge of 8% on the outstanding balance of the operating loan is included as a cash expense.

5. The cost of purchasing stock is based on information provided by surveying existing producers (Table 8).

6. Cow to Bull ratio will be maintained at no greater than 25:1.

7. Calving rate of 95% for cows and 85% for heifers with a weaning rate of 95% for both cows and heifers. Equal number of heifer and bull calves are born.

8. Forty acres of the quarter section are fenced in Year 1, with 40 acre pastures being fenced in Year 6, 7, and 8.

9. A used 1/2 ton pick-up truck is purchased in Year 1 for $10,000. Large chest freezers are purchased in Year 2, 5 and 7 for $700 each. A stock trailer is purchased in Year 3 for $5,000. A used tractor and loader is purchased in Year 4 for $9,000.

10. Elk are fed hay and grain for the winter and supplement year round.

11. Handling facilities are constructed in year 1. They are constructed in a manner that will allow for expansion as the herd size increases. A hydraulic squeeze is purchased in Year 4 for $6,000.

12. Market value of stock remains constant through the 10 years.

13. All sales and purchases of stock occur in the fourth quarter of the year, after velvetting and breeding. All velvet is sold in the year of production.

14. Slaughter bulls will be assumed to be at least 3 years of age. Hunt bulls will come from the velvet herd, and be bulls of at least 3 years of age. Any purchased velvet bulls will be assumed to be 2 years of age.

15. Breeding bulls will be purchased and sold as necessary to insure proper genetic progress. It is assumed that other than the purchase of the second breeding bull in Year 5, all replacement breeding bulls will be financed by the sale of the old breeding bull.

Table 11 is a summary of the projected cash flow and income statement for the elk enterprise.

Table 11: Projected Cash Flow & Income

	Year 1	Year 2	Year 3	Year 4	Year 5	Year 6	Year 7	Year 8	Year 9
Net Cash flow from Production	(2,906)	(26,602)	(31,710)	(38,965)	(57,265)	(52,575)	115,818	220,498	224,332
Accumulated Cash	(2,906)	(29,509)	(61,219)	(100,183)	(157,449)	(210,024)	(94,205)	126,293	350,624
Net Income	**($5,384)**	**$35,329**	**$78,295**	**$100,780**	**$133,537**	**$178,504**	**$235,322**	**$233,467**	**$245,592**

The Table 12 identifies the present value of the net income. It is important to note that in analyzing a proposed enterprise, the present value of future income is a significant measurement. If a proposed enterprise has a present value of net income that is less than the equity investment, it should be rejected. A present value of net income that is greater than the equity invested would reflect a net increase in equity over the term of the analysis.

Table 12: Present Value of Net Income

Present Value of Net Income	220,498
Less Initial Investment	126,293
Net Present Value of Income & Proprietor's Equity	**$233,467**

Table 13 shows the owners equity and the compound average rate of return on the initial investment in the enterprise for the first 10 years.

Table 13: Owner's Equity & Return on Initial Investment

Inital Equity Investment	$200,000	Average Compound Growth Rate
- end of Year 1	$ 194,616	(3%)
- end of Year 2	$ 229,945	7%
- end of Year 3	$ 308,240	16%
- end of Year 4	$ 409,021	20%
- end of Year 5	$ 542,558	22%
- end of Year 6	$ 721,062	24%
- end of year 7	$ 956,384	25%
- end of Year 8	$1,189,851	25%
- end of Year 9	$1,435,443	24%
- end of Year 10	$1,682,384	24%
Change in Owner Equity	$1,482,384	
Average Compound Rate of Growth	**24%**	

Calculations Re: Western Canadian Elk Farms in CDN Dollars

Table No. 1: Cow / Calf Operation Capital Investment Requirements

Item	#	Price/Unit	$ Invested	Useful Life	%	Depreciation $ per year	Interest $ per year
Livestock							
Yearling Heifers	95	7,000	665,000	20			$53,200
Bred Cows	5	12,000	60,000	15			$ 4,800
Bulls	5	18,000	90,000	15			$ 7,200
Total Costs			**$ 815,000**				**$65,200**
Land	**Acres**	**Lease/Acre**					
Pasture	360	$40	$14,400				$ 1,152
Total Costs			**$14,400**				**$ 1,152**
Improvments & Facilities							
Fencing (8 mi. @ $14,000/mi w/gates,water)			112,000	25	10	$11,200	$ 8,960
Facilities			73,000	25	10	$ 7,300	$ 5,840
Total Costs			**$ 185,000**			**$ 1,800**	**$14,800**
Equipment							
Stock Trailer			$ 8,000	7	30	$ 2,400	$ 640
Pick up truck			$ 20,000	5	30	$ 6,000	$ 1,600
Tractor & Loader			$ 25,000	10	30	$ 7,500	$ 2,000
Total Costs			**$ 53,000**			**$ 3,900**	**$ 4,240**
Total Fixed Costs			**$1,067,400**			**$ 5,700**	**$83,152**

Interest costs are the opportunity cost of using capital estimated at 8%

Financing Requirements	
Principle Amount	
Term (yrs.)	20
Interest %	9
Payment ($/yr.)	

Table 2: Cow / Calf Operation Production Plan & Annual Inventory

Inventory	Beginning	Year 1	Year 2	Year 3	Year 4	Year 5	Year 6
Heifer Calves Born		2	42	42	42	42	42
Cows	5	100	100	100	100	100	100
Heifer Calves Sold		0	40	41	41	41	41
Bull Calves Born		2	42	42	42	42	42
Bull Calves Sold		2	41	41	41	41	41
Herd Bulls	3	5	6	5	6	5	6
Adult Bulls Sold		0	1	0	1	0	1
Velvet Production (lbs)		50	60	80	120	200	200

Table No. 3: Elk Cow/Calf Operation Supplemental Feeding Costs

	Days on Feed	Lbs/Hd/Day	Total Fed - T.	$/Tonne	Cost /Head	# Head	Total Cost
Oats							
Cows	150	5	0.34	$ 125	$ 43	100	$ 4,261.36
Bulls	305	5	0.69	$ 125	$ 87	6	$ 519.89
Total			**1.03**				**$ 4,781**
Supplement							
Cows	365	1	0.17	$ 250	$ 41	100	$ 4,147.73
Bulls	365	2	0.33	$ 250	$ 83	6	$ 497.73
Total			**0.50**				**$ 4,645**
Hay							
Cows	180	20	1.64	$ 75	$123	100	$12,272.73
Bulls	180	25	2.05	$ 75	$153	6	$ 920.45
Total			**3.68**				**$13,193**
Total Feed Costs					**$213.40**	**106**	**$22,620**

Table No. 4: Cow / Calf Operation Estimated Revenues & Expenses

	$/Head	#Sold	Total
Revenues			
Heifer Calves	$ 9,000	40	$360,000
Bull Calves	$ 2,000	41	$ 82,000
Adult Bulls	$10,000	1	$ 10,000
Velvet at $60/lb		50	$ 2,000
Total Revenues			**$454,000**
	$/Head	**# Head**	
Operating Costs			
Veterinary & Medicine	$ 20.00	106	$ 2,120
Oats	$ 64.63		$ 4,781
Hay	$133.07		$ 13,193
Supplement	$ 62.22		$ 4,645
Land Lease			$ 14,400
Fence & Corral Repairs			$ 10,000
Marketing & Transportation			$ 12,000
Utilities			$ 1,800
Insurance			$ 6,000
Replace Breeding Bull			$ 12,000
Interest on Operating			$ 4,000
Total Operating Costs			**$ 84,940**
Fixed Costs			
Interest on Breeding Stock			$107,200
Interest on Land			$ 1,152
Interest on Fencing & Facilities			$ 14,800
Interest on Equipment			$ 4,240
Depreciation on Fencing & Facilities			$ 1,800
Depreciation on Equipment			$ 3,900
Land Taxes			$ 1,250
Total Fixed Costs			**$134,342**
Total Costs			**$219,282**
Net Profit			**$234,718**
Net Profit / Cow			**$ 2,347.18**

Table No. 5: Cow / Calf Operation Projected Cash Flows

	Year 1	Year 2	Year 3	Year 4	Year 5
Cash Outflow					
Breeding Stock	220,000				
Land	27,000				
Facilities	45,000				
Equipment	29,000				
Total Capital Investment	**$321,000**				
Veterinary & Medicine	395	395	395	395	395
Replace Breeding Bull		12,000		12,000	
Oats	814	814	814	814	814
Hay	980	980	980	980	980
Supplement	637	637	637	637	637
Fence & Corral Repairs	637	637	637	637	637
Marketing & Transportation	250	250	250	250	250
Utilities	150	150	150	150	150
Insurance	3,500	3,500	3,500	3,500	3,500
Interest on Operating	500	500	500	500	500
Land Taxes	250	250	250	250	250
Total Operating Expenses	**$ 8,113**	**$20,113**	**$ 8,113**	**$20,113**	**$ 8,113**
Personal Withdrawls	20,000	20,000	20,000	20,000	20,000
Annual Debt Payment	31,162	31,162	31,162	31,162	31,162
TOTAL CASH OUTFLOW	**$380,275**	**$71,275**	**$59,275**	**$71,275**	**$59,275**
Cash Inflow					
Borrowed Capital	200,000				
Equity Contribution	122,000				
Cash Sales					
Heifer Calves $10,000 / Head	60,000	60,000	60,000	60,000	60,000
Bull Calves $2,500 / Head	13,000	15,000	5,000	15,000	15,000
Cull Bulls $6,000 / Head		6,000	0	6,000	0
Velvet $80 / lb	2,000	2,000	2,000	2,000	2,000
Total Sales Revenue	**$ 77,000**	**$ 83,000**	**$ 77,000**	**$ 83,000**	**$ 77,000**
TOTAL CASH INFLOW	**$399,000**	**$ 83,000**	**$ 77,000**	**$ 83,000**	**$ 77,000**
Net Cashflow	**$ 18,725**	**$ 11,725**	**$ 17,725**	**$ 11,725**	**$ 17,725**
Cumulative Cash Flow	**$ 18,725**	**$ 30,450**	**$ 48,175**	**$ 59,900**	**$ 77,625**

Table No. 6: Commercial Velvet Operation - Capital Investment Requirements

Item	#	Price/Unit	$ Invested	Useful Life	Depreciation $ per year	Interest $ per year
Livestock						
Bull Calves	80	2,500	$ 200,000	15		$ 8,000
Total Costs			**$ 200,000**			**$ 8,000**
Land	**Acres**	**Lease/Acre**				
Pasture	160	$250	$ 40,000			$ 1,600
Total Costs			**$ 40,000**			**$ 1,600**
Improvments & Facilities						
Fencing (3 mi. @ $10,000/mi)			$ 30,000	20	$ 1,500	$ 1,200
Facilities			$ 20,000	20	$ 1,000	$ 800
Total Costs			**$ 50,000**		**$ 2,500**	**$ 2,000**
Equipment						
Stock Trailer			$ 4,000	10	$ 400	$ 160
Pick up truck			$ 10,000	5	$ 2,000	$ 400
Tractor & Loader			$ 15,000	10	$ 1,500	$ 600
Total Costs			**$ 29,000**		**$ 3,900**	**$ 1,160**
Totals			**$ 319,000**		**$ 6,400**	**$13,600**

Interest costs are the opportunity cost of using capital estimated at 8%

Financing Requirements	
Principle Amount	$ 200,000
Term (yrs.)	10
Interest %	9
Payment ($/yr.)	$ 31,162

Table 7: Velvet Operation Production Plan

Inventory	Beginning	Year 1	Year 2	Year 3	Year 4	Year 5	Year 6	Year 7
Velvet Bulls	80	80	80	80	80	80	80	80
Velvet Production (lbs/bull)		2	8	11	14	17	19	20
Velvet Price		$25	$45	$45	$65	$65	$75	$75
Velvet Revenues / bull		$50	$360	$495	$910	$1,105	$1,425	$1,500
Projected Velvet Revenues		$4,000	$28,800	$39,600	$72,800	$88,000	$114,000	120,000

Table No. 8: Velvet Operation Projected Cash Flows

	Year 1	Year 2	Year 3	Year 4	Year 5	Year 6	Year 7	Year 8	Year 9	Year 10
Cash Outflow										
Breeding Stock	200,000									
Land	88,000									
Facilities	50,000									
Equipment	29,000									
Total Capital	**$367,000**									
Veterinary & Medicine	900	900	900	900	900	900	900	900	900	900
Oats	3,453	3,453	3,453	3,453	3,453	3,453	3,453	3,453	3,453	3,453
Hay	16,200	16,200	16,200	16,200	16,200	16,200	16,200	16,200	16,200	16,200
Supplement	2,401	2,401	2,401	2,401	2,401	2,401	2,401	2,401	2,401	2,401
Fence & Corral Repairs	500	500	500	500	500	500	500	500	500	500
Marketing & Transportation	250	250	250	250	250	250	250	250	250	250
Utilities	150	150	150	150	150	150	150	150	150	150
Insurance	3,000	3,000	3,000	3,000	3,000	3,000	3,000	3,000	3,000	3,000
Interest on Operating	500	500	500	500	500	500	500	500	500	500
Land Taxes	250	250	250	250	250	250	250	250	250	250
Total Operating Expenses	**$ 27,604**	**$ 27,604**	**$ 27,604**	**$ 27,604**	**$ 27,604**	**$ 27,604**	**$ 27,604**	**$ 27,604**	**$ 27,604**	**$ 27,604**
Personal Withdrawls	20,000	20,000	20,000	20,000	20,000	20,000	20,000	20,000	20,000	20,000
Annual Debt Payment	31,162	31,162	31,162	31,162	31,162	31,162	31,162	31,162	31,162	31,162
TOTAL CASH OUTFLOW	**$445,766**	**$ 78,766**	**$ 78,766**	**$ 78,766**	**$ 78,766**	**$ 78,766**	**$ 78,766**	**$ 78,766**	**$ 78,766**	**$ 78,766**
Cash Inflow										
Borrowed Capital	200,000									
Equity Contribution	139,000									
Sales Revenue										
Velvet	4,000	28,800	39,600	72,800	88,400	114,000	120,000	126,000	132,000	138,000
Total Sales Revenue	**4,000**	**28,800**	**39,600**	**72,800**	**88,400**	**114,000**	**120,000**	**126,000**	**132,000**	**138,000**
TOTAL CASH INFLOW	**347,000**	**28,800**	**39,600**	**72,800**	**88,400**	**114,000**	**120,000**	**126,000**	**132,000**	**138,000**
Net Cash Revenue	**- $ 98,766**	**-$ 49,966**	**-$39,166**	**-$5,966**	**$ 9,634**	**$ 35,234**	**$ 41,234**	**$ 47,234**	**$ 53,234**	**$ 59,234**
Cumulative Cash Flow	**$138,331**	**$ 88,365**	**$ 49,199**	**$ 43,233**	**$ 52,867**	**$ 88,101**	**$129,335**	**$176,569**	**$229,803**	**$289,037**

Elk Economics
Bruce Friedel, M.Sc.

Cow/ calf operation's assumptions:

Buy Bred Cows

Wean 95% calf crop

Prices drop 5% per year

INVESTMENT	ELK:	FENCES:	SELL:	WEAN %	PRICES
250000	210000	40000	calves	0.95	(-5%/yr)

INVENTORY	2000	2001	2002	2003	2004	2005	2006	2007	2008	2009
Cows @ $8000	25	24	29	34	40	48	54	56	59	61
Bulls @ $10000	1	2	2	2	2	2	2	2	2	2
Yearling heifers	0	6	6	7	8	10	11	13	13	14
Weaning %	0.95	0.95	0.95	0.95	0.95	0.95	0.95	0.95	0.95	0.95
Weaned calves	24	23	28	33	38	45	51	53	56	57
SALES										
Heifer calves	6	6	7	8	10	11	13	13	14	14
$ / Heifer calf	5000	4750	4513	4287	4073	3869	3675	3492	3317	3151
Bull calves	12	11	14	16	19	23	26	22	22	22
$ / Bull calf	1500	1425	1354	1286	1222	1161	1103	1048	995	945
Bred cows	0	0	0	0	0	2	8	9	10	10
$ / cow	8000	7600	7220	6859	6516	6190	5881	5587	5307	5042
Deaths (2% / yr)	1	1	1	1	1	1	1	1	1	1
Cull bulls	0	1	0	1	0	1	0	1	0	1
$ / bull	0	0	0	4000	0	4000	0	4000	0	4000
Velvet (lbs)	25	50	50	50	50	50	50	50	50	50
$/lb	65	65	65	65	65	65	65	65	65	65
INCOME	49125	46570	53559	63187	65895	89703	125899	127132	124396	123757
Income ($/cow)	1965	1940	1826	1840	1628	1882	2326	2264	2122	2045
EXPENSES										
Breeding bulls	0	10000	0	10000	0	10000	0	10000	0	10000
Fence depr. (5%/yr)	1600	1600	1600	1600	1600	1600	1600	1600	1600	1600
Facilities depr. (10%/yr)	800	800	800	800	800	800	800	800	800	800
Hay ($50/T)	2339	2650	3119	3639	4258	4985	5661	5926	6175	6378
Oats ($100/T)	1148	1362	1589	1855	2169	2538	2885	3029	3155	3260
Supplement ($200/T)	1120	1330	1553	1816	2125	2490	2832	2975	3100	3204
Veterinary ($15/elk)	746	821	974	1139	1336	1568	1783	1865	1945	2009
Utilities ($5/elk)	249	274	325	380	445	523	594	622	648	670
Fuel & Machinery ($40/	1990	2190	2596	3037	3564	4182	4755	4973	5186	5357
Misc ($20/elk)	995	1095	1298	1519	1782	2091	2377	2487	2593	2678
COSTS	10986	22121	13854	25784	18079	30777	23288	34276	25201	35955
Cost/cow	439	922	472	751	447	646	430	611	430	594
NET MARGIN	38139	24449	39705	37403	47816	58926	102611	92856	99195	87802
Net / cow	1526	1019	1353	1089	1181	1236	1896	1654	1692	1451
% RETURN	0.14	0.08	0.12	0.10	0.12	0.13	0.21	0.19	0.20	0.18

TOTAL PROFITS	628901
Discount Rate (%)	10
NET PRESENT VALUE	257235

Notes: There are two parts to every investment decision!
Is it profitable? Look at the Net Present Value calculation. NPV reflects an increase (positive #) or decrease (negative #) in your wealth today
Can I do it? Examine your balance sheet, cash flow, debt payments, taxes, living costs, etc...

Need a loan?		
	Principle	100000
	Term (yrs)	10
	Interest (%)	9
	Payment ($/yr)	-15581

Elk Economics
Bruce Friedel, M.Sc.

Cow/ calf operation's assumptions:

Buy Bred Cows
Wean 85% calf crop
Prices drop 5% per year

INVESTMENT	ELK:	FENCES:	SELL:	WEAN %	PRICES
250000	210000	40000	calves	0.85	(-5%/yr)

INVENTORY	2000	2001	2002	2003	2004	2005	2006	2007	2008	2009
Cows @ $8000	25	24	29	33	38	45	50	50	50	50
Bulls @ $10000	1	2	2	2	2	2	2	2	2	2
Yearling heifers	0	5	5	6	7	8	9	11	11	11
Weaning %	0.85	0.85	0.85	0.85	0.85	0.85	0.85	0.85	0.85	0.85
Weaned calves	21	20	24	28	33	38	42	43	43	42
SALES										
Heifer calves	5	5	6	7	8	9	11	11	11	11
$ / Heifer calf	5000	4750	4513	4287	4073	3869	3675	3492	3317	3151
Bull calves	11	10	12	14	16	19	21	22	22	22
$ / Bull calf	1500	1425	1354	1286	1222	1161	1103	1048	995	945
Bred cows	0	0	0	0	0	2	8	9	10	10
$ / cow	8000	7600	7220	6859	6516	6190	5881	5587	5307	5042
Deaths (2% / yr)	1	1	1	1	1	1	1	1	1	1
Cull bulls	0	1	0	1	0	1	0	1	0	1
$ / bull	0	0	0	4000	0	4000	0	4000	0	4000
Velvet (lbs)	25	50	50	50	50	50	50	50	50	50
$/lb	65	65	65	65	65	65	65	65	65	65
INCOME	44125	42010	47323	55567	56516	78303	112431	117687	113722	111802
Income ($/cow)	1765	1750	1647	1676	1469	1756	2261	2353	2258	2246
EXPENSES										
Breeding bulls	0	10000	0	10000	0	10000	0	10000	0	10000
Fence depr. (5%/yr)	1600	1600	1600	1600	1600	1600	1600	1600	1600	1600
Facilities depr. (10%/yr)	800	800	800	800	800	800	800	800	800	800
Hay ($50/T)	2279	2558	2962	3405	3924	4524	5044	5129	5164	5116
Oats ($100/T)	1103	1296	1489	1712	1971	2271	2534	2586	2603	2581
Supplement ($200/T)	1102	1290	1483	1709	1970	2272	2538	2590	2608	2585
Veterinary ($15/elk)	709	776	904	1041	1203	1390	1552	1576	1587	1572
Utilities ($5/elk)	236	259	301	347	401	463	517	525	529	524
Fuel & Machinery ($40/	1890	2069	2410	2777	3208	3708	4139	4204	4233	4192
Misc ($20/elk)	945	1034	1205	1389	1604	1854	2069	2102	2116	2096
COSTS	10663	21682	13154	24781	16681	28883	20794	31113	21240	31065
Cost/cow	427	903	458	748	434	648	418	622	422	624
NET MARGIN	33462	20328	34169	30786	39835	49420	91637	86574	92482	80737
Net / cow	1338	847	1189	929	1036	1108	1843	1731	1836	1622
% RETURN	0.12	0.07	0.10	0.09	0.10	0.12	0.21	0.20	0.22	0.20

TOTAL PROFITS	559431
Discount Rate (%)	10
NET PRESENT VALUE	184720

Notes: There are two parts to every investment decision!
Is it profitable? Look at the Net Present Value calculation. NPV reflects an increase (positive #) or decrease (negative #) in your wealth today
Can I do it? Examine your balance sheet, cash flow, debt payments, taxes, living costs, etc...

Need a loan?		
	Principle	100000
	Term (yrs)	10
	Interest (%)	9
	Payment ($/yr)	-15581

Elk Economics
Bruce Friedel, M.Sc.

Cow/ calf operation's assumptions:

Buy Bred Cows
Wean 85% calf crop
Prices drop 10% per year

INVESTMENT	ELK:	FENCES:	SELL:	WEAN %	PRICES
250000	210000	40000	calves	0.85	(- 10%/yr)

INVENTORY	2000	2001	2002	2003	2004	2005	2006	2007	2008	2009
Cows @ $8000	25	24	29	33	38	45	50	50	50	50
Bulls @ $10000	1	2	2	2	2	2	2	2	2	2
Yearling heifers	0	5	5	6	7	8	9	11	11	11
Weaning %	0.85	0.85	0.85	0.85	0.85	0.85	0.85	0.85	0.85	0.85
Weaned calves	21	20	24	28	33	38	42	43	43	42
SALES										
Heifer calves	5	5	6	7	8	9	11	11	11	11
$ / Heifer calf	5000	4500	4050	3645	3281	2952	2657	2391	2152	1937
Bull calves	11	10	12	14	16	19	21	22	22	22
$ / Bull calf	1500	1350	1215	1094	984	886	797	717	646	581
Bred cows	0	0	0	0	0	2	8	9	10	10
$ / cow	8000	7200	6480	5832	5249	4724	4252	3826	3444	3099
Deaths (2% / yr)	1	1	1	1	1	1	1	1	1	1
Cull bulls	0	1	0	1	0	1	0	1	0	1
$ / bull	0	0	0	4000	0	4000	0	4000	0	4000
Velvet (lbs)	25	50	50	50	50	50	50	50	50	50
$/lb	65	65	65	65	65	65	65	65	65	65
INCOME	44125	39970	42806	48332	46157	61472	82184	82889	74931	71519
Income ($/cow)	1765	1665	1490	1458	1200	1378	1653	1657	1488	1437
EXPENSES										
Breeding bulls	0	10000	0	10000	0	10000	0	10000	0	10000
Fence depr. (5%/yr)	1600	1600	1600	1600	1600	1600	1600	1600	1600	1600
Facilities depr. (10%/yr)	800	800	800	800	800	800	800	800	800	800
Hay ($50/T)	2279	2558	2962	3405	3924	4524	5044	5129	5164	5116
Oats ($100/T)	1103	1296	1489	1712	1971	2271	2534	2586	2603	2581
Supplement ($200/T)	1102	1290	1483	1709	1970	2272	2538	2590	2608	2585
Veterinary ($15/elk)	709	776	904	1041	1203	1390	1552	1576	1587	1572
Utilities ($5/elk)	236	259	301	347	401	463	517	525	529	524
Fuel & Machinery ($40/	1890	2069	2410	2777	3208	3708	4139	4204	4233	4192
Misc ($20/elk)	945	1034	1205	1389	1604	1854	2069	2102	2116	2096
COSTS	10663	21682	13154	24781	16681	28883	20794	31113	21240	31065
Cost/cow	427	903	458	748	434	648	418	622	422	624
NET MARGIN	33462	18288	29652	23552	29476	32589	61390	51776	53691	40454
Net / cow	1338	762	1032	710	766	731	1235	1035	1066	813
% RETURN	0.12	0.06	0.10	0.08	0.09	0.10	0.18	0.17	0.19	0.15

TOTAL PROFITS	374330
Discount Rate (%)	10
NET PRESENT VALUE	44784

Notes: There are two parts to every investment decision!
Is it profitable? Look at the Net Present Value calculation. NPV reflects an increase (positive #) or decrease (negative #) in your wealth today
Can I do it? Examine your balance sheet, cash flow, debt payments, taxes, living costs, etc...

Need a loan?	Principle	100000
	Term (yrs)	10
	Interest (%)	9
	Payment ($/yr)	-15581

Elk Economics
Bruce Friedel, M.Sc.

Cow/ calf operation's assumptions:

Buy Bred Cows
Wean 80% calf crop
Sell bred cows rather than calves

Prices stable

INVESTMENT	ELK:	FENCES:	SELL:	WEAN %	PRICES
250000	210000	40000	bred cows	0.8	stable

INVENTORY	2000	2001	2002	2003	2004	2005	2006	2007	2008	2009
Cows @ $8000	25	24	33	35	39	43	49	55	63	72
Bulls @ $10000	1	2	2	2	2	2	2	2	2	2
Yearling heifers	0	10	10	13	14	16	17	20	22	25
Weaning %	0.80	0.80	0.80	0.80	0.80	0.80	0.80	0.80	0.80	0.80
Weaned calves	20	19	27	28	32	35	39	44	50	58
SALES										
Heifer calves	0	0	0	0	0	0	0	0	0	0
$ / Heifer calf	5000	5000	5000	5000	5000	5000	5000	5000	5000	5000
Bull calves	10	10	13	14	16	17	20	22	25	29
$ / Bull calf	1500	1500	1500	1500	1500	1500	1500	1500	1500	1500
Bred cows	0	0	7	8	9	9	10	10	11	11
$ / cow	8000	8000	8000	8000	8000	8000	8000	8000	8000	8000
Deaths (2% / yr)	1	1	1	1	1	1	1	1	2	2
Cull bulls	0	1	0	1	0	1	0	1	0	1
$ / bull	0	0	0	4000	0	4000	0	4000	0	4000
Velvet (lbs)	25	50	50	50	50	50	50	50	50	50
$ / lb	65	65	65	65	65	65	65	65	65	65
INCOME	16625	17650	79242	92287	98903	105277	112628	120243	129099	138675
Income ($/cow)	665	735	2378	2632	2509	2427	2300	2187	2047	1916
EXPENSES										
Breeding bulls	0	10000	0	10000	0	10000	0	10000	0	10000
Fence depr. (5%/yr)	1600	1600	1600	1600	1600	1600	1600	1600	1600	1600
Facilities depr. (10%/yr)	800	800	800	800	800	800	800	800	800	800
Hay ($50/T)	2249	2783	3569	3922	4337	4774	5344	5987	6819	7800
Oats ($100/T)	1080	1444	1815	2021	2227	2454	2742	3072	3493	3995
Supplement ($200/T)	1093	1452	1832	2041	2251	2482	2776	3112	3542	4053
Veterinary ($15/elk)	690	828	1074	1177	1305	1438	1612	1808	2063	2363
Utilities ($5/elk)	230	276	358	392	435	479	537	603	688	788
Fuel & Machinery ($40/	1840	2208	2863	3138	3479	3834	4299	4823	5502	6300
Misc ($20/elk)	920	1104	1432	1569	1740	1917	2150	2411	2751	3150
COSTS	10502	22495	15342	26659	18174	29777	21860	34215	27257	40848
Cost/cow	420	937	460	760	461	686	446	622	432	564
NET MARGIN	6124	-4845	63900	65628	80729	75499	90768	86027	101842	97828
Net / cow	245	-202	1918	1872	2048	1741	1854	1564	1614	1352
% RETURN	0.02	-0.01	0.14	0.13	0.14	0.12	0.13	0.11	0.12	0.10

TOTAL PROFITS	663500
Discount Rate (%)	10
NET PRESENT VALUE	461583

Notes: There are two parts to every investment decision!
Is it profitable? Look at the Net Present Value calculation. NPV reflects an increase (positive #) or decrease (negative #) in your wealth today.
Can I do it? Examine your balance sheet, cash flow, debt payments, taxes, living costs, etc...

Need a loan?		
	Principle	100000
	Term (yrs)	10
	Interest (%)	9
	Payment ($/yr)	-15581

Elk Economics
Bruce Friedel, M.Sc.

Velvet bull operation's assumptions:

Buy Bull Calves

Velvet Antler Prices Stable

Livestock prices drop 5% per year

INVESTMENT	ELK:	FENCES:	SELL:	WEAN %	PRICES
250000	210000	40000	velvet antler and bulls	0.85	as above

INVENTORY	2000	2001	2002	2003	2004	2005	2006	2007	2008	2009
Cows @ $8000	0	0	0	0	0	0	0	0	0	0
Bulls @ $1500	140	140	140	140	140	140	140	140	140	140
Yearling heifers	0	0	0	0	0	0	0	0	0	0
Weaning %	0.85	0.85	0.85	0.85	0.85	0.85	0.85	0.85	0.85	0.85
Weaned calves	0	0	0	0	0	0	0	0	0	0
SALES										
Heifer calves	0	0	0	0	0	0	0	0	0	0
$ / Heifer calf	5000	4750	4513	4287	4073	3869	3675	3492	3317	3151
Breeding Bulls	0	0	1	4	4	4	4	4	4	4
$ / Bull	1500	2500	5000	8000	7600	7220	6859	6516	6190	5881
Bred cows	0	0	0	0	0	0	0	0	0	0
$ / cow	8000	7600	7220	6859	6516	6190	5881	5587	5307	5042
Deaths (3% / yr)	4	4	4	4	4	4	4	4	4	4
Cull and Hunt Bulls	0	0	14	14	14	14	14	14	14	14
$ / bull	0	0	2500	2500	2500	2500	2500	2500	2500	2500
Velvet (lbs)	300	1260	1680	1960	2240	2520	2800	3080	3080	3080
$ / lb	55	55	55	55	55	55	55	55	55	55
INCOME	16500	69300	132400	176525	190245	204049	217933	231892	230524	229224
Income ($/bull)	118	495	946	1261	1359	1457	1557	1656	1647	1637
EXPENSES										
Replacement bull calves	6300	6300	27300	27375	27375	27375	27375	27375	27375	27375
Fence depr. (5%/yr)	1600	1600	1600	1600	1600	1600	1600	1600	1600	1600
Facilities depr. (10%/yr)	800	800	800	800	800	800	800	800	800	800
Hay ($50/T)	11340	11340	11340	11340	11340	11340	11340	11340	11340	11340
Oats ($100/T)	6300	6300	6300	6300	6300	6300	6300	6300	6300	6300
Supplement ($200/T)	5110	5110	5110	5110	5110	5110	5110	5110	5110	5110
Veterinary ($15/elk)	2100	2100	2100	2100	2100	2100	2100	2100	2100	2100
Utilities ($5/elk)	700	700	700	700	700	700	700	700	700	700
Fuel & Machinery ($40/elk)	5600	5600	5600	5600	5600	5600	5600	5600	5600	5600
Misc ($20/elk)	2800	2800	2800	2800	2800	2800	2800	2800	2800	2800
COSTS	42650	42650	63650	63725	63725	63725	63725	63725	63725	63725
Cost/bull	305	305	455	455	455	455	455	455	455	455
NET MARGIN	-26150	26650	68750	112800	126520	140324	154208	168167	166799	165499
Net / bull	-187	190	491	806	904	1002	1101	1201	1191	1182
% RETURN	-0.02	0.02	0.06	0.10	0.11	0.12	0.13	0.14	0.14	0.14

TOTAL PROFITS	1103567
Discount Rate (%)	10
NET PRESENT VALUE	326843

Notes: There are two parts to every investment decision!
Is it profitable? Look at the Net Present Value calculation. NPV reflects an increase (positive #) or decrease (negative #) in your wealth today
Can I do it? Examine your balance sheet, cash flow, debt payments, taxes, living costs, etc...

Need a loan?	Principle	100000	200000
	Term (yrs)	10	10
	Interest (%)	9	9
	Payment ($/yr)	-15581	-32000

Elk Economics
Bruce Friedel, M.Sc.

Velvet bull operation's assumptions:

Buy Bull Calves

Velvet Antler Prices drop 10% per year

Livestock prices drop 5% per year

INVESTMENT	ELK:	FENCES:	SELL:	WEAN %	PRICES
250000	210000	40000	velvet antler and bulls	0.85	as above

INVENTORY	2000	2001	2002	2003	2004	2005	2006	2007	2008	2009
Cows @ $8000	0	0	0	0	0	0	0	0	0	0
Bulls @ $1500	140	140	140	140	140	140	140	140	140	140
Yearling heifers	0	0	0	0	0	0	0	0	0	0
Weaning %	0.85	0.85	0.85	0.85	0.85	0.85	0.85	0.85	0.85	0.85
Weaned calves	0	0	0	0	0	0	0	0	0	0
SALES										
Heifer calves	0	0	0	0	0	0	0	0	0	0
$ / Heifer calf	5000	4750	4513	4287	4073	3869	3675	3492	3317	3151
Breeding Bulls	0	0	1	4	4	4	4	4	4	4
$ / Bull	1500	2500	5000	8000	7600	7220	6859	6516	6190	5881
Bred cows	0	0	0	0	0	0	0	0	0	0
$ / cow	8000	7600	7220	6859	6516	6190	5881	5587	5307	5042
Deaths (3% / yr)	4	4	4	4	4	4	4	4	4	4
Cull and Hunt Bulls	0	0	14	14	14	14	14	14	14	14
$ / bull	0	0	2500	2500	2500	2500	2500	2500	2500	2500
Velvet (lbs)	300	1260	1680	1960	2240	2520	2800	3080	3080	3080
$ / lb	55	50	45	40	36	32	29	26	24	21
INCOME	16500	62370	114844	147311	147877	147291	145775	143516	134045	125453
Income ($/bull)	118	446	820	1052	1056	1052	1041	1025	957	896
EXPENSES										
Replacement bull calves	6300	6300	27300	27375	27375	27375	27375	27375	27375	27375
Fence depreciation (5%/yr)	1600	1600	1600	1600	1600	1600	1600	1600	1600	1600
Facilities depr. (10%/yr)	800	800	800	800	800	800	800	800	800	800
Hay ($50/T)	11340	11340	11340	11340	11340	11340	11340	11340	11340	11340
Oats ($100/T)	6300	6300	6300	6300	6300	6300	6300	6300	6300	6300
Supplement ($200/T)	5110	5110	5110	5110	5110	5110	5110	5110	5110	5110
Veterinary ($15/elk)	2100	2100	2100	2100	2100	2100	2100	2100	2100	2100
Utilities ($5/elk)	700	700	700	700	700	700	700	700	700	700
Fuel & Machinery ($40/elk)	5600	5600	5600	5600	5600	5600	5600	5600	5600	5600
Misc ($20/elk)	2800	2800	2800	2800	2800	2800	2800	2800	2800	2800
COSTS	42650	42650	63650	63725	63725	63725	63725	63725	63725	63725
Cost/bull	305	305	455	455	455	455	455	455	455	455
NET MARGIN	-26150	19720	51194	83586	84152	83566	82050	79791	70320	61728
Net / bull	-187	141	366	597	601	597	586	570	502	441
% RETURN	-0.02	0.02	0.04	0.07	0.07	0.07	0.07	0.07	0.06	0.05

TOTAL PROFITS	589957
Discount Rate (%)	10
NET PRESENT VALUE	70447

Notes: There are two parts to every investment decision!
Is it profitable? Look at the Net Present Value calculation. NPV reflects an increase (positive #) or decrease (negative #) in your wealth today
Can I do it? Examine your balance sheet, cash flow, debt payments, taxes, living costs, etc...

Need a loan?			
	Principle	100000	200000
	Term (yrs)	10	10
	Interest (%)	9	9
	Payment ($/yr)	-15581	-32000

Elk Economics

Bruce Friedel, M.Sc.
780 826 6475

Velvet bull operation's assumptions:

Buy Bull Calves

Velvet Antler Prices up 10% per year

Livestock prices drop 5% per year

INVESTMENT	ELK:	FENCES:	SELL:	WEAN %	PRICES
250000	210000	40000	velvet antler and bulls	0.85	as above

INVENTORY	2000	2001	2002	2003	2004	2005	2006	2007	2008	2009
Cows @ $8000	0	0	0	0	0	0	0	0	0	0
Bulls @ $1500	140	140	140	140	140	140	140	140	140	140
Yearling heifers	0	0	0	0	0	0	0	0	0	0
Weaning %	0.85	0.85	0.85	0.85	0.85	0.85	0.85	0.85	0.85	0.85
Weaned calves	0	0	0	0	0	0	0	0	0	0
SALES										
Heifer calves	0	0	0	0	0	0	0	0	0	0
$ / Heifer calf	5000	4750	4513	4287	4073	3869	3675	3492	3317	3151
Breeding Bulls	0	0	1	4	4	4	4	4	4	4
$ / Bull	1500	2500	5000	8000	7600	7220	6859	6516	6190	5881
Bred cows	0	0	0	0	0	0	0	0	0	0
$ / cow	8000	7600	7220	6859	6516	6190	5881	5587	5307	5042
Deaths (3% / yr)	4	4	4	4	4	4	4	4	4	4
Cull and Hunt Bulls	0	0	14	14	14	14	14	14	14	14
$ / bull	0	0	2500	2500	2500	2500	2500	2500	2500	2500
Velvet (lbs)	300	1260	1680	1960	2240	2520	2800	3080	3080	3080
$ / lb	55	58	61	64	67	70	74	77	81	85
INCOME	16500	72765	141871	193517	216795	242342	270308	300855	311405	322619
Income ($/bull)	118	520	1013	1382	1549	1731	1931	2149	2224	2304
EXPENSES										
Replacement bull calves	6300	6300	27300	27375	27375	27375	27375	27375	27375	27375
Fence depreciation (5%/yr)	1600	1600	1600	1600	1600	1600	1600	1600	1600	1600
Facilities depr. (10%/yr)	800	800	800	800	800	800	800	800	800	800
Hay ($50/T)	11340	11340	11340	11340	11340	11340	11340	11340	11340	11340
Oats ($100/T)	6300	6300	6300	6300	6300	6300	6300	6300	6300	6300
Supplement ($200/T)	5110	5110	5110	5110	5110	5110	5110	5110	5110	5110
Veterinary ($15/elk)	2100	2100	2100	2100	2100	2100	2100	2100	2100	2100
Utilities ($5/elk)	700	700	700	700	700	700	700	700	700	700
Fuel & Machinery ($40/elk)	5600	5600	5600	5600	5600	5600	5600	5600	5600	5600
Misc ($20/elk)	2800	2800	2800	2800	2800	2800	2800	2800	2800	2800
COSTS	42650	42650	63650	63725	63725	63725	63725	63725	63725	63725
Cost/bull	305	305	455	455	455	455	455	455	455	455
NET MARGIN	-26150	30115	78221	129792	153070	178617	206583	237130	247680	258894
Net / bull	-187	215	559	927	1093	1276	1476	1694	1769	1849
% RETURN	-0.02	0.03	0.07	0.11	0.13	0.15	0.18	0.20	0.21	0.22

TOTAL PROFITS	1493952
Discount Rate (%)	10
NET PRESENT VALUE	515884

Notes: There are two parts to every investment decision!
Is it profitable? Look at the Net Present Value calculation. NPV reflects an increase (positive #) or decrease (negative #) in your wealth today
Can I do it? Examine your balance sheet, cash flow, debt payments, taxes, living costs, etc...

Need a loan?			
	Principle	100000	200000
	Term (yrs)	10	10
	Interest (%)	9	9
	Payment ($/yr)	-15581	-32000